Advances in Plant Physiology of Abiotic Stresses

Advances in Plant Physiology of Abiotic Stresses

Editors

Sara Álvarez
José Ramón Acosta-Motos

MDPI • Basel • Beijing • Wuhan • Barcelona • Belgrade • Manchester • Tokyo • Cluj • Tianjin

Editors

Sara Álvarez
Instituto Tecnológico Agrario
de Castilla y León (ITACYL)
Spain

José Ramón Acosta-Motos
Universidad Católica San
Antonio de Murcia (UCAM)
Spain

Editorial Office
MDPI
St. Alban-Anlage 66
4052 Basel, Switzerland

This is a reprint of articles from the Special Issue published online in the open access journal *Agronomy* (ISSN 2073-4395) (available at: https://www.mdpi.com/journal/agronomy/special_issues/physiology_stresses).

For citation purposes, cite each article independently as indicated on the article page online and as indicated below:

LastName, A.A.; LastName, B.B.; LastName, C.C. Article Title. *Journal Name* **Year**, *Volume Number*, Page Range.

ISBN 978-3-0365-5999-5 (Hbk)
ISBN 978-3-0365-6000-7 (PDF)

Contents

About the Editors . . . . **vii**

Preface to "Advances in Plant Physiology of Abiotic Stresses" . . . . **ix**

Sara Álvarez and José Ramón Acosta-Motos
Miscellaneous Sets of Abiotic Stresses and Plant Strategies to Cope with Them
Reprinted from: *Agronomy* **2022**, *12*, 2727, doi:10.3390/agronomy12112727 . . . . **1**

Agnieszka Pawełek, Joanna Wyszkowska, Daniele Cecchetti, Mergi Daba Dinka, Krzysztof Przybylski and Adriana Szmidt-Jaworska
The Physiological and Biochemical Response of Field Bean (*Vicia faba* L. (partim)) to Electromagnetic Field Exposure Is Influenced by Seed Age, Light Conditions, and Growth Media
Reprinted from: *Agronomy* **2022**, *12*, 2161, doi:10.3390/agronomy12092161 . . . . **7**

Larisa I. Fedoreyeva, Elena M. Lazareva, Olga V. Shelepova, Ekaterina N. Baranova and Neonila V. Kononenko
Salt-Induced Autophagy and Programmed Cell Death in Wheat
Reprinted from: *Agronomy* **2022**, *12*, 1909, doi:10.3390/agronomy12081909 . . . . **31**

Mónica Pineda and Matilde Barón
Health Status of Oilseed Rape Plants Grown under Potential Future Climatic Conditions Assessed by Invasive and Non-Invasive Techniques
Reprinted from: *Agronomy* **2022**, *12*, 1845, doi:10.3390/agronomy12081845 . . . . **51**

María R. Conesa, Wenceslao Conejero, Juan Vera and Mª Carmen Ruiz-Sánchez
Root Reserves Ascertain Postharvest Sensitivity to Water Deficit of Nectarine Trees
Reprinted from: *Agronomy* **2022**, *12*, 1805, doi:10.3390/agronomy12081805 . . . . **71**

La Hoang Anh, Nguyen Van Quan, Vu Quang Lam, Akiyoshi Takami, Tran Dang Khanh and Tran Dang Xuan
Rice Momilactones and Phenolics: Expression of Relevant Biosynthetic Genes in Response to UV and Chilling Stresses
Reprinted from: *Agronomy* **2022**, *12*, 1731, doi:10.3390/agronomy12081731 . . . . **91**

Lingxin Xu, Hong Chen, Tingting Zhang, Yanan Deng, Junxin Yan and Lei Wang
Salicylic Acid Improves the Salt Tolerance Capacity of *Saponaria officinalis* by Modulating Its Photosynthetic Rate, Osmoprotectants, Antioxidant Levels, and Ion Homeostasis
Reprinted from: *Agronomy* **2022**, *12*, 1443, doi:10.3390/agronomy12061443 . . . . **107**

Lei Qin, Chengyuan Li, Dongbin Li, Jiayan Wang, Li Yang, Aili Qu and Qingfei Wu
Physiological, Metabolic and Transcriptional Responses of Basil (*Ocimum basilicum* Linn. var. *pilosum* (Willd.) Benth.) to Heat Stress
Reprinted from: *Agronomy* **2022**, *12*, 1434, doi:10.3390/agronomy12061434 . . . . **121**

Ramiro Alonso-Salinas, José Ramón Acosta-Motos, Estrella Núñez-Delicado, José Antonio Gabaldón and Santiago López-Miranda
Combined Effect of Potassium Permanganate and Ultraviolet Light as Ethylene Scavengers on Post-Harvest Quality of Peach at Optimal and Stressful Temperatures
Reprinted from: *Agronomy* **2022**, *12*, 616, doi:10.3390/agronomy12030616 . . . . **139**

Daniel Bañón, Beatriz Lorente, Sebastián Bañón, María Fernanda Ortuño, María Jesús Sánchez-Blanco and Juan José Alarcón
Control of Substrate Water Availability Using Soil Sensors and Effects of Water Deficit on the Morphology and Physiology of Potted *Hebe andersonii*
Reprinted from: *Agronomy* **2022**, *12*, 206, doi:10.3390/agronomy12010206 . . . . **155**

Virginia Birlanga, José Ramón Acosta-Motos and José Manuel Pérez-Pérez
Mitigation of Calcium-Related Disorders in Soilless Production Systems
Reprinted from: *Agronomy* **2022**, *12*, 644, doi:10.3390/agronomy12030644 . . . . **175**

Ademola Emmanuel Adetunji, Tomi Lois Adetunji, Boby Varghese, Sershen and Norman W. Pammenter
Oxidative Stress, Ageing and Methods of Seed Invigoration: An Overview and Perspectives
Reprinted from: *Agronomy* **2021**, *11*, 2369, doi:10.3390/agronomy11122369 . . . . **193**

Chao Han, Junna Chen, Zemao Liu, Hong Chen, Fangyuan Yu and Wanwen Yu
Morphological and Physiological Responses of *Melia azedarach* Seedlings of Different Provenances to Drought Stress
Reprinted from: *Agronomy* **2022**, *12*, 1461, doi:10.3390/agronomy12061461 . . . . **221**

About the Editors

Sara Álvarez

Sara Álvarez has published 40 indexed papers in 16 different journals (19 Q1), appearing as first author in 15 of them, in 6 as last author and in 8 as corresponding author; my total number of citations, according to Scopus, is 1295 (h-index Scopus: 23). In addition, He has written 1 book chapters and 16 scientific-technical reports; have participated in 63 international and national congresses; and have given 23 oral communications and 3 invited talks. He have participated in 35 courses, seminars, webinars, etc., related to my research. He is a member of a jury of a doctoral thesis, have refereed 89 papers in SCI journals; and have been a guest editor of 5 Special Issues. He has participated in 20 projects to date, five of which have been financed by European funds; He was the principal investigator of a FEADER project and earned three contracts of special relevance from various companies. Finally, He was awarded for the best communication in the VII Congreso Ibérico de Agroingeniería y Ciencias Horticolas in 2013.

José Ramón Acosta-Motos

José Ramón Acosta-Motos has published 42 articles in 16 different journals (27Q1; 3Q2; 1Q3), appearing as first author in 14 of them and as last author in 3; my H-index and total citations according to Google Scholar are 13 and 1637, respectively, and 11 and 1160, respectively, according to Scopus. He has published three reviews, appearing as first author in two of them. In addition, one of these reviews was awarded as the best publication of the year 2017 in the journal Agronomy. He has also published five books and two book chapters, appearing in six of them as first author. He is the co-director of two PhD programs and the member of the jury of two PhDs. He has participated in 32 international and national congresses, 13 as first author and 2 as last author, and have presented 5 oral communications. He has participated in 46 participations courses, seminars, webinars, etc., related to my research; 19 organizing events related to R&D&I; and 12 projects to date, five of which have been financed by European funds and the remaining seven by national funds.

Preface to "Advances in Plant Physiology of Abiotic Stresses"

This Special Issue is focused on "Advances in Plant Physiology of Abiotic Stresses". Several authors have contributed novel research, reviews, and opinion pieces to this Special Issue, covering a large number of topics associated with how plants respond to physical and chemical stresses associated with abiotic stress. In addition, most of these articles are framed within the context of the current effects of climate change. Abiotic stress, environmental stress, salinity, and extreme temperature are some of the keywords most used in the accepted manuscripts.

Sara Álvarez and José Ramón Acosta-Motos
Editors

agronomy

MDPI

Editorial

Miscellaneous Sets of Abiotic Stresses and Plant Strategies to Cope with Them

Sara Álvarez [1,*] and José Ramón Acosta-Motos [2,3,*]

1 Instituto Tecnológico Agrario de Castilla y León (ITACYL), 47071 Valladolid, Spain
2 Campus de los Jerónimos, Universidad Católica San Antonio de Murcia (UCAM), 30107 Guadalupe, Spain
3 Group of Fruit Biotechnology, Department of Fruit Breeding, CEBAS-CSIC, P.O. Box 164, 30100 Murcia, Spain
* Correspondence: alvmarsa@itacyl.es (S.Á.); jracosta@ucam.edu (J.R.A.-M.)

Citation: Álvarez, S.; Acosta-Motos, J.R. Miscellaneous Sets of Abiotic Stresses and Plant Strategies to Cope with Them. *Agronomy* **2022**, *12*, 2727. https://doi.org/10.3390/agronomy12112727

Received: 17 October 2022
Accepted: 1 November 2022
Published: 3 November 2022

Plant stress can be defined as any adverse situation or agent that can damage or block the metabolism, growth or development of a plant. The responses of plants can vary according to the frequency and intensity of the stress, as well as the growth phase of the plant under stress.

Throughout their life cycle, plants are subjected to a wide variety of conditions or potential stress factors. One of the main characteristics of plants is that they do not have the ability to travel, and therefore, they remain fixed to the substrate and they are planted through their root systems. In other words, adult individuals do not have the capacity to displace, but they are able to move and respond to stimuli such as light (phototropism) or the effect of gravity (gravitropism). This means that they are unable to escape from unfavourable conditions such as abiotic stress caused by non-living agents, which are differentiated from living agents such as herbivores, fungi, bacteria or viruses which cause biotic stress.

Depending on the kind of causative agent involved, abiotic stress may be divided into physical or chemical types of stress. Physical stress includes salinity (due to its osmotic component), water shortage, temperature extremes, excessive or insufficient irradiation, anaerobiosis caused by waterlogging or flooding, mechanical stress caused by wind or excessive soil compaction and stress induced by wounds or injuries. Chemical stress can be due to salinity (due to its ionic or toxic components), a lack of mineral nutrients and environmental contaminants. In addition to the particular damage caused by each of these stresses, they all facilitate the overproduction of reactive oxygen species (ROS) such as superoxide anions, hydrogen peroxide or hydroxyl radicals associated with oxidative stress. Plants have developed a series of adaptive responses against abiotic stresses and oxidative metabolisms to cope with the environment in which they live. More specifically, they have developed resistance mechanisms, which include avoidance mechanisms and tolerance mechanisms.

Throughout this Special Issue, some of the stresses mentioned above, as well as the resistance mechanisms implemented at different stages of the vegetative and reproductive cycles of plants (seed, seedling, adult plant or fruit), will be discussed, including: a review regarding oxidative stress, ageing and methods of seed invigoration; a study on how to control the availability of water in substrate and water deficit responses in variegata, a type of ornamental plant; an analysis of the postharvest quality of peach fruit under optimal and stressful temperatures; a review concerning nutritional disorders related to the macronutrient calcium and strategies to avoid them, such as soilless production systems; the salt stress resistance of soapwort, a native Chinese plant; responses to heat stress in basil plants; drought stress responses of Chinese *Melia azedarach* of different origins; responses to UV and cold stress in rice plants associated with the expression of genes associated with the production of phytoalexins; an electron microscopy analysis of histological changes in wheat plants due to salinity; an analysis of the resilience of oilseed grown in greenhouses under stress conditions which simulate the future climate change scenario, using a variety of techniques; and a study of the different responses of field bean to the application of electromagnetic fields [1–12].

Adetunji et al. focus on the description of different strategies to enable seed invigoration. The maintenance of the quality of seeds during long processes of the conservation of plant genetic resources is key to avoid the expected food crises due to climate change and the increase in the global population. It is therefore imperative to effectively implement these invigoration techniques without neglecting other equally important agronomic practices, such as plant cultivation, fertiliser supply and pest and disease monitoring. In general, these techniques fall into one of two groups: (1) the priming or pre-hydration of seeds in a solution to improve yields after harvest, or (2) reinvigoration after storage, which often consists of soaking seeds recovered from storage in a solution. Seed priming methods are further divided into traditional methods (e.g., hydropriming, osmopriming, redox priming or priming with biostimulants) and innovative methods (e.g., nanopriming, magnetopriming and priming with other physical agents) [1].

Bañón et al. sought to increase farmers' awareness of the need to avoid excessive water use, which can damage plants due to water and salt stress. Moreover, these bad practices not only result in damage to plants, but they can also result in leaching into soil, which is harmful to the ecosystem and leads to increases in production costs. With these premises, the authors conducted a study under greenhouse conditions regarding the morphological and physiological responses of *Hebe andersonii* to three volumetric water contents in substrate (49%, 39% and 32%). In addition, due to the type of irrigation applied, which consisted of adding small volumes of water, they avoided drainage through the pots, and by using dielectric sensors, they monitored the humidity of the substrate and thus ensured that the amount of water provided was correct for each treatment. By way of conclusions, the authors demonstrated better growth and water use efficiency in variegata when humidity (and therefore water supply to the substrate) is low and there is no drainage [2].

Alonso-Salinas et al. carried out a study framed within the search for solutions to food wastage, specifically by extending the shelf life of a climacteric fruit such as peach. The concept that allows us to differentiate between climacteric and non-climacteric fruits is fundamental in important aspects such as harvesting, marketing, and postharvest preservation systems. Climacteric fruits are able to continue ripening once separated from the plant, assuming they have reached a level of development characterised by its maximum size (physiological maturity). Non-climacteric fruit, on the other hand, only ripen on the plant, and irreversibly stop their ripening process once separated from the plant. The physiological processes related to the ripening of climacteric fruits are characterised by an initial autocatalytic production of ethylene accompanied by an increase in the production rate of carbon dioxide, which accelerates the respiration of these fruits and therefore their ripening, which may culminate in the premature deterioration of the product, which has to be avoided. The authors implemented a highly efficient system of ethylene elimination using machines that combine the use of two types of eliminators of this phytohormone: by oxidation, namely potassium permanganate incorporated in filters and patented by the company ("Nuevas Tecnologías Agroalimentarias" (KEEPCOOL), patent No. 2548787 (2016)), and ultraviolet radiation. The use of this system is shown to be efficient both at low temperatures and at high temperatures, for the conservation in peach, extending its shelf life [3].

In a review, Birlanga et al. summarised the physiological mechanisms involved in the leaf necrosis of lettuce leaves known as tipburn, and the apical rot of tomato fruit, known as blossom end rot. Both physiopathologies are due to the deficient translocation of calcium from the root system to the leaves and fruit, but their effects are aggravated by abiotic stresses related to climate change, such as drought, salinity and temperature extremes. Subsequently, the advantages and disadvantages of the different soilless cultivation systems were also assessed, as they allow for the use of remote sensors and their automation for the precise monitoring and control of nutritional balance throughout the crop cycle. Finally, they described a set of solutions to deal with the physiopathologies discussed above. At the level of a soilless cultivation system, the authors concluded that it is very important to know the tools that currently are available for more exhaustive control of the environmental conditions, as well as the precise physiological control of crops. The use of intelligent management practices and their digitalisation allow us to monitor not only the environmental conditions inside the greenhouse, but also possible critical states at the plant level, evapotranspiration, water and nutrient consumption and even pathogen detection. At the nutritional level, these solutions include the use of organic fertilisers, the exogenous application of calcium or the use of nanoparticles to achieve more precise nutrition and a reduction in the amount of nutrients supplied. At the physiological level, it is important to analyse the behaviour of certain phytohormones in the vascular system of a plant and their influence during its growing cycle. At the genetic level, the use of genomic tools has allowed the identification of quantitative trait loci (QTL) in several populations of recombinant inbred lines and the subsequent development of molecular markers linked to these QTL [4].

Xu et al. used a plant named *Saponaria officinalis*, which is native from China, as one of the best ways to improve the salinity level of soil, as more than 70% of the soil in the north-eastern area in China is currently affected by salinity. In this way, it is possible to avoid the use of physical and chemical methods, which are expensive and can also lead to the secondary salinisation of soils. In addition, they also built their work on the known evidence of the positive role of salicylic acid (SA) in regulating resistance to salt stress in plants. Considering that the exogenous effect of SA on *S. officinalis* subjected to salt stress has not been studied so far, the authors analysed the resistance of this plant to salt stress in pot trials by using different concentrations of NaCl and spraying different concentrations of SA on leaves. The results indicated that SA enhances the salt resistance capacity of *S. officinalis* by regulating its photosynthetic rate, osmoprotectant levels, antioxidant concentrations, and ionic homeostasis. In spite of this, the efficacy of SA was not linearly related to its concentration. Taken together, the authors' insights suggest that the use of 0.6 mmol L^{-1} of SA under salt stress conditions could be effective in reducing the injury caused in *S. officinalis* by saline soils [5].

Qin et al., due to the observed increase in heat stress as a consequence of global warming and its negative effect on plant growth and yield, decided to perform a novel assay on the morphological, metabolic and transcriptional responses of basil—a medicinal and edible plant with important nutritional and economic value—to heat stress. The results revealed that heat stress led to severe oxidative damage and a decrease in the photosynthesis of basil. Metabonomic screening indicated that, compared to the control group, 29 differentially accumulated metabolites (DAMs) were found after one day of heat treatment, and 37 DAMs were identified after three days of heat treatment. Furthermore, a transcriptomic analysis revealed that 15,066 and 15,445 differentially expressed genes (DEGs) were present after one and three days of heat treatment, respectively. Of these, 11,183 DEGs were common response genes under one and three days of heat treatment, including 5437 down-regulated and 6746 up-regulated DEGs. All DEGs were significantly enriched in various KEGG (Kyoto Encyclopedia of Genes and Genomes) pathways. Taken together, all of these responses generated useful information which improved our knowledge regarding plants' response mechanisms to heat stress, which in turn could be useful for crop breeding [6].

Han et al. studied the mechanisms of drought resistance in *Melia azedarach*, a tree native to China, by selecting eight provenances as research subjects and applying four levels of drought stress as treatments. The overall results, after the analysis of individual parameters such as seedling height, root–shoot ratio, soil diameter, relative water content, transpiration rate, gas exchange parameters, chlorophyll, malondialdehyde and superoxide dismutase contents, and linear combinations of all the variables studied by means of principal component analysis, revealed that the drought resistance of Henan Shihe and Jiangxi Xihu was stronger, while the drought resistance of Guangdong Luogang and Hubei

Shayang was weaker. The authors, based on the results obtained, could select provenances with strong or weak drought resistance for transcriptome sequencing in order to screen for drought-resistant genes and perform an extensive study at the molecular level [7].

Hoang Anh et al. conducted the first study demonstrating that momilactones A (MA) and B (MB) contribute, in addition to their well-known role in protecting rice crops against pathogens, to the resistance of rice to environmental stresses, including ultraviolet radiation and chilling conditions. Although momilactones with valuable biological activities have been identified in recent decades, the catalytic steps in the biosynthetic pathway of these compounds are poorly understood. This is especially true when the genes encoding the hydrolysis of MA to MB in rice are unknown. The discovery of these unclear concepts may result in new research approaches to enhance rice's resistance to adverse conditions. In addition, further research regarding the existence of new secondary compounds and relevant gene expressions also offers the potential to achieve other valuable results. For instance, phenolics and momilactones show potential for the treatment of several human diseases. Consequently, the concurrent accumulation of phenolics and momilactones in rice may hold significant potential for medicinal and pharmaceutical applications. The authors hope that this new study will contribute to the sustainable development goals of guaranteeing healthy lifestyles and ending world poverty and starvation, particularly in developing countries [8].

Fedoreyeva et al. analysed the responses to the salinity of wheat of a sensitive variety such as Zolotaya and a tolerant variety such as Orenburgskaya 22. In the sensitive variety, a higher accumulation of phytotoxic ions was accompanied by a higher level of ROS production. At a 150 mM NaCl concentration, there was an increase in the expression level of *TOR*, which is a negative controller of autophagosome generation (their formation contributes to protect a plant against salinity). The level of *TOR* expression in Zolotaya was higher in roots and leaves than in Orenburgskaya 22. The build-up of ROS generation resulted in autophagy and programmed cell death (PCD). PCD biomarkers indicated DNA ruptures in nuclei and chromosomes in metaphase, superficial phosphatidylserine localisation at the cellular level, and cytochrome c liberation in the cytoplasm, suggesting a mitochondrial route for cell death during salinity events. Based on findings from electron microscopy, the incidence of mitophagy in the root and leaf cells of wheat under salinity conditions was demonstrated [9].

Pineda and Baron analysed the behaviour of oilseed rape grown under environmental conditions that simulated severe and intermediate climate changes, establishing the current climatic conditions as the control treatment, with the principal goal being to assess the effect of climate change on the health status of this crop of agronomic relevance. Two approaches were applied (invasive and non-invasive techniques). Invasive quantitative methods relate to the absorbance of biochemical compounds. Non-invasive methods, such as the use of thermal, multi-colour fluorescence and hyperspectral reflectance sensors, are focused on the spectral properties of plants. The results show changes in lipid peroxidation, altered pigment content, photosynthesis, transpiration and secondary metabolite synthesis, which are more pronounced in severe treatments. In addition, the new metric, diseased broccoli index 3, could reveal the extent of lipid peroxidation in leaves, while the climatic stress index for brassicas showed early symptoms of stress caused by climate change in oilseed rape plants [10].

Conesa et al. investigated the postharvest susceptibility of early-ripening nectarines to water stress. Apart from a well-irrigated treatment (T-0), three water shortage treatments (by retention of irrigation) were implemented: T-1: early postharvest from June to July, T-2: late postharvest from August to September, and T-3: water stress during the entire postharvest period, from June to September. Throughout the study, soil water content (θ_v) and midday stem water potential (Ψ_{stem}) were determined. During winter dormancy, L-arginine, starch and phosphorus contained in the roots were studied. At harvest, the yield, fruit quality and metabolites were measured. The most interesting treatment, according to the results obtained, was T-2, which showed the highest percentage of cracked fruit.

Additionally, significantly lower values in leaves (−2.3 MPa) were observed in T-2. In a similar way, lower values of L-arginine and phosphorus were obtained in the roots of the T-2 trees, as compared to the T-0 trees. This allowed the authors to conclude that, despite the fact that the early postharvest stage is key for the implementation of regulated deficit irrigation approaches, the late postharvest period was also a period sensitive to water restriction, as the build-up of winter root reserves was depleted, resulting in limited yields [11].

Pawełek et al. analysed the exposure of young and old bean seeds to electromagnetic fields (EMFs) (50 Hz, 7 mT) and studied seed germination and seedling growth in different situations. The results obtained showed the stimulation of germination and early root growth of old seeds, sown in Petri dishes in continuous darkness, and the inhibition of the germination of young seeds sown in pots under long day conditions. The root growth of two-week-old seedlings of young seeds sown in pots was stimulated by an EMF treatment, while stem growth was suppressed. In conclusion, the results indicated that the EMF (50 Hz, 7 mT) priming of bean seeds might be a critical factor that influences germination, early growth, and cell processes, and it may have a positive influence on the capability of bean plants to grow and develop in more stressful conditions at later stages [12].

Presently, due to the palpable effects of anthropogenic climate change, some of the abiotic stresses plants suffer from are being aggravated. Abiotic stresses that are more closely related to climatic change are drought, salinity and extreme temperatures. Human activities are the main cause of the global warming observed since the mid-20th century. The latest intergovernmental panel on climate change (IPCC) report has estimated that 23% of total greenhouse gas emissions generated between 2007 and 2016 mainly came from agriculture, forestry and other anthropogenic uses of land resources. Concern about global climate change and the search for measures that minimise its environmental impact has brought about new strategies for the use and consumption of natural resources. The modernisation of agricultural practices, as well as the digitisation of farming activities and the development of new varieties that grow more efficiently and are better adapted to adverse conditions such as climate change, are the main focus of farmers' efforts to achieve sustainable agriculture.

Author Contributions: Both authors contributed equally to the writing of this editorial. All authors have read and agreed to the published version of the manuscript.

Acknowledgments: The second author would like to thank the UCAM-Santander Chair in Agri-Food Entrepreneurship for its financial support and Mario Fon for his help with English editing.

Conflicts of Interest: The authors declare no conflict of interest.

References

1. Adetunji, A.E.; Adetunji, T.L.; Varghese, B.; Sershen; Pammenter, N.W. Oxidative Stress, Ageing and Methods of Seed Invigoration: An Overview and Perspectives. *Agronomy* **2021**, *11*, 2369. [CrossRef]
2. Bañón, D.; Lorente, B.; Bañón, S.; Ortuño, M.F.; Sánchez-Blanco, M.J.; Alarcón, J.J. Control of Substrate Water Availability Using Soil Sensors and Effects of Water Deficit on the Morphology and Physiology of Potted *Hebe andersonii*. *Agronomy* **2022**, *12*, 206. [CrossRef]
3. Alonso-Salinas, R.; Acosta-Motos, J.R.; Núñez-Delicado, E.; Gabaldón, J.A.; López-Miranda, S. Combined Effect of Potassium Permanganate and Ultraviolet Light as Ethylene Scavengers on Post-Harvest Quality of Peach at Optimal and Stressful Temperatures. *Agronomy* **2022**, *12*, 616. [CrossRef]
4. Birlanga, V.; Acosta-Motos, J.R.; Pérez-Pérez, J.M. Mitigation of Calcium-Related Disorders in Soilless Production Systems. *Agronomy* **2022**, *12*, 644. [CrossRef]
5. Xu, L.; Chen, H.; Zhang, T.; Deng, Y.; Yan, J.; Wang, L. Salicylic Acid Improves the Salt Tolerance Capacity of *Saponaria officinalis* by Modulating Its Photosynthetic Rate, Osmoprotectants, Antioxidant Levels, and Ion Homeostasis. *Agronomy* **2022**, *12*, 1443. [CrossRef]
6. Qin, L.; Li, C.; Li, D.; Wang, J.; Yang, L.; Qu, A.; Wu, Q. Physiological, Metabolic and Transcriptional Responses of Basil (*Ocimum basilicum* Linn. var. *pilosum* (Willd.) Benth.) to Heat Stress. *Agronomy* **2022**, *12*, 1434. [CrossRef]
7. Han, C.; Chen, J.; Liu, Z.; Chen, H.; Yu, F.; Yu, W. Morphological and Physiological Responses of *Melia azedarach* Seedlings of Different Provenances to Drought Stress. *Agronomy* **2022**, *12*, 1461. [CrossRef]

8. Hoang Anh, L.; Van Quan, N.; Quang Lam, V.; Takami, A.; Dang Khanh, T.; Dang Xuan, T. Rice Momilactones and Phenolics: Expression of Relevant Biosynthetic Genes in Response to UV and Chilling Stresses. *Agronomy* **2022**, *12*, 1731. [CrossRef]
9. Fedoreyeva, L.I.; Lazareva, E.M.; Shelepova, O.V.; Baranova, E.N.; Kononenko, N.V. Salt-Induced Autophagy and Programmed Cell Death in Wheat. *Agronomy* **2022**, *12*, 1909. [CrossRef]
10. Pineda, M.; Barón, M. Health Status of Oilseed Rape Plants Grown under Potential Future Climatic Conditions Assessed by Invasive and Non-Invasive Techniques. *Agronomy* **2022**, *12*, 1845. [CrossRef]
11. Conesa, M.R.; Conejero, W.; Vera, J.; Ruiz-Sánchez, M.C. Root Reserves Ascertain Postharvest Sensitivity to Water Deficit of Nectarine Trees. *Agronomy* **2022**, *12*, 1805. [CrossRef]
12. Pawełek, A.; Wyszkowska, J.; Cecchetti, D.; Dinka, M.D.; Przybylski, K.; Szmidt-Jaworska, A. The Physiological and Biochemical Response of Field Bean (*Vicia faba* L. (partim)) to Electromagnetic Field Exposure Is Influenced by Seed Age, Light Conditions, and Growth Media. *Agronomy* **2022**, *12*, 2161. [CrossRef]

agronomy

MDPI

Article

The Physiological and Biochemical Response of Field Bean (*Vicia faba* L. (partim)) to Electromagnetic Field Exposure Is Influenced by Seed Age, Light Conditions, and Growth Media

Agnieszka Pawełek [1,*], Joanna Wyszkowska [2], Daniele Cecchetti [1], Mergi Daba Dinka [3], Krzysztof Przybylski [1] and Adriana Szmidt-Jaworska [1]

1 Department of Plant Physiology and Biotechnology, Faculty of Biological and Veterinary Sciences, Nicolaus Copernicus University in Toruń, Lwowska 1, 87-100 Toruń, Poland
2 Department of Animal Physiology and Neurobiology, Faculty of Biological and Veterinary Sciences, Nicolaus Copernicus University in Toruń, Lwowska 1, 87-100 Toruń, Poland
3 Department of Ecology and Biogeography, Faculty of Biological and Veterinary Sciences, Nicolaus Copernicus University in Toruń, Lwowska 1, 87-100 Toruń, Poland
* Correspondence: apawelek@umk.pl

Abstract: Research interest into the exposure of plants to magnetic fields (MF), including electromagnetic fields (EMF), has increased recently but results often vary depending on factors such as plant species and treatment dose. In this study, we exposed young (one year) and old (four years) field bean (*Vicia faba* L. (partim)) seeds to EMF (50 Hz, 7 mT) and observed seed germination and seedling growth under different conditions (growth media and light). The results indicated a stimulation by EMF of germination and early root growth of Petri dish-sown old seeds in continuous darkness and inhibition of germination of the pot-sown young seeds under long-day conditions. Root growth of two-week-old seedlings from pot-sown young seeds was stimulated by EMF treatment while their stem growth was inhibited. Some selected biochemical traits were examined, showing specific changes in membrane integrity, amylase activity, H_2O_2 levels, photosynthetic pigments, and content of the main groups of phytohormones, depending on seed age. The results indicate that priming of field bean seeds with EMF (50 Hz, 7 mT) could be a eustress factor that influences germination, early growth, and cellular activities and could positively influence the ability of field bean plants to grow and develop in more stressful conditions at later stages.

Keywords: electromagnetic field; seed priming; eustress; seed aging; germination; phytohormones

Citation: Pawełek, A.; Wyszkowska, J.; Cecchetti, D.; Dinka, M.D.; Przybylski, K.; Szmidt-Jaworska, A. The Physiological and Biochemical Response of Field Bean (*Vicia faba* L. (partim)) to Electromagnetic Field Exposure Is Influenced by Seed Age, Light Conditions, and Growth Media. *Agronomy* **2022**, *12*, 2161. https://doi.org/10.3390/agronomy12092161

Academic Editors: José Ramón Acosta-Motos and Sara Álvarez

Received: 7 August 2022
Accepted: 7 September 2022
Published: 11 September 2022

Publisher's Note: MDPI stays neutral with regard to jurisdictional claims in published maps and institutional affiliations.

1. Introduction

The quality of seeds and other planting materials is an essential component of successful crop production. The ability of seeds to develop into healthy seedlings and subsequently produce high yields influences decisions on crop production methods. In the face of adverse environmental conditions, farmers increase the use of inputs such as fertilizers and water to ensure substantial crop yield [1,2]. The quality of seeds and crop yield is expected to be significantly reduced by abiotic stress factors, such as drought and high temperatures resulting from climate change [3]. Seed aging is an important issue in crop production, which is associated with different internal (morphological, physiological, and genetic) and external (storage temperature and humidity) factors. Old seeds that are more likely to have low germinability or vigor are often discarded. These practices tend to increase both the economic and environmental costs to crop production [4].

Legumes (Fabaceae family) serve as a fundamental, worldwide source of high-quality food and feed, as well as help to sustain soil health during intensified crop production [5]. The limited spread of the cultivation of legumes is attributed to the reduction and instability in yield and its vulnerability to biotic and abiotic stress factors [6]. Therefore, it is essential to combine both genetic and agronomic techniques to produce high yields of legumes, especially under changing environmental conditions [7]. Field bean (*Vicia faba* L. var. minor; *Vicia faba* L. partim) is one of the major leguminous crops cultivated for animal feed, green forage, hay, silage, or green manure and also serves as an alternative to transgenic soybean. Due to its ability to fix nitrogen, field bean is increasingly becoming relevant in organic production systems to limit the use of mineral fertilizer inputs [8,9].

Seed priming aims to stimulate the physiological, biochemical, and metabolic activity of seeds in order to increase germinability, improve crop yield under unfavorable conditions, and reduce the adverse environmental footprint of crop production. Conventional priming techniques include hydropriming, osmopriming, biopriming, and chemopriming [10]. In modern seed improvement techniques, utilizing physical agents to improve crop production provides some advantages over conventional methods, particularly, the use of chemical substances [11]. Among the physical stimulation methods, the use of magnetic fields (MF), which may be in the form of static magnetic fields (SMF) or electromagnetic fields (EMF), is considered an eco-friendly, relatively cheap, and non-invasive technique with proven beneficial effects in plant production [12,13]. Seed priming with different MFs (SMF and EMFs) is assumed to enhance plant vigor by influencing a plethora of biochemical and molecular processes, such as changes in reactive oxygen species (ROS) production [14] and modulation of phytohormone balance [15,16].

EMF presence in the environment, particularly from man-made sources, has increased tremendously [17] and is regarded as a silent stressor with a great impact on developmental patterns in living systems, including plants [18]. A stressor may be regarded as either a eustress, a type of physiological priming with positive effects on the performance of a living organism, or a distress, which is a continuing factor whose high dose causes negative effects [19]. The treatment of plant materials, including seeds, with different doses of MFs, has shown positive results in a number of crop plants [12]. Pea (*Pisum sativum*) seeds treated with MF (30 mT, 85 mT) expressed faster water uptake and enhanced germination [20], as well as faster early growth [16]. In the case of broad beans (*Vicia faba*), the same MF application (30 mT, 85 mT) also had a positive effect on seed emergence and crop yield during field cultivation but the efficiency of this treatment was dependent on weather conditions [21]. Soybean (*Glycine max*) seeds exposed to MF (10 Hz, 1.5 μT) produce plants with a greater number of leaves, pods, and seeds and also improved the length of the pods and weight of seeds [22]. Beneficial effects of the pre-sowing treatment of red clover (*Trifolium pratense*) seeds on their agronomic performance have been reported where seed treatment with EMF (5.28 MHz, 0.74 mT) significantly increased the number of root nodules [23] and changed the amount of flavonoids in the root exudates important for communication with nitrogen-fixing rhizobacteria [24]. Treatment of seeds with different doses of MF has also been shown to alleviate the harmful effects of various abiotic stresses. The enhancement of germination rate and seedling growth under different salinity levels was reported for magneto-primed seeds of chickpea (*Cicer arietinum*) (with SMF of 100 mT) [25], as well as maize (*Zea mays*) and soybean (with SMF of 200 mT) [26].

On the other hand, not all studies resulted in positive outcomes of agronomic importance. Aguilar et al. [27] reported that the growth response of maize seeds to EMF (60 Hz, 20–100 mT) treatment was strongly cultivar-dependent, showing positive, neutral, or even negative effects compared to the controls. The pre-sowing exposure of pea seeds to EMF (60–180 mT, 50 Hz) did not affect the chlorophyll content of one-month-old plants despite the stimulation of their stems and roots [28]. The treatment of spring wheat (*Triticum aestivum*) seeds with MF (30 mT) also did not produce any effect on yield [29]. Cakmak et al. [30] reported that the exposure of seeds of bean (*Phaseolus vulgaris*) and wheat to SMF (7 mT) significantly stimulated dry biomass accumulation of wheat but not of beans. The existing reports suggest that the positive response to MF treatment may only occur under specific conditions dictated by plant species and growth environment [13]. Moreover, a wide range of MFs are used in seed priming and variations in their duration and intensity could change a positive effect on the treated plant to either a negative or no effect [31].

These inconsistencies in reported studies serve as a drawback to the application of MFs as means of priming seeds. Therefore, further investigation is needed to determine the optimum conditions under which exposure of seeds to MFs could be used as sustainable means of enhancing crop production.

This study aims to determine how priming the seeds of field bean (*Vicia faba* L. (partim)) of different ages (1-year-olds vs. 4-year-olds), with EMF (50 Hz, 7 mT), affects their germination and growth in different media and under two distinct light conditions. Additionally, the mechanism of EMF action in field bean tissues is investigated by analyzing some biochemical traits, including membrane integrity, α-amylase activity, H_2O_2 levels, photosynthetic pigment content, and changes in phytohormones controlling growth (indole-3-acetic acid, IAA; abscisic acid, ABA; gibberellins, GAs) and stress responses (jasmonic acid, JA; salicylic acid, SA) in plants. The outcome of these studies will, thus, contribute to understanding better the effect of MF exposure on plants under different conditions.

2. Materials and Methods

2.1. Plant Material and Cultivation Conditions

Two groups of seeds of field bean, *Vicia faba* L. (partim) of Polish variety Fernando, were used in this study: (1) young seeds of one year old, and (2) old seeds of four years old. Both groups of seeds were harvested in the Kuyavian-Pomeranian region in Poland and stored in similar controlled conditions of temperature (approx. 15 °C), humidity (approx. 35%), and light.

2.2. Exposure to Electromagnetic Field

Before sowing in Petri dishes and in pots, dry seeds were exposed for 24 h to a sinusoidal electromagnetic field (50 Hz, 7 mT), generated in a coil of 0.1 m in radius (Elektronika i Elektromedycyna Sp. J., Otwock, Poland). A detailed description of the exposure system (Figure 1) is presented in previously published papers [32,33]. The control groups (without EMF exposure) were put in a sham exposure setup and were affected by only the local geomagnetic field. The EMF was measured before each experiment with a Gauss meter (Model GM2, AlphaLab, Inc., UT, USA). The non-homogeneity of EMF within the area where the seeds in falcon tubes were kept was less than 4%. The temperature during all experiments was set to 24 ± 1 °C and monitored using thermocouples.

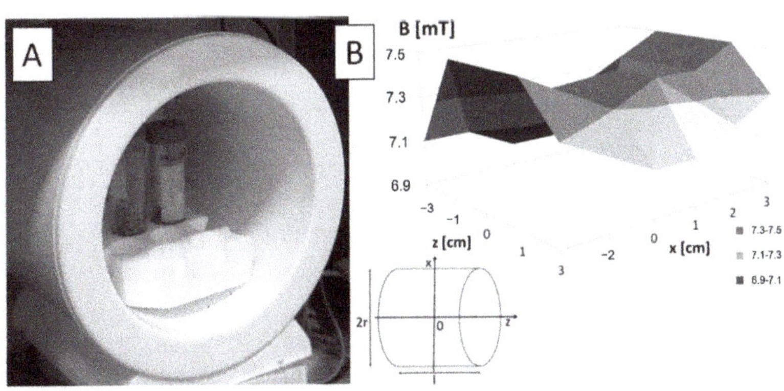

Figure 1. Exposure system. (**A**) Plant material in the coil. (**B**) The average magnetic flux density distribution inside the solenoid along the Z and X axes. Inset shows the coordinate system. 2r—diameter of the solenoid, B—magnetic flux density vector, l –coil length.

2.3. Cultivation Conditions

Immediately after exposure to EMF, the dry seeds were either germinated in Petri dishes for six days or in plastic pots (11 × 11 × 21.5 cm) filled with a universal substrate

(Substral, Warszawa, Poland) for two weeks. For the tests in Petri dishes, seeds were first surface-sterilized for five minutes using a mixture of 30% hydrogen peroxide and 96% ethanol (1/1 (v/v)), then washed 10 times for 30 s in sterile distilled water. Afterwards, the seeds were sown on the 9 cm Petri dishes lined with sterile filter paper moistened with 8 mL of sterile distilled water. For the Petri dish tests, seed germination and seedling growth were carried out in cultivation rooms at 21 ± 0.5 °C under two conditions: (1) long-day constituting 15 h of light and 9 h of darkness and (2) continuous darkness. The study involving seed germination and plant growth in substrate-filled pots was only carried out under long-day conditions. The light source produced a photosynthetic photon flux of 30 μmol $\times$ m^{-2} $\times$ s, and the humidity was 60%. In the Petri dish tests, three replicates of 50 seeds each were used with each Petri dish holding 10 seeds, while in the pot tests, five replicates of 10 seeds each were used.

2.4. Seed Germination Analysis

The germination counts of seeds sown in Petri dishes and in the pots were performed daily. Germination kinetics of young and old field bean seeds was presented on graphs and expressed as the total proportion (%) of germinated seeds on a particular day from sowing. Additionally, five different germination parameters were evaluated using the following formulas reported by Ranal and Santana [34] (parameters 1, 3, 4), Kader [35] (parameter 2), and Coolbear et al. [36] (parameter 5):

$$\text{Germinability, G} = 100\ (N/S) \tag{1}$$

where: N is the number of total germinated seeds at end of the counts; S is the number of initial seeds used;

$$\text{Germination index, GI} = (6 \times N_1) + (5 \times N_2) + \ldots + (1 \times N_6) \tag{2}$$

where: $N_1, N_2 \ldots N_6$ are the number of germinated seeds on the first, second, and subsequent days, respectively; 6, 5, ... , 1 are the weights given to the days of germination;

$$\text{Mean germination time, MGT} = (N_1T_1 + N_2T_2 + \ldots + N_nT_n)/(N_1 + N_2 + \ldots + N_n) \tag{3}$$

where: N is the number of seeds germinated on each day; T is the time point in days;

$$\begin{gathered}\text{Coefficient of the velocity of germination,}\\ \text{CVG} = 100\ [(N_1 + N_2 + \ldots + N_n)/(N_1T_1 + N_2T_2 + \ldots + N_nT_n)]\end{gathered} \tag{4}$$

where: N is the number of seeds germinated each day; T is the number of days from sowing corresponding to N;

$$\begin{gathered}\text{Median response (time to reach 50\% of final germination),}\\ \text{t50} = Ti + [(((N + 1)/2) - Ni)\ (Tj - Ti)]/(Nj - Ni)\end{gathered} \tag{5}$$

where N is the final number of seeds germinated and Ni and Nj are the total number of germinated seeds in adjacent counts at time Ti and Tj, respectively, when $Ni < (N + 1)/2 < Nj$.

2.5. Measurement of Seedling Morphological Parameters

The physiological traits of the seedlings (root and epicotyl/stem length; fresh and dry mass) grown in Petri dishes and in the pots were determined with a millimetric ruler on the 6th and 14th day after sowing, respectively. Additionally, the seedlings growing in Petri dishes, under long-day conditions and in continuous darkness, were analyzed separately for seedlings with fully emerged roots and epicotyls, and those with only protruded roots. The seedlings were oven-dried for 48 h at 70 °C to prepare for dry mass estimation.

2.6. Plant Material Collections

The plant materials collected were germinating seeds from the Petri dish tests and plant roots and leaves from the pot tests. They were frozen in liquid nitrogen and stored at −80 °C until further use. In the Petri dish tests, the germinating seeds were extracted

24 h after sowing, and this was repeated every 24 h up to the sixth day and used in the analyses described in Sections 2.8 and 2.9. On the other hand, the plants growing in the pots were carefully removed from the pots 14 days after sowing. Their roots and leaves were separated, cleaned, and used for the analyses described in Sections 2.10 and 2.11.

2.7. Examination of Seeds Water Uptake and Membrane Integrity

Water uptake (endpoint analysis) and membrane integrity of seeds (control and EMF-treated) were determined in four replicates with each replicate having 25 seeds. The dry seeds were weighed and soaked in 25 mL of deionized water at 23 °C for 24 h in darkness. They were then removed from the water, blotted dry, and weighed again. The change in weight due to imbibition was expressed as the amount of water absorbed per seed dry weight which was calculated by the following formula:

$$\text{Water uptake [\%]} = ((\text{fresh weight of seed} - \text{dry weight of seed}) \times 100)/(\text{dry weight of seed})$$

The membrane integrity of seeds was measured based on their ion leakage. After soaking in deionized water for 24 h, the seed leachate was decanted off in a clean beaker and the electrical conductance of the leachate (mS/cm) was measured at room temperature using a digital conductivity meter (Elmetron CX-105, Zabrze, Poland).

2.8. α-Amylase Assay

The activity of α-amylase was quantified using a slightly modified 3,5-dinitrosalicylic acid (DNS) method [37]. Young and old field bean seeds germinating in Petri dishes under long-day conditions were used for this analysis and were collected every 24 h for 6 days beginning from the first 24 h after sowing. They were then homogenized in liquid nitrogen and 100 mg of each sample tissue was extracted with 2 mL of chilled distilled water, followed by centrifugation at 4500× *g* for 15 min. Next, 1 mL of crude enzyme extract was mixed with 0.5 mL of phosphate buffer (0.5 M; pH 6.9) and the reaction was initiated by adding 1 mL of the 1% starch solution as a substrate, incubated for 5 min at 37 °C, and terminated by adding 1% DNS (0.5 mL). Afterwards, the samples were incubated at 100 °C for 5 min. After cooling the samples on ice, 2.5 mL of distilled water was added and the amount of reducing sugar released was measured using a spectrophotometer (UV-160 1PC, Shimadzu, Kyoto, Japan) at 540 nm with maltose as the reducing sugar standard. The activity of α-amylase was calculated from a standard curve and expressed as mg of maltose per g of fresh weight (FW).

2.9. Hydrogen Peroxide Measurement

Hydrogen peroxide (H_2O_2) content was assayed in young and old field bean seeds germinated in Petri dishes under long-day conditions. The seeds were collected every 24 h for 6 days (0–6th day) and ground in liquid nitrogen. The 100 mg samples were homogenized with 0.5 mL of 0.1% (*w*/*v*) trichloroacetic acid in an ice bath with shaking, followed by centrifugation at 16,000× *g* for 10 min at 4 °C. Afterwards, 0.3 mL of supernatant was mixed with 0.3 mL of 0.1 M sodium phosphate buffer (pH 7.6) and 0.6 mL of 1 M KI. The samples were incubated in darkness for 1 h at room temperature. Next, the absorbance of the solution was measured at 390 nm in a spectrophotometer (UV-160 1PC, Shimadzu, Kyoto, Japan) and the content of H_2O_2 was determined using a standard curve of 0–20 μM H_2O_2.

2.10. Determination of Photosynthetic Pigments

Chlorophylls (a, b, and total) and total carotenoids (xanthophylls + b-carotene) concentrations were determined from leaf materials (100 mg fresh weight) of two-week-old plants growing in pots, then ground in a pre-chilled mortar and extracted in 1 mL of 80% acetone overnight in the dark at 4 °C. Afterwards, the samples were centrifuged at 12,000× *g* for 10 min at 4 °C and the supernatant was collected and diluted in cold 80% acetone. The absorbance of the extract was measured at 664, 647, and 470 nm using a spectrophotometer (UV-160 1PC, Shimadzu, Kyoto, Japan) and the pigment concentrations were calculated according to Lichtenthaler [38]. Three biological repetitions were performed and the data are presented as mean ± standard error (SE).

2.11. *Quantification of Phytohormones by Liquid Chromatography–Tandem Mass Spectrometry (LC-MS/MS)*

Mass spectrometry combined with liquid chromatography (LC-MS/MS) and the QuEChERS-based extraction method [39] were used to determine the concentrations of endogenous indole-3-acetic acid (IAA), abscisic acid (ABA), gibberellins (GA_1, GA_3), salicylic acid (SA), and jasmonic acid (JA). For this analysis, the roots and leaves of two-week-old pot-grown plants were ground in liquid nitrogen, and each sample (100 mg) was extracted overnight at 8 °C with shaking in a buffer containing 80% acetonitrile, 5% formic acid (FA), 15% water, 1mM butylhydroxytoluene (BHT), and stable isotope-labeled internal standards (5 ng/mL d2IAA; 5 ng/mL d6ABA; 10 ng/mL d2GA_1; 15 ng/mL d2GA_3; 10 ng/mL d4SA; 10 ng/mL d5JA). Afterwards, a salt mixture (magnesium sulfate heptahydrate and sodium chloride, 1/3 [m/m]) was added to the samples, vigorously mixed for 3 min, and centrifuged at 10,000× *g* for 10 min. The obtained supernatant was purified by adding sodium sulfate anhydrous and mixed vigorously for 1 min, followed by centrifugation (10,000× *g* for 10 min). Collected supernatants were dried with nitrogen gas, dissolved in 1 M FA (1 mL), and subjected to solid phase extraction (SPE) using polymer-based columns (Discovery® DSC-18 SPE Tube, Supelco, Darmstadt, Germany). The DSC-18 columns were activated and conditioned by using 100% methanol and 1M FA, respectively. The column-loaded samples were washed with 1 M FA, 1 M FA with 20% (*v*/*v*) methanol, and eluted with 80% methanol (*v*/*v*). Next, samples were lyophilized in CentriVap Centrifugal Concentrator (Labconco Corporation, Kansas City, MO, USA), resuspended in 100 µL of 35% methanol (*v*/*v*), and collected in glass vials. The concentrations of phytohormones were determined using LC-MS/MS Nexera UHPLC and LCMS-8045 integrated system (Shimadzu Corporation, Kyoto, Japan). Chromatographic separation of samples was performed on a reversed-phase C18 column (150 × 2.1 mm, 2.6 µm, Kinetex®, Phenomenex Inc., Torrance, CA, USA). Water with 0.1% FA (*v*/*v*) (A) and methanol with 0.1% FA (*v*/*v*) (B) were used as the mobile phase. The separation was carried out on a linear gradient of 40–90% (*v*/*v*) methanol for 7 min at 30 °C. The flow rate and injection volume were 0.4 mL/min and 10 µL, respectively. In mass spectrometry, the samples were subjected to negative and positive electrospray ionization (ESI) (4 kV voltage) and ions were fragmented by collision-induced dissociation (CID). Analysis of individual phytohormones was based on multiple reactions monitoring (MRM) with the LabSolutions workstation for LCMS-8045 (Shimadzu Corporation, Kyoto, Japan). Three biological repetitions were performed and the data are presented as mean ± standard error (SE).

2.12. *Statistical Analysis*

All data were tested for normality and homogeneity, and the level of significance was set at $p < 0.05$. The results were given as mean ± standard error of the mean (SE). The data obtained from the germination kinetics, other germination parameters (G, GI, MGT, CVG, t50), and the morphometric traits of field bean seedlings growing in Petri dishes and in pots, were analyzed with a one-way ANOVA test using the R package version 4.0.4 (R Core Team, Vienna, Austria) [40].

The Mann-Whitney test was applied for the analysis of water uptake, electrolyte leakage, and photosynthetic pigments. An unpaired *t*-test with Welch's correction was used to determine whether exposure to EMF had an effect on the activity of α-amylases and the amount of hydrogen peroxide. Similarly, the phytohormones were analyzed by applying the Students' unpaired *t*-test. These data were analyzed with the SPSS 25.0 package (IBM Inc., Armonk, NY, USA).

3. Results

3.1. *Effect of Seed Age and EMF Exposure on Germination Kinetics of Petri Dish-Sown and Pot-Sown Field Bean Seeds*

The germination kinetics of the control groups was first examined in order to observe the effects of field bean seed age on the germination process. In the Petri dish studies, young seeds germinated for 6 days at a significantly higher rate than old seeds under both the long-day conditions ($p < 0.01$) and in continuous darkness ($p < 0.05$) (Figure 2A,B). However, the germination kinetics of the control group of pot-sown seeds (young and old) under long-day conditions observed for 14 days, was not influenced by seed age (Figure 2C).

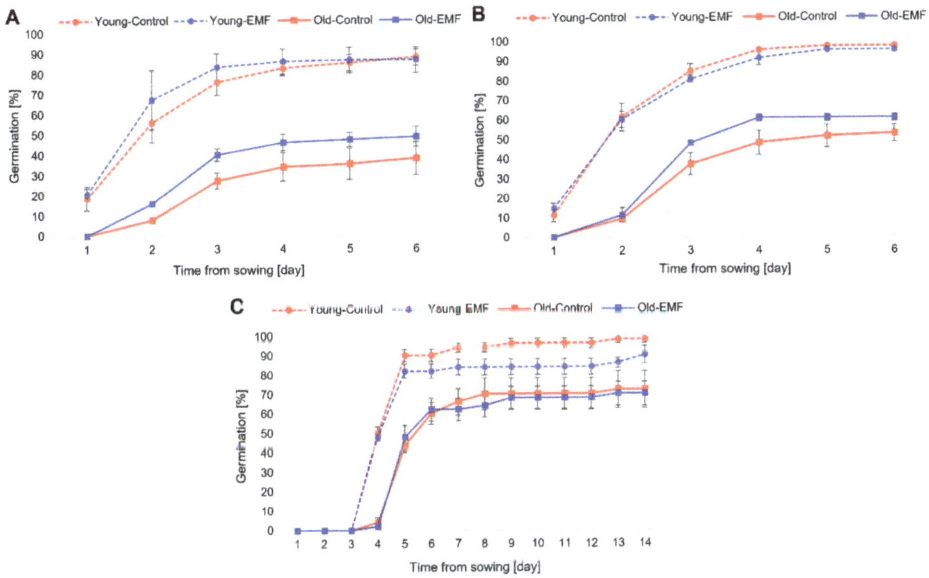

Figure 2. Germination kinetics of young and old field bean seeds sown in Petri dishes under long-day conditions (**A**), in Petri dishes in continuous darkness (**B**), and in pots under long-day conditions (**C**). The points represent mean values ± standard error (SE). Petri dish and pot tests were replicated three times (n = 50 per replicate) and five times (n = 10 per replicate), respectively. The levels of significant differences ($p < 0.05$; $p < 0.01$) between particular experimental groups are indicated in the text in the Results section.

When analyzing the results for the effect of EMF treatment in all the variants of the study, the germination kinetics in continuous darkness of Petri dish-sown old seeds exposed to EMF significantly improved ($p < 0.05$) compared to the untreated control (Figure 2B). However, in the other experimental variants (Petri dish-sown young and old seeds under long-day conditions; Petri dish-sown young seeds in continuous darkness; and pot-sown young and old seeds under long-day conditions), EMF exposure did not affect the germination kinetics (Figure 2A–C).

Among the EMF-treated groups, Petri dish-sown young seeds under long-day conditions germinated at a significantly higher rate ($p < 0.05$) compared to old seeds in the same conditions (Figure 2A), thus indicating again the moderating effect of seed age on germination kinetics under certain conditions. On the other hand, seed age did not affect the germination kinetics of the EMF-treated groups of Petri dish-sown seeds in continuous darkness and pot-sown seeds under long-day conditions (Figure 2B,C).

3.2. Analysis of Germination Parameters (G, GI, MGT, CVG, t50)

In addition to the analysis of germination kinetics (Figure 2A–C), other parameters (G, GI, MGT, CVG, and t50) were evaluated to determine the germination changes in both the EMF-treated and control groups of young and old seeds.

Within the control groups, seed age was found to significantly affect these germination parameters in all of the studied media and light conditions. The germinability (G) of old seeds was significantly lower than that of young seeds by the following margins within the control groups: (a) 57% less in Petri dishes under long-day conditions; (b) 46% less in Petri dishes in continuous darkness; and (c) 27% less in pots under long-day conditions (Tables 1 and 2). Further comparison of the control groups of young and old seeds germinated in Petri dishes under long-day conditions showed that the realized values of GI and CVG of the old seeds were significantly lower by 65% and 25%, respectively, while their MGT and t50 values were significantly higher than young seeds by 33% and 41%, respectively (Table 1). Similarly, old seeds germinated in Petri dishes in continuous darkness obtained GI and CVG values significantly lower by 56% and 26%, respectively and MGT and t50 values significantly higher by 35% and 48%, respectively when compared to young seeds in the control groups (Table 1). When comparing the germination parameters of the control groups of pot-sown young and old seeds, the GI and CVG parameters of the old seeds were significantly lower by 33% and 15%, respectively, while their t50 value was significantly higher by 16% (Table 2). These marked differences between the germination parameters of young and old seeds in the control groups clearly indicate that old seeds without any exposure to EMF germinate at a much lower rate than young seeds in all the analyzed growth media and light conditions.

Table 1. The effects of EMF on germination parameters of field bean seeds cultivated in Petri dishes for 6 days under long-day conditions and in continuous darkness.

	Long-Day Conditions				Continuous Darkness			
	Young—Control	Young—EMF	Old—Control	Old—EMF	Young—Control	Young—EMF	Old—Control	Old—EMF
G	88 ± 4	86.66 ± 6.36	38 ± 8.08 $^{\&}$	48.66 ± 4.81 $^{\#\#}$	97.33 ± 0.66	95.33 ± 0.66	52.66 ± 4.37 $^{\&}$	60.66 ± 1.76 $^{\#\#\#\#}$
GI	203.33 ± 0.82	215.33 ± 1.27	71.33 ± 1.51 $^{\&}$	99 ± 0.81 $^{\#\#}$	223.67 ± 0.37	218.67 ± 0.27	99.33 ± 1.12 $^{\&}$	120.67 ± 0.17 $^{\#\#\#\#}$
MGT	2.38 ± 0.1	2.04 ± 0.05	3.17 ± 0.1 $^{\&}$	2.92 ± 0.05 $^{\#\#}$	2.41 ± 0.06	2.41 ± 0.07	3.25 ± 0.08 $^{\&}$	3.02 ± 0.04 $^{\#\#}$
CVG	42.41 ± 0.43	49.12 ± 0.26	31.77 ± 0.34 $^{\&}$	34.33 ± 1.68 $^{\#\#}$	41.68 ± 0.23	41.63 ± 0.28	30.9 ± 0.23 $^{\&}$	33.17 ± 0.13 $^{\#}$
t50	1.8 ± 0.15	1.64 ± 0.18	2.53 ± 0.16 $^{\&}$	2.39 ± 0.03 $^{\#}$	1.8 ± 0.12	1.77 ± 0.07	2.67 ± 0.09 $^{\&}$	2.53 ± 0.05 $^{\#\#\#}$

Note. The presented values are the average of three independent experiments ± SE. "&" indicates significant differences between control groups ($^{\&}$ $p < 0.05$); "#" indicates significant differences between EMF-treated groups ($^{\#}$ $p < 0.05$, $^{\#\#}$ $p < 0.01$, $^{\#\#\#}$ $p < 0.001$, $^{\#\#\#\#}$ $p < 0.0001$).

Table 2. The effects of EMF on germination parameters of field bean seeds cultivated for 14 days in pots under long-day conditions.

	Young—Control	Young—EMF	Old—Control	Old—EMF
G	98 ± 2	90 ± 4.47	72 ± 9.69 $^{\&}$	70 ± 6.32 $^{\#}$
GI	99.8 ± 2.4	89.2 ± 3.84 *	66.8 ± 7.78 $^{\&}$	65 ± 5.67 $^{\#\#}$
MGT	4.81 ± 0.14	5.06 ± 0.27	5.74 ± 0.43	5.69 ± 0.19
CVG	20.83 ± 0.58	19.98 ± 1.11	17.76 ± 1.11 $^{\&}$	17.65 ± 0.6
t50	4.07 ± 0.08	4.03 ± 0.09	4.73 ± 0.2 $^{\&}$	4.51 ± 0.24

Note. The presented values are the average of five independent repetitions ± SE. The symbol (*) indicates significant differences between EMF-treated and control groups (* $p < 0.05$); "&" indicates significant differences between control groups ($^{\&}$ $p < 0.05$); "#" indicates significant differences between EMF-treated groups ($^{\#}$ $p < 0.05$, $^{\#\#}$ $p < 0.01$).

Young seeds exposed to EMF and germinated in pots obtained a GI value, which is 11% lower than that of the untreated control (Table 2). However, the analysis of germination parameters in the remaining experimental variants (Petri dish-sown young and old seeds under long-day conditions and in continuous darkness, and pot-sown old seeds) showed that pre-sowing exposure of seeds to EMF did not affect their germination parameters (Tables 1 and 2).

The EMF-treated groups of young and old seeds were analyzed to determine the influence of seed age similar to the analysis of the control groups. The comparison of the germination parameters within these groups follows a similar trend to the analysis within the control groups (apart from differences between CVG and t50 parameters for pot-sown seeds) (Tables 1 and 2), which confirms that the germination rate reduces with increasing seed age.

The above analysis of the various germination parameters indicates that the germination process of field beans is negatively affected by seed age irrespective of the light conditions and growth media. On the contrary, pre-sowing exposure of field bean seed to EMF is found to affect germination differently and depends on seed age, light conditions, and growth media.

3.3. Morphometric Analysis of Seedling's Growth in Control Groups

The growth parameters were measured in 6-day-old and 14-day-old seedlings growing in Petri dishes and in substrate-filled pots, respectively. Different changes in the morphometric parameters were observed depending on seed age, light conditions, and growth media.

To assess seed age's influence on early seedling growth, the control groups in different growth media were analyzed. The control group of old seeds germinated in Petri dishes under long-day conditions did not develop seedlings with fully emerged roots and epicotyls, in contrast to the young seeds (Table 3). Analysis within the control group for differences in delayed growth (seedlings with only roots protruded) showed that old seeds germinated in Petri dishes under long-day conditions had shorter root lengths (–22%) than roots from young seeds.

Further assessment of the control groups showed that seedlings from old seeds growing in Petri dishes in continuous darkness had reduced growth parameters compared to young seeds (Table 4) at the following rates: 27% shorter root length of fully emerged seedlings, 14% less seedling fresh weight, 29% shorter root length of seedlings with only roots protruded, and 25% less emerged embryo fresh weight.

These results show that seedling growth in Petri dishes under long-day conditions and in continuous darkness was slower in seedlings from old seeds, and may be related to the similar trend of reduced rate of germination of old seeds compared to young seeds. On the contrary, analysis of the control groups of seedlings growing in substrate-filled pots shows that there are no differences in growth parameters between seedlings from young and old seeds (Table 5).

Table 3. The effects of EMF on growth parameters of field bean seedlings cultivated in Petri dishes for 6 days under long-day conditions.

Seedling Growth Parameters		Long-Day Conditions			
		Young—Control	Young—EMF	Old—Control	Old—EMF
Fully emerged roots and epicotyls	Epicotyl length (mm)	7.48 ± 0.37	8.27 ± 0.66	-	6.5 ± 0.5
	Root length (mm)	17.45 ± 1.22	14.85 ± 1.05	-	11.5 ± 0.5
	Seedling fresh weight (mg)	157.70 ± 8.15	142.77 ± 7.97	-	145.5 ± 7.5
	Seedling dry weight (mg)	30.33 ± 4.81	47 ± 22.55	-	26 ± 0.1
Only roots protruded	Root length (mm)	8.38 ± 0.53	8.87 ± 0.58	6.56 ± 0.38 &	7.13 ± 0.3 #
	Emerged embryo fresh weight (mg)	76.42 ± 3.4	83.04 ± 3.18	67.21 ± 4.14	75.97± 3.55

Note. The presented values are the average of three independent repetitions ± SE. "&" indicates significant differences between control groups (& $p < 0.05$); "#" indicates significant differences between EMF-treated groups (# $p < 0.05$).

Table 4. The effects of EMF on growth parameters of field bean seedlings cultivated in Petri dishes for 6 days in continuous darkness.

Seedling Growth Parameters		Continuous Darkness			
		Young—Control	Young—EMF	Old—Control	Old—EMF
Fully emerged roots and epicotyls	Epicotyl length (mm)	9.61 ± 0.42	11 ± 0.54 *	9.22 ± 0.74	10 ± 0.62
	Root length (mm)	16.83 ± 1.06	15.81 ± 0.97	12.22 ± 0.85 &	11.35 ± 0.79 ##
	Seedling fresh weight (mg)	172.55 ± 5.25	179.35 ± 7.53	147.83 ± 7.31 &	142.96 ± 6.88 ##
	Seedling dry weight (mg)	32.66 ± 6.12	29.33 ± 7.13	17.33 ± 2.85	20.33 ± 0.88
Only roots protruded	Root length (mm)	9.11 ± 0.46	9.72 ± 0.46	6.44 ± 0.31 &&&&	7.84 ± 0.25 ***,##
	Emerged embryo fresh weight (mg)	104.6 ± 3.47	108.62 ± 3.22	78.67 ± 3.98 &&&&	95.72 ± 3.25 **,##

Note. The presented values are the average of three independent repetitions ± SE. The symbol (*) indicates significant differences between EMF-treated and control groups (* $p < 0.05$, ** $p < 0.01$, *** $p < 0.001$); "&" indicates significant differences between control groups (& $p < 0.05$, &&&& $p < 0.0001$); "#" indicates significant differences between EMF-treated groups (## $p < 0.01$).

Table 5. The effects of EMF on growth parameters of field bean seedlings cultivated for 14 days in pots under long-day conditions.

Seedling Growth Parameters	Young—Control	Young—EMF	Old—Control	Old—EMF
Stem length (cm)	31.02 ± 0.83	27.95 ± 1.23 *	29.24 ± 1.06	26.79 ± 1.39
Root length (cm)	20.69 ± 0.54	23.43 ± 0.93 **	22.28 ± 0.62	21.5 ± 1.08
Stem fresh weight (g)	3.68 ± 0.14	3.33 ± 0.17	3.6 ± 0.25	3.05 ± 0.20
Root fresh weight (g)	1.87 ± 0.09	1.66 ± 0.09	1.61 ± 0.10	1.61 ± 0.13
Stem dry weight (mg)	232.90 ± 9.19	205.19 ± 14.53	216.06 ± 60.79	190.78 ± 16.46
Root dry weight (mg)	115.41 ± 8.85	102.64 ± 5.96	112.70 ± 27.04	99.88 ± 7.44

Note. The presented values are the average of five independent repetitions ± SE. The symbol (*) indicates significant differences between EMF-treated and control groups (* $p < 0.05$, ** $p < 0.01$).

3.4. EMF Treatment Effect on Growth Parameters

Pre-sowing EMF treatment of both young and old seeds germinated in Petri dishes under long-day conditions had no statistically significant effect on the analyzed growth parameters of their respective seedlings compared to the untreated control group (Table 3). However, it can be hypothesized that EMF treatment of old seeds could be responsible for the enhanced growth of seedlings in Petri dishes under long-day conditions to the point of the appearance of fully emerged roots and epicotyls, a phenomenon which is entirely absent in seedlings from untreated samples of old seeds, although the statistical significance of this effect could not be determined (Table 3).

Contrary to the results of seedling growth in Petri dishes under long-day conditions, some selected parameters of seedling growth in Petri dishes in continuous darkness were significantly stimulated by the pre-sowing treatment of both young and old seeds. EMF treatment of young seeds enhanced the epicotyl growth of their seedlings by 14%. Compared to their untreated controls, the growth of seedlings from EMF-treated old seeds was enhanced as follows: 22% longer roots in seedlings with only protruded roots (delayed growth) and 22% more fresh weight of emerged embryos. This enhancement of seedling growth from old seeds by EMF exposure (Table 4) is associated with the observed improvement of germination kinetics in EMF-treated old seeds germinating in Petri dishes in continuous darkness (Figure 2B). This shows that the early stages of plant growth may be more prone to stimulation by seed exposure to EMF, especially if the growth process is delayed, as was observed for the old seeds.

Furthermore, the analysis of the growth parameters of 14-day-old field bean plants germinated in substrate-filled pots under long-day conditions from young seeds indicated that pre-sowing EMF treatment of the seeds significantly stimulated the root length by 13% but inhibited stem length by 10% compared to the untreated control (Table 5). Other morphometric traits of field bean plants growing in the pots were not affected by seed age or exposure to EMF.

The comparison of growth parameters within EMF-treated groups follows a similar trend to the analysis within control groups (Tables 3 and 4), confirming that the seedling growth rate decreases with seed aging.

3.5. Assessment of Water Uptake and Membrane Integrity of Field Bean Seeds

To check whether the observed effects of seed age and EMF exposure on germination and seedling growth are related to changes in membrane permeability of young and old seeds of field beans, parameters of water uptake and membrane integrity were examined.

Among the control groups, there were no differences in water uptake for young and old seeds (Figure 3A). EMF treatment also did not affect the water uptake of young and old seeds, compared to the untreated controls. However, among the EMF-treated groups, young seeds absorbed significantly more water (+5%) than old seeds (Figure 3A).

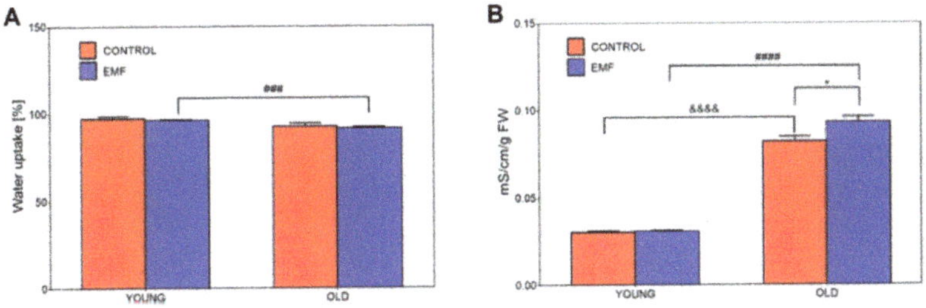

Figure 3. Changes in water uptake (**A**) and electrolyte leakage (**B**) in EMF-treated and untreated young and old seeds of field bean. Data are the means ± SE (n = 4). The symbol (*) indicates significant differences between EMF-treated and control groups (* $p < 0.05$); "&" indicates significant differences between control groups (&&&& $p < 0.0001$); "#" indicates significant differences between EMF-treated groups (### $p < 0.001$, #### $p < 0.0001$).

When membrane integrity was analyzed with respect to seed age, it was observed that in the control groups, the electrolyte leakage in old seeds was significantly higher (+172%) than in young seeds (Figure 3B). Moreover, EMF treatment significantly increased the electrolyte leakage by 13% in old seeds compared to their untreated controls. In the case of young seeds, EMF exposure did not affect their electrolyte leakage compared to their untreated controls. Similarly, old seeds in the EMF-treated groups had significantly higher (+204%) values of electrolyte leakage compared to treated young seeds (Figure 3B). These results show that the integrity of cellular membranes of seed tissues was negatively affected by seed age and EMF exposure.

3.6. Amylolytic Activity and H_2O_2 Content in Seeds of Field Bean

For the analyses of α-amylases activity and H_2O_2 levels, the Petri dish-sown seeds germinating under long-day conditions were selected due to the fact that EMF-treated old seeds were able to germinate in Petri dishes under long-day conditions, to the point of the appearance of seedlings with fully emerged roots and epicotyls, which are completely absent in the untreated old seeds (Table 3). Sarraf et al. [12] have reported that amylases and reactive oxygen species (ROS) are known factors affected by seed exposure to different doses of MF. Thus, our objective was to determine whether the observed changes in seedling growth from old seeds after EMF exposure, which could not be statistically determined, would be reflected in stimulations at the cellular level, specifically, in the enzymatic activity of amylases, which hydrolyze starch reserves in seeds, and in the level of H_2O_2, which belongs to ROS molecules.

In the germinating young and old seeds of field beans, the α-amylase activity showed specific patterns of changes in the control groups and EMF-treated groups (Figure 4A). In the control groups, α-amylase activity in old seeds on the 2nd, 5th, and 6th days of germination was significantly higher (+38%, +41%, +43%), compared to young seeds.

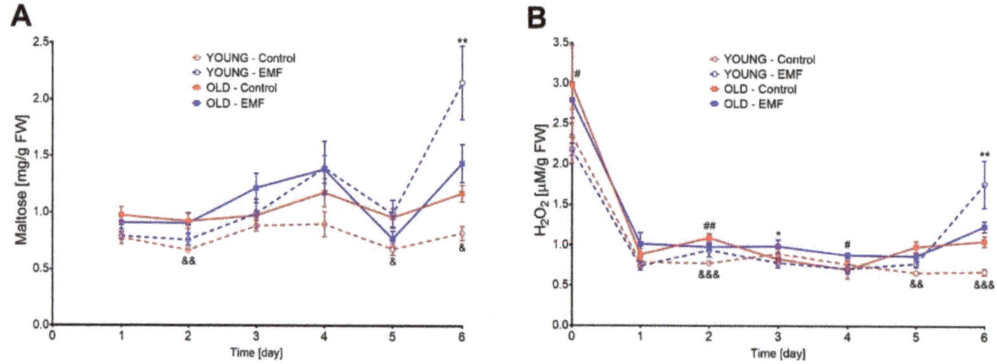

Figure 4. The activity of α-amylases (**A**) and amount of hydrogen peroxide (H_2O_2) molecules (**B**) in young and old field bean seeds treated with EMF and their controls, germinating in Petri dishes for 6 days under long-day conditions. Data are the means ± SE (n = 3). The symbol (*) indicates significant differences between EMF-treated and control groups (* $p < 0.05$, ** $p < 0.01$); "&" indicates significant differences between control groups (& $p < 0.05$, && $p < 0.01$, &&& $p < 0.001$); "#" indicates significant differences between EMF-treated group (# $p < 0.05$, ## $p < 0.01$).

EMF treatment significantly stimulated, by 162%, the activity of α-amylases in young seeds on the last (6th) day of germination, compared to their untreated controls. In old seeds, the activity of α-amylases was not affected by EMF exposure. This suggests that the observed appearance of seedlings with fully emerged roots and epicotyls under long-day conditions after exposure of old seeds to EMF, was not directly related to the stimulation of α-amylase activity by the EMF factor.

In contrast to the control groups, where α-amylase activity was observed to be significantly higher in old seeds than in young seeds, analysis of the EMF-treated groups showed that there were no differences in the α-amylase activity in young and old seeds. This revealed a strong stimulatory effect of EMF exposure in young seeds in the latter days of germination.

To reveal the influence of seed age on H_2O_2 content in field bean seeds, the control groups were analyzed and showed that on the 2nd, 5th, and 6th days of germination, the amount of H_2O_2 in old seeds was significantly higher (by +41%, +49%, and +57%), compared to young seeds (Figure 4B). This elevated H_2O_2 level in the old seeds may be responsible for the observed higher amylase activity in old seeds compared to young seeds in control groups.

The H_2O_2 levels were 162% higher in EMF-treated young seeds than in their controls on the 6th day of germination, while treated old seeds had 19% higher H_2O_2 levels than their untreated controls on the 3rd day of germination (Figure 4B). The strong stimulation of H_2O_2 levels in young seeds by EMF exposure recorded on the 6th day of germination is positively associated with the observed strong enhancement of α-amylase activity by EMF treatment of young seeds, also on the 6th day of germination. This again shows that during the germination process of field beans, α-amylase activity may be positively regulated by higher levels of H_2O_2.

Among the EMF-treated groups, the H_2O_2 content of old dry seeds (before sowing) and germinating old seeds on the 2nd and 4th days were found to be 28%, 4%, and 24% more than in the young seeds, respectively (Figure 4B). This increase in the H_2O_2 level, also revealed in the control groups, can be a consequence of metabolic changes in cells of old seeds during the senescence process. It is also worth pointing out that in both the control and EMF-treated groups, dry seeds (young and old) before sowing had about three times higher the amount of H_2O_2 in the imbibed seeds during the next five days of germination, showing a strong dependence of H_2O_2 level on the stage of the germination process.

3.7. Effect of Seed Age and EMF Exposure on Photosynthetic Pigments Content

Plant productivity depends on photosynthesis, and the level of photosynthetic pigments can be an indication of changes in plant physiology, sometimes due to the actions of different stress factors [41]. In this study, the contents of chlorophylls (a, b, and total) and total carotenoids (xanthophylls + b-carotene) were examined in two-week-old field bean plants growing in substrate-filled pots. The analysis was performed in the control groups, as well as in the EMF-treated groups, to assess the dependence of seed age and the potential physical eustress factor due to EMF on photosynthetic pigments content.

Among the control groups, there were no differences in the contents of the chlorophylls and total carotenoids in the leaves of plants growing from young and old seeds (Figure 5A,B), indicating that seed age does not influence the content of photosynthetic pigments in field bean plants.

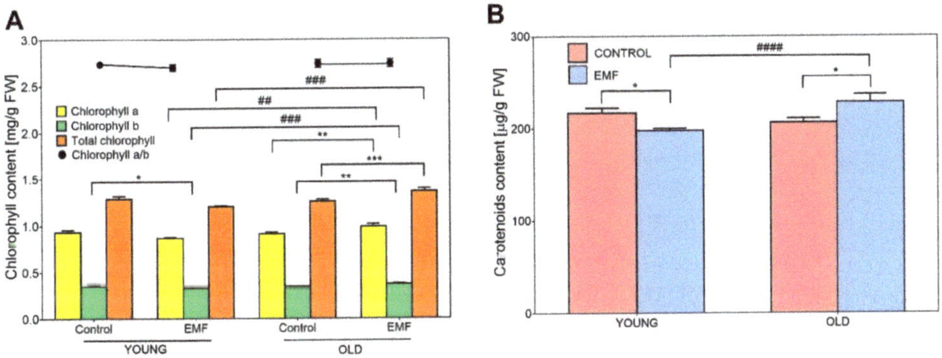

Figure 5. Photosynthetic pigments content of chlorophylls (**A**) and carotenoids (**B**) in leaves of two-week-old plants of field beans growing in pots from young and old seeds treated with EMF, and their controls. Data are the means ± SE (n = 3). The symbol (*) indicates significant differences between EMF-treated and control groups (* $p < 0.05$, ** $p < 0.01$ and *** $p < 0.001$); "#" indicates significant differences between EMF-treated groups (## $p < 0.01$, ### $p < 0.001$, #### $p < 0.0001$).

Leaves of plants grown from EMF-treated young seeds had significantly lower levels of chlorophyll b (−5%) and carotenoids (−9%) than in their untreated controls (Figure 5A,B). This reduction in the pigment content in leaves of plants from EMF-treated young seeds can be related to the observed inhibition of growth of above-ground organs (stems) from the EMF-treated young seeds (Table 5). In contrast, leaves of plants growing from EMF-treated old seeds had a significant increase in the content of all photosynthetic pigments studied compared to their untreated controls: chlorophyll a (+8%), chlorophyll b (+6%), total chlorophyll (+8%), and total carotenoids (+11%) (Figure 5A,B).

Among the EMF-treated groups, changes in pigment content specific to seed age were observed. In these groups, leaves of plants growing from EMF-treated old seeds express significantly higher levels of chlorophyll a (+14%), chlorophyll b (+11%), total chlorophyll (+13%), and carotenoids (+15%), compared to EMF-treated young seeds (Figure 5A,B), highlighting the strong stimulatory effect of EMF exposure on photosynthetic pigments in plants growing from old seeds.

Moreover, in all the experimental variants, the chlorophyll a/b ratio did not change and therefore was not affected by seed aging or EMF exposure (Figure 5A).

3.8. Influence of Seed Age and EMF Exposure on Phytohormone Levels in Seedlings Growing in Substrate-Filled Pots

Phytohormones are essential signaling elements that control many aspects of plant growth and development, as well as their response to environmental stress [42]. This association of phytohormones to nearly all fundamental biological processes makes them a good candidate for consideration during testing and engineering stress tolerance in agronomically important crops [43].

To check if seed age and pre-sowing exposure of seeds to EMF can influence longer-term biochemical traits in field bean plants, the levels of six selected phytohormones (IAA, ABA, GA_1, GA_3, SA, JA) were analyzed quantitatively in the roots and leaves of two-week-old plants growing from EMF-treated young and old seeds as well as their untreated controls. Additionally, we sought to determine whether the changes observed in the growth parameters (Table 5) and the content of photosynthetic pigments (Figure 5) attributed to pre-sowing exposure of seeds to EMF will be related to changes in the levels of the main phytohormones controlling growth (IAA, ABA, GAs) and stress response (SA, JA).

In the case of IAA, there were no differences in the auxin levels in the roots and leaves of plants growing from young and old seeds in the control groups. This indicates that seed age did not affect the IAA amount (Figure 6A). Concerning the influence of EMF exposure, the changes in IAA levels in plants from treated seeds were observed solely in underground organs (Figure 6A). The IAA amount in the roots of plants growing from EMF-treated young seeds was significantly lower (−28%) than in the untreated control. This reduction in IAA amount in the roots of treated plants is negatively associated with the observed stimulation of root growth in plants growing from EMF-treated young seeds (Table 5). Contrary to the changes in the roots of plants growing from EMF-treated young seeds, the roots of plants from EMF-treated old seeds did not express changes in IAA levels, compared to their untreated controls (Figure 6A). However, among the EMF-treated groups, changes related to seed age were detected. In these treatment groups, roots of plants growing from old seeds expressed significantly higher levels (+82%) of IAA compared to roots growing from young seeds (Figure 6A).

The ABA level changed significantly among the control groups, revealing that this hormone content was affected by seed age. In these groups, the ABA amount in the roots of plants growing from young seeds was higher compared to the roots of plants growing from old seeds (Figure 6B).

EMF-treated young seeds produced plants with significantly reduced ABA levels in their roots (−21%) and leaves (−20%) compared to their untreated controls (Figure 6B). Similar to the results from the IAA analysis, the observed reduction in ABA level in roots is negatively associated with the enhanced root growth of plants growing from EMF-treated young seeds (Table 5). Concerning the effects of EMF exposure on photosynthetic pigments, the observed reduction in ABA amount in leaves corresponds with the decrease in chlorophyll b and total carotenoids in leaves of plants growing from EMF-treated young seeds (Figure 5). The roots and leaves of plants growing from EMF-treated old seeds had no significant changes in ABA levels compared to their untreated controls (Figure 6B). Moreover, among the EMF-treated groups, the ABA level in leaves of plants growing from treated young seeds was lower (−52%) than in plants from treated old seeds, showing that in these groups the ABA level was also affected by seed age.

The analysis of the level of bioactive gibberellins showed that among the control groups, there was no difference in the GA_1 levels in the roots and leaves of plants growing from young and old seeds (Figure 6C). However, the level of GA_1 in leaves of plants growing from EMF-treated young seeds was significantly reduced by 58%, compared with their untreated control (Figure 6C). This reduction in GA_1 level in leaves was associated with the decrease in stem length and content of particular photosynthetic pigments in leaves due to pre-sowing exposure of seeds to EMF (Table 5, Figure 5). No other significant effects of EMF exposure on the levels of GA_1 in the organs of plants growing from treated young and old seeds compared to their untreated controls were detected. Concerning the effect of seed age among the EMF-treated groups, the leaves of plants growing from treated young seeds contained less GA_1 (−51%) than the leaves of plants growing from treated old seeds (Figure 6C).

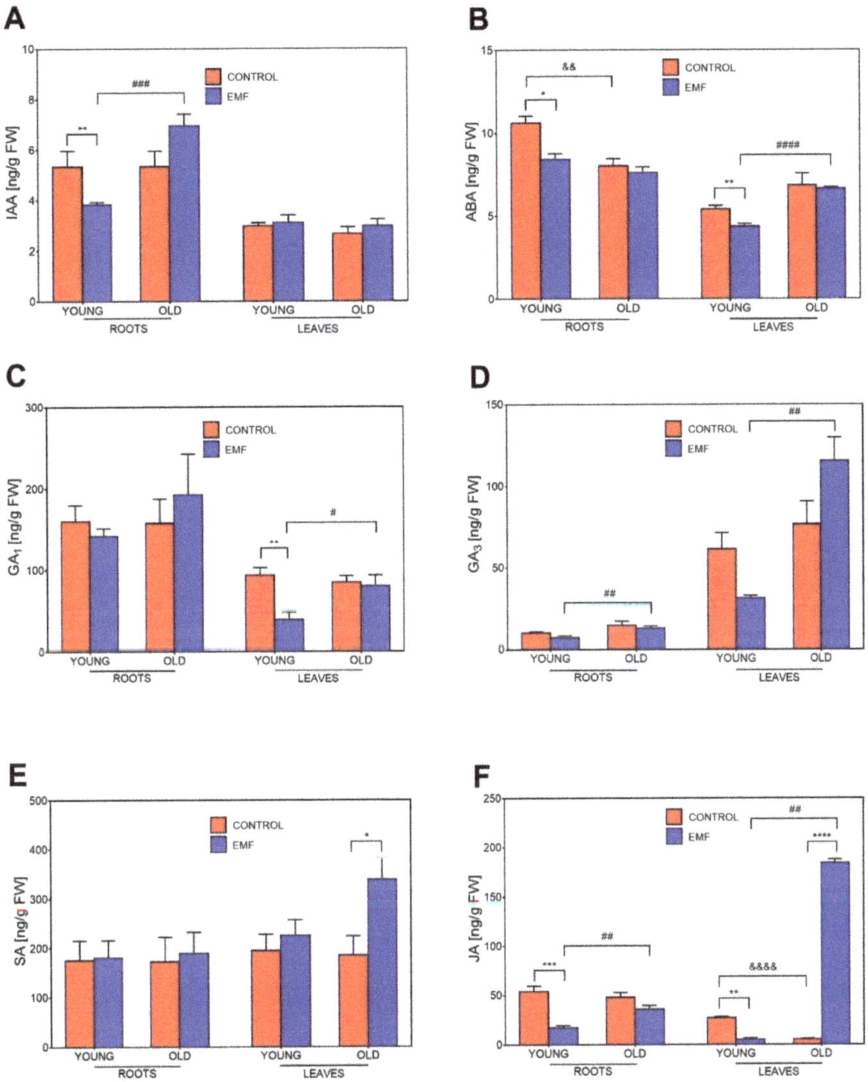

Figure 6. The amount of IAA (**A**), ABA (**B**), GA_1 (**C**), GA_3 (**D**), SA (**E**) and JA (**F**) phytohormones in roots and leaves of two-week-old plants of field bean growing in pots from young and old seeds treated with EMF, and their controls. Data are the means ± SE (n = 3). The symbol (*) indicates significant differences between EMF-treated and control groups ($^{*}\ p < 0.05$, $^{**}\ p < 0.01$, $^{***}\ p < 0.001$, $^{****}\ p < 0.0001$); "&" indicates significant differences between control groups ($^{\&\&}\ p < 0.01$, $^{\&\&\&\&}\ p < 0.0001$); "#" indicates significant differences between EMF-treated groups ($^{\#}\ p < 0.05$, $^{\#\#}\ p < 0.01$, $^{\#\#\#}\ p < 0.001$, $^{\#\#\#\#}\ p < 0.0001$).

Among the control groups, the level of GA_3 in the roots and leaves was not affected by seed age (Figure 6D). Furthermore, the pre-sowing exposure of young and old seeds to EMF did not lead to any changes in the GA_3 levels in the roots and leaves of plants compared to their untreated controls. However, when the EMF-treated groups were analyzed, the age of the treated seeds had an effect on their GA_3 levels. The GA_3 levels in the roots and leaves of plants growing from treated old seeds were respectively, 78% and 266% higher than those of plants growing from treated young seeds (Figure 6D).

The SA level in roots and leaves in the control groups was not affected by seed age. Pre-sowing exposure of seeds to EMF led to a significant increase (+83%) in SA levels in the leaves of plants growing from old seeds (Figure 6E), which is associated with the higher content of the photosynthetic pigments in leaves of plants growing from the EMF-treated old seeds (Figure 5). No other specific changes in SA levels attributable to the pre-sowing exposure of seeds to EMF or seed age were detected (Figure 6E).

Significant changes in the amount of JA were found among the control groups of young and old seeds, indicating the influence of seed age on the JA level. In the control groups, leaves of plants growing from old seeds had an 80% lower JA level compared to leaves growing from young seeds (Figure 6F).

The JA levels in the roots and leaves of plants growing from EMF-treated young seeds were, respectively, 68% and 81% lower than in their untreated controls (Figure 6F). These results indicate a possible difference in the role of JA in the growth regulation of roots and above-ground organs since the observed reduction in JA levels attributed to EMF corresponds with root growth stimulation and the inhibition of stem growth and photosynthetic pigment content in plants growing from young treated seeds (Table 5, Figure 5). In contrast, the level of JA in the leaves of plants growing from old seeds treated with EMF was strongly stimulated (34-fold increase) compared to their untreated control. This particular change in JA level was associated with the increase in photosynthetic pigment content in leaves growing from EMF-treated old seeds (Figure 5). Among the EMF-treated groups, particular seed age-specific changes in JA levels were also detected. In these groups, the roots and leaves of plants growing from treated old seeds had significantly higher levels of JA (two-fold and 36-fold, respectively) than the organs of plants growing from treated young seeds.

4. Discussion

Plant response to MFs is regarded as a non-linear phenomenon since all living organisms, including plants, can be considered as non-linear systems which pass through different stages of developmental processes [44]. Studies on the effect of MF treatments on plant growth and development have produced varying results of stimulation, inhibition, and sometimes no effect, which indicates a complex mechanism of MF action in plant cells and the likely occurrence of different interactions between MF response and endogenous rhythms in plant cells [31].

Our current study revealed different effects of seed exposure to EMF on germination and early growth of field beans, which were dependent on seed age, light conditions, and growth media. In control groups, the germination kinetics and germination parameters of old seeds were significantly reduced compared to young seeds in all the experimental variants. On the contrary, EMF exposure significantly improved germination kinetics, but only for old seeds germinating in Petri dishes in continuous darkness (Figure 2B). However, the germination rate of EMF-treated young seeds in the substrate-filled pots was significantly reduced. In all the experimental variants, the only instance where EMF treatment led to an improvement in germination kinetics, as well as subsequent seedling growth, was during the germination and growth in Petri dishes of treated old seeds in continuous darkness (Table 4). Although the pre-sowing exposure to EMF inhibited the germination rate of young seeds in substrate-filled pots, the root length of two-week-old seedlings growing from EMF-treated young seeds was stimulated. However, the growth of stems derived from these EMF-treated young seeds in the substrate-filled pots was inhibited (Table 5). These results strongly point to the dependence of EMF effects on the physiological and developmental stage of the plant material, as well as the growth (environmental) conditions. Similar results were obtained by Ivankov et al. [23] where treatment of red clover seeds with EMF (5.28 MHz, 0.74 mT) stimulated five-week-old seedlings but only in the morphometric parameters of the roots and not in the above-ground parts. In the case of sunflower (*Helianthus annus*) seeds, the same dose of EMF exposure (5.28 MHz, 0.74 mT) resulted in a higher germination rate and an increase in leaf weight of two-week-old seedlings sown in a substrate, while other growth parameters remained unaffected by the EMF treatment [45]. Furthermore, the treatment of tomato (*Solanum lycopersicum*) seeds with EMF (3 Hz, 12.5 mT) led to the inhibition of stem length in field conditions, despite the stimulation of other growth parameters such as the number of leaves and flowers [46]. There are several other reports showing inconsistencies in the influence of MF exposure on particular physiological traits of plants, which suggests the possibility of improving plant growth without altering the germination rate of seeds. The EMF (50 Hz, 15 mT) treatment of durum wheat (*Triticum durum*) seeds caused no influence on the dormant seed germination process but positively affected the fresh weight of seedlings [47]. A study involving the application of EMF (5.28 MHz, 0.74 mT) on seeds of two perennial woody plants (*Rhododendron smirnowii, Morus nigra*) showed that the most adverse effect of EMF on the germination of seeds could be followed by an increase in the growth of leaves during the later developmental stages of the plants [48].

An earlier study showed that roots seem to be more sensitive to MF exposure than shoots and that the stimulation or inhibition of root growth by MF depends on plant species and their physiological state [49]. In addition, the roots of Arabidopsis seedlings were reported to be the most sensitive organs to MF exposure, with the observed stimulation in root growth attributed to an increase in cell number [50]. In our study, the observed stimulation of root length of field bean plants growing in the substrate-filled pots from EMF-treated young seeds can be of particular importance for crop production in field conditions. Similar results of significant improvement in the main root characteristics (length, surface area, and volume) due to MF exposure have been found in cotton (*Gossypium hirsutum*) (3 Hz, 12.5 mT) [51], chickpea (100 mT) [25], and maize (50–250 mT) [52]. This improvement in root growth highlights the potential of MF treatment as a useful priming method to improve crop resistance to drought stress.

Under certain stress conditions, roots are found to accelerate growth to absorb more water and nutrients for survival, which in turn increases the root-to-shoot ratio and this is attributed to the relatively greater negative consequence of the stress factor on the above-ground organs [53]. These observations support the hypothesis that particular doses of EMF exposure on some plants could be a eustress factor that affects plant growth and development. Mildaziene et al. [45] have reported that proteome analysis of two-week-old sunflower seedlings after exposure to EMF (5.28 MHz, 0.74 mT) shows a eustress-like response localized mainly in their chloroplasts.

In our Petri dish experiments, EMF effects on germination and growth were also dependent on light conditions. This is similar to the results obtained by Novitskii et al. [54], where the examination of the composition and content of lipids in five-day-old radish (*Raphanus sativus*) seedlings growing from seeds exposed to MF (50 Hz, 500 μT) showed existing differences in the mechanisms of MF action in light and darkness. In addition, EMF (50 Hz, 7 mT) treatment of winter wheat seeds stimulated, to a higher degree, germination and early growth in continuous darkness, compared to continuous light conditions [15]. Currently, the mechanism of this light-dependent response to MF is not fully understood, but existing reports indicate the involvement of cryptochromes in this response [55].

In old seeds, many cellular processes could change as a consequence of the aging process, including membrane damage, loss of enzymatic activity, and many oxidative damages to lipids, proteins, and DNA [4]. This, in turn, can influence specific signaling pathways in cells of seeds of different ages and cause distinct, sometimes stronger, seed responses to MF exposure [48]. In our study, the germination kinetics of only old seeds was stimulated by exposure to EMF (50 Hz, 7 mT), which later led to accelerated early root growth (Figure 2B, Table 4). Bilalis et al. [51] found that in the case of cotton seeds, EMF (3 Hz, 12.5 mT) exposure was an effective priming technique for plants growing in pots under field conditions, especially for old seeds with reduced germinability. When six-year-old pea seeds were exposed to MF (100 mT), seed vigor was enhanced, and this improvement in the quality of the aged seed was mediated by changes in free radicals by the antioxidant defense system and protein oxidation [56]. Additionally, the treatment of aged broccoli (*Brassica oleracea*) seeds with MF (60 Hz, 3.6 mT) to improve their germination led to different outcomes (positive, negative, and no effects) depending on the length of their aging process [57].

Many reports have indicated an increase in water uptake and improvement in seed coat membrane integrity—expressed in lower electrolyte leakage—after seed exposure to different doses of MFs, which in turn is often positively associated with an increase in germination speed [16,52,58]. In our study, EMF (50 Hz, 7 mT) exposure did not change the water uptake of field bean seeds, but reduced the already impaired membrane integrity of old seeds, and had no effect on the membrane integrity of the young seeds (Figure 3). In other studies where wheat seeds were magneto-primed, the water uptake of seeds was reduced after treatment with SMF (30 mT) but did not change after EMF (10 Hz) exposure [59]. Moreover, the treatment of two genotypes of chickpea seeds with SMF (100 mT) caused an increase in water uptake in only one genotype, while in the second (native) genotype, MF priming did not cause a significant change in seed water uptake, suggesting that the rate of water uptake by seeds may depend mainly on the internal water potential of the seeds [25]. Thus, the influence of MF on membrane permeability can depend on seed type (structure of seed coat and type of storage material) and MF characteristics.

In our study, biochemical analysis of the field bean seeds showed that in control groups, α-amylase activity in old seeds was higher than in young seeds, and was also associated with a higher level of H_2O_2 in old seeds. Moreover, EMF exposure (50 Hz, 7 mT) caused a significant increase in α-amylase activity on the last day of germination of young seeds, which again was associated with a large rise in H_2O_2 level in those seeds at the same point. Similar positive associations of H_2O_2 concentration with amylase activity were reported in barley (*Hordeum vulgare*) grains for β-amylase [60] and α-amylase [61]. Furthermore, the increased amylase activity at the later stages of germination due to MF exposure was also found in these seeds: chickpea treated with SMF of 100 mT [25]; millet (*Setaria italica*) treated with EMF of 10 Hz, 0.03 mT [62]; and faba bean (*Vicia faba*) treated with MFs of 35 mT and 80 mT seeds [63]. In the case of chickpea, the amylase activity after MF exposure was found to vary depending on plant genotype [25]. Additionally, the higher content of H_2O_2 after exposure of seeds to MF was associated with the reported improvement in the germination of MF-treated tomato (100 mT) [14] and cucumber (*Cucumis sativus*) (100–250 mT) [58]. Apart from the beneficial role of H_2O_2 in MF signaling, this molecule can also have a detrimental effect during seed aging, leading to a lowering of the germination rate caused by factors including membrane integrity loss [64]. In our study, the concentration of H_2O_2 was highest in dry seeds (young and old) compared to imbibed seeds. Moreover, within the control groups, old seeds of field beans contained more H_2O_2 compared to young seeds and this was also associated with higher electrolyte leakage and a reduced germination rate in the old seeds. These results are supported by existing reports showing that H_2O_2 accumulates in dry seeds during the after-ripening process and with seed aging during prolonged storage [64].

The content of photosynthetic pigments is a good marker of plant health and productivity, and a higher composition of chlorophylls and carotenoids can enhance photosynthesis and plant growth [65]. Existing evidence show that the effects of MF exposure on photosynthetic pigments depend on plant species and MF parameters. In this study, EMF treatment of seeds led to the photosynthetic pigment content increasing in the leaves of seedlings grown from old seeds while decreasing in seedlings from young seeds (Figure 5). Vashisth and Joshi [52] have reported a higher chlorophyll content in plants growing from MF-treated seeds of maize (50–250 mT). However, in other studies, MF exposure produced inhibitory effects or no effects on photosynthetic pigment content. EMF (50 Hz, 65 μT) decreased the content of chlorophylls a, b, and carotenoids in barley seedlings [66], while in pea seedlings, the chlorophyll content was not significantly affected after treatment of seeds with EMF (50 Hz, 60–180 mT) [28]. Similar to our results, it was reported that the exposure of durum wheat seeds to EMF (50 Hz, 15 mT) did not affect the chlorophyll a and b ratios [47].

Currently, only a few reports exist presenting hormonal changes after MF exposure and they often refer to hormonal analysis in seeds [14,23,45]. The existing knowledge base of the influence of EMF exposure on longer-term traits like phytohormone balance in growing plants is, thus, negligible. Our results show that in the case of the main growth hormones (IAA, ABA, and GAs), more changes due to exposure to EMF were observed in the roots and leaves of plants growing from young seeds. Plant hormones typical for stress response (SA and JA) were affected by EMF exposure the most in the leaves of plants growing from old seeds (Figure 6E,F). Differences in hormonal changes in roots and leaves due to seed age were detected mostly by comparing within the EMF-treated groups. In most of these cases, the levels of analyzed hormones were higher in organs of plants growing from old seeds than in organs of plants growing from young seeds (Figure 6). Khan et al. [67] have reported that the levels of IAA, ABA, SA, and JA increase in certain senescing plant organs. Some age-dependent functions of gibberellins have also been reported, indicating the role of GAs in the delay of nodule senescence in peas [68].

Regarding changes in IAA levels, our results showed that the reduction in the IAA amount in roots of plants growing from EMF-treated young seeds was associated with the stimulation of root growth from those seeds. Auxins are reported to inhibit primary root growth [69] and this supports our results. The stimulation of root growth associated with a reduction in IAA level was also reported in wheat seedlings growing from seeds exposed to EMF (50 Hz, 7 mT) [15]. However, another study indicated that exposing pea seeds to MF (30 mT and 85 mT) instead caused a significant increase in IAA levels in 6-day-old stems and roots [16]. Similarly, the treatment of 3-day-old Arabidopsis seedlings with SMF (600 mT) increased the level of auxin in the MF-treated root tip via enhanced expression of auxin influx transporter, AUX1, and decreased the expression of auxin efflux transporter, PIN3 [50]. Thus, the effect of MF treatment on IAA level in growing plants seems to be strongly dependent on MF parameters.

ABA is considered a general inhibitor of plant growth [70]. Lowering the ABA level in seeds after MF exposure is regarded as a priming mechanism to stimulate germination [15,45]. In this study, we observed a reduction in ABA level in the roots of plants growing from EMF-treated young seeds which was also associated with the stimulation of root growth from those seeds. Moreover, changes in ABA and IAA levels observed in roots of plants growing from EMF-treated young seeds were similar, which indicates possible interactions between ABA and IAA signaling pathways during root growth of field beans. Such interactions are known to occur during Arabidopsis root growth under different abiotic stress conditions [42].

Gibberellins (GAs) can enhance organ growth by stimulating cell elongation and division [43]. GA_1, GA_3, GA_4, and GA_7 are the most common biologically active forms of gibberellins found in higher plants [71]. In our study, the GA_1 level was reduced in leaves of plants growing from EMF-treated young seeds. Achard et al. [72] have shown that under salt stress, the levels of bioactive gibberellins reduce, possibly through the signaling of ABA and that the reduction in the growth of Arabidopsis due to the gibberellin pathways is beneficial and enhances the survival of plants. Furthermore, results in our study indicate that leaves of plants growing from EMF-treated old seeds contained more GA_1, chlorophylls, and carotenoids than leaves of plants growing from EMF-treated young seeds (Figure 6 C). This indicates the possible involvement of GA_1 in the regulation of photosynthetic processes. Iftikhar et al. [73] have shown that exogenous application of GA positively affects chlorophyll content in wheat. It has also been revealed that the IAA/GA_1 ratio reflects changes in growth parameters of soybean plants under specific photosynthetic conditions [74].

In our experiments, the GA_3 levels were not affected by the pre-sowing exposure of seeds to EMF. However, Anand et al. [14] found that the GA_3 level was higher in MF-treated (100 mT) seeds of tomato. Similarly, Podleśny et al. [16] reported that the GA_3 levels in seeds of pea, along with roots and stems of their 6-day-old seedlings, increased after pre-sowing exposure of the seeds to MF (30 mT and 85 mT).

SA and JA are known to participate in defense reactions to many biotic and abiotic stressors [43]. SA content was previously investigated in dry seeds of red clover and sunflower, where exposure to EMF (5.28 MHz, 0.74 mT) resulted in SA content reduction in the red clover [23,24] but an elevation in the sunflower [45]. However, as far as we know, the effect of EMF exposure on JA content in plants has not yet been analyzed. In our study, JA level was inhibited in roots of plants growing from EMF-treated young seeds, while at the same time, the growth of those roots was stimulated. These results are supported by the fact that JA is known to inhibit root growth [75] and thus, lowering JA levels can have a stimulatory effect on root development.

The exposure of old seeds to EMF in our study led to a significant increase in SA and JA levels in the leaves of plants growing from such seeds, which resembles stressor-specific changes. SA is essential in regulating defense processes, such as hypersensitive response and systemic acquired resistance, while JA regulates biotic and abiotic stress responses [67]. Thus, the increase in SA and JA levels suggests that EMF may boost plant defense systems, preparing field bean plants for future stress conditions. One consequence of such a boost in the plant defense system could be the increase in photosynthetic pigment content, which was observed in this study for leaves of plants growing from EMF-treated old seeds. Kaya and Doganlar [76] found that tobacco (*Nicotiana tabacum*) plants treated with endogenous JA produce more chlorophyll and carotenoids, which helps plants to alleviate the negative effects of herbicide stressors. *Brassica oleracea* plants grown from seeds treated with various concentrations of JA, have also been shown to enhance photosynthetic efficiency and chlorophyll fluorescence [77].

Despite a significant increase in the level of specific stress hormones (SA and JA), the complex hormonal changes observed in the roots and leaves of plants growing from EMF-treated seeds suggests that EMF treatment of 50 Hz, 7 mT does not cause the typical stress response in field bean organs. However, the modifications in hormonal balance detected in our studies suggest a response of the plants to low-intensity stress, i.e., eustress, especially in the case of plants growing from old seeds. In recent years, it has become evident that plant hormones do not act only in a linear pathway, but also produce numerous and often complicated interactions [43,78]. Therefore, more detailed studies are necessary to elucidate the role of phytohormones in the mechanism of MF action in plant tissues and to lead to the development of more resilient crops.

5. Conclusions

We analyzed the effectiveness of the pre-sowing exposure of field bean seeds of different ages to EMF (50 Hz, 7 mT) as an alternative physical seed treatment to improve germination and growth. In control groups, old seeds expressed a reduced germination rate and slower growth compared to young seeds. After EMF treatment, different effects on seed germination, growth parameters, and biochemical traits, depending on seed age, light condition, and growth media were obtained. EMF treatment stimulated only the germination of old seeds in Petri dishes in continuous darkness and this was followed by enhanced early growth of their roots. In the studies using a universal substrate in pots, EMF exposure led to a lower rate of germination of young seeds but had no significant effect on the germination rate of old seeds. Our studies showed that membrane integrity was only affected (negatively) by EMF treatment in old seeds. Moreover, increased α-amylase activity was associated with higher H_2O_2 levels in the control group of old seeds and group of EMF-treated young seeds. Furthermore, the response to EMF exposure was assessed by measuring the morphological and biochemical parameters of two-week-old plants, which can affect field bean competition under field conditions. The improved root growth of plants growing from EMF-treated young seeds, corresponding to reduced IAA, ABA, and JA levels, suggests that magnetically treated field bean seeds may grow into plants with a better uptake of water under rainfed (un-irrigated) or even drought conditions. Additionally, the stimulation of photosynthetic pigment content in leaves growing from

EMF-treated old seeds was associated with increased levels of SA and JA in those organs and this could help field bean plants to produce bigger biomass during field cultivation.

We conclude that the observed germination and growth effects, as well as the hormonal changes due to the pre-sowing exposure of seeds to EMF (50 Hz, 7 mT), point to a field bean response to low-intensity stress (i.e., eustress). Although pre-sowing EMF treatment did not lead to direct stimulation of germination of young and old seeds in the substrate medium, this priming method produced some positive long-term effects by improving root growth and chlorophyll content. Thus, the results of our studies indicate that pre-sowing exposure of seeds of field beans to EMF (50 Hz, 7 mT) has the potential to be an alternative method of improving their production. The effectiveness of this seed priming method to enhance long-term growth and development of field beans could also be further explored in field experiments.

Author Contributions: Conceptualization, A.P.; methodology, A.P. and J.W.; investigation, A.P., D.C., J.W., M.D.D. and K.P.; writing—original draft preparation, A.P.; writing—review and editing, A.P., J.W., D.C., M.D.D., K.P. and A.S.-J.; visualization, A.P. and J.W.; supervision, A.P. and A.S.-J. All authors have read and agreed to the published version of the manuscript.

Funding: This research was funded by 1. The EU Program Knowledge, Education, Development and the Polish National Center for Research and Development ("Universitas Copernicana Thoruniensis in Futuro", project no. POWER.03.05.00.00-Z302/17-00); 2. INCOOP competition "Excellence Initiative—Research University" at the Nicolaus Copernicus University in Toruń.

Institutional Review Board Statement: Not applicable.

Informed Consent Statement: Not applicable.

Data Availability Statement: Relevant data are contained within the article.

Acknowledgments: We express our gratitude to Grażyna Czeszewska-Rosiak for her technical support during the laboratory work.

Conflicts of Interest: The authors declare no conflict of interest.

References

1. Robert, M.; Thomas, A.; Bergez, J.-E. Processes of Adaptation in Farm Decision-Making Models. A Review. *Agron. Sustain. Dev.* **2016**, *36*, 64. [CrossRef]
2. Laik, R.; Kumara, B.H.; Pramanick, B.; Singh, S.K.; Nidhi; Alhomrani, M.; Gaber, A.; Hossain, A. Labile Soil Organic Matter Pools Are Influenced by 45 Years of Applied Farmyard Manure and Mineral Nitrogen in the Wheat—Pearl Millet Cropping System in the Sub-Tropical Condition. *Agronomy* **2021**, *11*, 2190. [CrossRef]
3. Hampton, J.G.; Conner, A.J.; Boelt, B.; Chastain, T.G.; Rolston, P. Climate Change: Seed Production and Options for Adaptation. *Agriculture* **2016**, *6*, 33. [CrossRef]
4. Zhang, K.; Zhang, Y.; Sun, J.; Meng, J.; Tao, J. Deterioration of Orthodox Seeds during Ageing: Influencing Factors, Physiological Alterations and the Role of Reactive Oxygen Species. *Plant Physiol. Biochem.* **2021**, *158*, 475–485. [CrossRef]
5. Kumar, M.; Mitra, S.; Mazumdar, S.P.; Majumdar, B.; Saha, A.R.; Singh, S.R.; Pramanick, B.; Gaber, A.; Alsanie, W.F.; Hossain, A. Improvement of Soil Health and System Productivity through Crop Diversification and Residue Incorporation under Jute-Based Different Cropping Systems. *Agronomy* **2021**, *11*, 1622. [CrossRef]
6. Stagnari, F.; Maggio, A.; Galieni, A.; Pisante, M. Multiple Benefits of Legumes for Agriculture Sustainability: An Overview. *Chem. Biol. Technol. Agric.* **2017**, *4*, 2. [CrossRef]
7. Dutta, A.; Trivedi, A.; Nath, C.P.; Gupta, D.S.; Hazra, K.K. A Comprehensive Review on Grain Legumes as Climate-smart Crops: Challenges and Prospects. *Environ. Chall.* **2022**, *7*, 100479. [CrossRef]
8. Mariotti, M.; Andreuccetti, V.; Arduini, I.; Minieri, S.; Pampana, S. Field Bean for Forage and Grain in Short-Season Rainfed Mediterranean Conditions. *Ital. J. Agron.* **2018**, *13*, 208–215. [CrossRef]
9. Sulas, L.; Roggero, P.P.; Canu, S.; Seddaiu, G. Potential Nitrogen Source from Field Bean for Rainfed Mediterranean Cropping Systems. *Agron. J.* **2013**, *105*, 1735–1742. [CrossRef]
10. Devika, O.S.; Singh, S.; Sarkar, D.; Barnwal, P.; Suman, J.; Rakshit, A. Seed Priming: A Potential Supplement in Integrated Resource Management Under Fragile Intensive Ecosystems. *Front. Sustain. Food Syst.* **2021**, *5*, 654001. [CrossRef]
11. Araújo, S.S.; Paparella, S.; Dondi, D.; Bentivoglio, A.; Carbonera, D.; Balestrazzi, A. Physical Methods for Seed Invigoration: Advantages and Challenges in Seed Technology. *Front. Plant Sci.* **2016**, *7*, 646. [CrossRef]
12. Sarraf, M.; Kataria, S.; Taimourya, H.; Santos, L.O.; Menegatti, R.D.; Jain, M.; Ihtisham, M.; Liu, S. Magnetic Field (MF) Applications in Plants: An Overview. *Plants* **2020**, *9*, 1139. [CrossRef]

13. Pietruszewski, S.; Muszyński, S.; Dziwulska, A. Electromagnetic Fields and Electromagnetic Radiation as Non-Invasive External Stimulants for Seeds (Selected Methods and Responses). *Int. Agrophys.* **2007**, *21*, 95–100.
14. Anand, A.; Kumari, A.; Thakur, M.; Koul, A. Hydrogen Peroxide Signaling Integrates with Phytohormones during the Germination of Magnetoprimed Tomato Seeds. *Sci. Rep.* **2019**, *9*, 8814. [CrossRef]
15. Cecchetti, D.; Pawełek, A.; Wyszkowska, J.; Antoszewski, M.; Szmidt-Jaworska, A. Treatment of Winter Wheat (*Triticum aestivum* L.) Seeds with Electromagnetic Field Influences Germination and Phytohormone Balance Depending on Seed Size. *Agronomy* **2022**, *12*, 1423. [CrossRef]
16. Podleśny, J.; Podleśna, A.; Gładyszewska, B.; Bojarszczuk, J. Effect of Pre-Sowing Magnetic Field Treatment on Enzymes and Phytohormones in Pea (*Pisum sativum* L.) Seeds and Seedlings. *Agronomy* **2021**, *11*, 494. [CrossRef]
17. Vian, A.; Davies, E.; Gendraud, M.; Bonnet, P. Plant Responses to High Frequency Electromagnetic Fields. *BioMed Res. Int.* **2016**, *2016*, 1830262. [CrossRef]
18. Ribeiro-Oliveira, J.P. Electromagnetism and Plant Development: A New Unknown in a Known World. *Theor. Exp. Plant Physiol.* **2019**, *31*, 423–427. [CrossRef]
19. Vázquez-Hernández, M.C.; Parola-Contreras, I.; Montoya-Gómez, L.M.; Torres-Pacheco, I.; Schwarz, D.; Guevara-González, R.G. Eustressors: Chemical and Physical Stress Factors Used to Enhance Vegetables Production. *Sci. Hortic.* **2019**, *250*, 223–229. [CrossRef]
20. Podleśny, J.; Misiak, L.E.; Podleśna, A.; Pietruszewski, S. Concentration of Free Radicals in Pea Seeds after Pre-Sowing Treatment with Magnetic Field. *Int. Agrophys.* **2005**, *19*, 243–249.
21. Podleśny, J.; Pietruszewski, S.; Podleśna, A. Efficiency of the Magnetic Treatment of Broad Bean Seeds Cultivated under Experimental Plot Conditions. *Int. Agrophys.* **2004**, *18*, 65–71.
22. Radhakrishnan, R.; Ranjitha Kumari, B.D. Pulsed Magnetic Field: A Contemporary Approach Offers to Enhance Plant Growth and Yield of Soybean. *Plant Physiol. Biochem.* **2012**, *51*, 139–144. [CrossRef]
23. Ivankov, A.; Zukiene, R.; Nauciene, Z.; Degutyte-Fomins, L.; Filatova, I.; Lyushkevich, V.; Mildaziene, V. The Effects of Red Clover Seed Treatment with Cold Plasma and Electromagnetic Field on Germination and Seedling Growth Are Dependent on Seed Color. *Appl. Sci.* **2021**, *11*, 4676. [CrossRef]
24. Mildaziene, V.; Ivankov, A.; Pauzaite, G.; Naucienė, Z.; Zukiene, R.; Degutyte-Fomins, L.; Pukalskas, A.; Venskutonis, P.R.; Filatova, I.; Lyushkevich, V. Seed Treatment with Cold Plasma and Electromagnetic Field Induces Changes in Red Clover Root Growth Dynamics, Flavonoid Exudation, and Activates Nodulation. *Plasma Processes Polym.* **2021**, *18*, 2000160. [CrossRef]
25. Thomas, S.; Anand, A.; Chinnusamy, V.; Dahuja, A.; Basu, S. Magnetopriming Circumvents the Effect of Salinity Stress on Germination in Chickpea Seeds. *Acta Physiol. Plant.* **2013**, *35*, 3401–3411. [CrossRef]
26. Kataria, S.; Baghel, L.; Guruprasad, K.N. Pre-Treatment of Seeds with Static Magnetic Field Improves Germination and Early Growth Characteristics under Salt Stress in Maize and Soybean. *Biocatal. Agric. Biotechnol.* **2017**, *10*, 83–90. [CrossRef]
27. Aguilar, C.H.; Dominguez-Pacheco, A.; Carballo, A.; Cruz-Orea, A.; Ivanov, R.; Luis, J.; Pastor, J. Alternating Magnetic Field Irradiation Effects on Three Genotype Maize Seed Field Performance. *Acta Agroph.* **2009**, *14*, 7–17.
28. Iqbal, M.; Haq, Z.U.; Jamil, Y.; Ahmad, M.R. Effect of Presowing Magnetic Treatment on Properties of Pea. *Int. Agrophys.* **2012**, *26*, 25–31. [CrossRef]
29. Bujak, K.; Frant, M. Influence of pre-sowing seed stimulation with magnetic field on spring wheat yielding. *Acta Agroph.* **2009**, *14*, 19–29.
30. Cakmak, T.; Dumlupinar, R.; Erdal, S. Acceleration of Germination and Early Growth of Wheat and Bean Seedlings Grown under Various Magnetic Field and Osmotic Conditions. *Bioelectromagnetics* **2010**, *31*, 120–129. [CrossRef]
31. Teixeira da Silva, J.A.; Dobránszki, J. Magnetic Fields: How Is Plant Growth and Development Impacted? *Protoplasma* **2016**, *253*, 231–248. [CrossRef] [PubMed]
32. Trawiński, T.; Szczygieł, M.; Wyszkowska, J.; Kluszczyński, K. Analysis of Magnetic Field Distribution and Mechanical Vibration of Magnetic Field Exciter under Different Voltage Supply. In *Information Technologies in Biomedicine*; Piętka, E., Kawa, J., Eds.; Springer: Berlin/Heidelberg, Germany, 2010; pp. 613–622.
33. Bieńkowski, P.; Wyszkowska, J. Technical aspects of exposure to magnetic fields of extremely low frequencies (ELF) in biomedical research. *Med. Pr.* **2015**, *66*, 185–197. [CrossRef] [PubMed]
34. Ranal, M.A.; de Santana, D.G. How and Why to Measure the Germination Process? *Braz. J. Bot.* **2006**, *29*, 1–11. [CrossRef]
35. Kader, M. A Comparison of Seed Germination Calculation Formulae and the Associated Interpretation of Resulting Data. *J. Proc. R. Soc. N. S. W.* **2005**, *138*, 65–75.
36. Coolbear, P.; Francis, A.; Grierson, D. The Effect of Low Temperature Pre-Sowing Treatment on the Germination Performance and Membrane Integrity of Artificially Aged Tomato Seeds. *J. Exp. Bot.* **1984**, *35*, 1609–1617. [CrossRef]
37. Miller, G.L. Use of Dinitrosalicylic Acid Reagent for Determination of Reducing Sugar. *Anal. Chem.* **1959**, *31*, 426–428. [CrossRef]
38. Lichtenthaler, H.K. Chlorophylls and Carotenoids: Pigments of Photosynthetic Biomembranes. In *Methods in Enzymology*; Plant Cell Membranes; Academic Press: San Diego, CA, USA, 1987; Volume 148, pp. 350–382.
39. Pu, C.-H.; Lin, S.-K.; Chuang, W.-C.; Shyu, T.-H. Modified QuEChERS Method for 24 Plant Growth Regulators in Grapes Using LC-MS/MS. *J. Food Drug Anal.* **2018**, *26*, 637–648. [CrossRef]
40. R Core Team. *R: A Language and Environment for Statistical Computing*; R Foundation for Statistical Computing: Vienna, Austria, 2020.

41. Sherin, G.; Aswathi, K.P.R.; Puthur, J.T. Photosynthetic Functions in Plants Subjected to Stresses Are Positively Influenced by Priming. *Plant Stress* **2022**, *4*, 100079. [CrossRef]
42. Waadt, R.; Seller, C.A.; Hsu, P.-K.; Takahashi, Y.; Munemasa, S.; Schroeder, J.I. Plant Hormone Regulation of Abiotic Stress Responses. *Nat. Rev. Mol. Cell Biol.* **2022**, *23*, 1–15. [CrossRef]
43. Salvi, P.; Manna, M.; Kaur, H.; Thakur, T.; Gandass, N.; Bhatt, D.; Muthamilarasan, M. Phytohormone Signaling and Crosstalk in Regulating Drought Stress Response in Plants. *Plant Cell Rep.* **2021**, *40*, 1305–1329. [CrossRef]
44. Ruzic, R.; Jerman, I.; Gogala, N. Effects of Weak Low-Frequency Magnetic Fields on Spruce Seed Germination under Acid Conditions. *Can. J. For. Res.* **1998**, *28*, 609–616. [CrossRef]
45. Mildažienė, V.; Aleknavičiūtė, V.; Žūkienė, R.; Paužaitė, G.; Naučienė, Z.; Filatova, I.; Lyushkevich, V.; Haimi, P.; Tamošiūnė, I.; Baniulis, D. Treatment of Common Sunflower (*Helianthus annus* L.) Seeds with Radio-Frequency Electromagnetic Field and Cold Plasma Induces Changes in Seed Phytohormone Balance, Seedling Development and Leaf Protein Expression. *Sci. Rep.* **2019**, *9*, 6437. [CrossRef]
46. Efthimiadou, A.; Katsenios, N.; Karkanis, A.; Papastylianou, P.; Triantafyllidis, V.; Travlos, I.; Bilalis, D.J. Effects of Presowing Pulsed Electromagnetic Treatment of Tomato Seed on Growth, Yield, and Lycopene Content. *Sci. World J.* **2014**, *2014*, 369745. [CrossRef]
47. Muszyński, S.; Gagoś, M.; Pietruszewski, S. Short-Term Pre-Germination Exposure to ELF Magnetic Field Does Not Influence Seedling Growth in Durum Wheat (Triticum Durum). *Pol. J. Environ. Stud.* **2009**, *18*, 1065–1072.
48. Mildaziene, V.; Pauzaite, G.; Malakauskiene, A.; Zukiene, R.; Nauciene, Z.; Filatova, I.; Azharonok, V.; Lyushkevich, V. Response of Perennial Woody Plants to Seed Treatment by Electromagnetic Field and Low-Temperature Plasma. *Bioelectromagnetics* **2016**, *37*, 536–548. [CrossRef]
49. Peñuelas, J.; Llusià, J.; Martínez, B.; Fontcuberta, J. Diamagnetic Susceptibility and Root Growth Responses to Magnetic Fields in Lens Culinaris, Glycine Soja, and Triticum Aestivum. *Electromagn. Biol. Med.* **2004**, *23*, 97–112. [CrossRef]
50. Jin, Y.; Guo, W.; Hu, X.; Liu, M.; Xu, X.; Hu, F.; Lan, Y.; Lv, C.; Fang, Y.; Liu, M.; et al. Static Magnetic Field Regulates Arabidopsis Root Growth via Auxin Signaling. *Sci. Rep.* **2019**, *9*, 14384. [CrossRef]
51. Bilalis, D.J.; Katsenios, N.; Efthimiadou, A.; Karkanis, A.; Efthimiadis, P. Investigation of Pulsed Electromagnetic Field as a Novel Organic Pre-Sowing Method on Germination and Initial Growth Stages of Cotton. *Electromagn. Biol. Med.* **2012**, *31*, 143–150. [CrossRef]
52. Vashisth, A.; Joshi, D.K. Growth Characteristics of Maize Seeds Exposed to Magnetic Field. *Bioelectromagnetics* **2017**, *38*, 151–157. [CrossRef]
53. Xu, W.; Cui, K.; Xu, A.; Nie, L.; Huang, J.; Peng, S. Drought Stress Condition Increases Root to Shoot Ratio via Alteration of Carbohydrate Partitioning and Enzymatic Activity in Rice Seedlings. *Acta Physiol. Plant.* **2015**, *37*, 9. [CrossRef]
54. Novitskii, Y.I.; Novitskaya, G.V.; Serdyukov, Y.A. Lipid Utilization in Radish Seedlings as Affected by Weak Horizontal Extremely Low Frequency Magnetic Field. *Bioelectromagnetics* **2014**, *35*, 91–99. [CrossRef]
55. Vanderstraeten, J.; Gailly, P.; Malkemper, E.P. Low-Light Dependence of the Magnetic Field Effect on Cryptochromes: Possible Relevance to Plant Ecology. *Front. Plant Sci.* **2018**, *9*, 121. [CrossRef]
56. Bhardwaj, J.; Anand, A.; Pandita, V.K.; Nagarajan, S. Pulsed Magnetic Field Improves Seed Quality of Aged Green Pea Seeds by Homeostasis of Free Radical Content. *J. Food Sci. Technol.* **2016**, *53*, 3969–3977. [CrossRef]
57. Martínez, F.R.; Pacheco, A.D.; Aguilar, C.H.; Pardo, G.P.; Ortiz, E.M. Effects of Magnetic Field Irradiation on Broccoli Seed with Accelerated Aging. *Acta Agroph.* **2014**, *21*, 63–73.
58. Bhardwaj, J.; Anand, A.; Nagarajan, S. Biochemical and Biophysical Changes Associated with Magnetopriming in Germinating Cucumber Seeds. *Plant Physiol. Biochem.* **2012**, *57*, 67–73. [CrossRef]
59. Payez, A.; Ghanati, F.; Behmanesh, M.; Abdolmaleki, P.; Hajnorouzi, A.; Rajabbeigi, E. Increase of Seed Germination, Growth and Membrane Integrity of Wheat Seedlings by Exposure to Static and a 10-KHz Electromagnetic Field. *Electromagn. Biol. Med.* **2013**, *32*, 417–429. [CrossRef]
60. Wei, K.; Jin, X.; Chen, X.; Wu, F.; Zhou, W.; Qiu, B.; Qiu, L.; Wang, X.; Li, C.; Zhang, G. The Effect of H_2O_2 and Abscisic Acid (ABA) Interaction on β-Amylase Activity under Osmotic Stress during Grain Development in Barley. *Plant Physiol. Biochem.* **2009**, *47*, 778–784. [CrossRef]
61. Ishibashi, Y.; Tawaratsumida, T.; Kondo, K.; Kasa, S.; Sakamoto, M.; Aoki, N.; Zheng, S.-H.; Yuasa, T.; Iwaya-Inoue, M. Reactive Oxygen Species Are Involved in Gibberellin/Abscisic Acid Signaling in Barley Aleurone Cells. *Plant Physiol.* **2012**, *158*, 1705–1714. [CrossRef]
62. Ramesh, B.; Kavitha, G.; Gokiladevi, S.; Balachandar, R.K.; Kavitha, K.; Gengadharan, A.C.; Puvanakrishnan, R. Effect of Extremely Low Power Time-Varying Electromagnetic Field on Germination and Other Characteristics in Foxtail Millet (Setaria Italica) Seeds. *Bioelectromagnetics* **2020**, *41*, 526–539. [CrossRef]
63. Podleśna, A.; Bojarszczuk, J.; Podleśny, J. Effect of Pre-Sowing Magnetic Field Treatment on Some Biochemical and Physiological Processes in Faba Bean (*Vicia faba* L. Spp. Minor). *J. Plant Growth Regul.* **2019**, *38*, 1153–1160. [CrossRef]
64. Wojtyla, Ł.; Lechowska, K.; Kubala, S.; Garnczarska, M. Different Modes of Hydrogen Peroxide Action During Seed Germination. *Front. Plant Sci.* **2016**, *7*, 66. [CrossRef] [PubMed]
65. Rochalska, M. Influence of Frequent Magnetic Field on Chlorophyll Content in Leaves of Sugar Beet Plants. *Nukleonika* **2005**, *50*, S25–S28.

66. Pazur, A.; Rassadina, V.; Dandler, J.; Zoller, J. Growth of Etiolated Barley Plants in Weak Static and 50 Hz Electromagnetic Fields Tuned to Calcium Ion Cyclotron Resonance. *Biomagn. Res. Technol.* **2006**, *4*, 1. [CrossRef] [PubMed]
67. Khan, M.; Rozhon, W.; Poppenberger, B. The Role of Hormones in the Aging of Plants–A Mini-Review. *Gerontology* **2014**, *60*, 49–55. [CrossRef] [PubMed]
68. Serova, T.A.; Tsyganova, A.V.; Tikhonovich, I.A.; Tsyganov, V.E. Gibberellins Inhibit Nodule Senescence and Stimulate Nodule Meristem Bifurcation in Pea (*Pisum sativum* L.). *Front. Plant Sci.* **2019**, *10*, 285. [CrossRef] [PubMed]
69. Qin, H.; Huang, R. Auxin Controlled by Ethylene Steers Root Development. *Int. J. Mol. Sci.* **2018**, *19*, 3656. [CrossRef] [PubMed]
70. Humplík, J.F.; Bergougnoux, V.; Van Volkenburgh, E. To Stimulate or Inhibit? That Is the Question for the Function of Abscisic Acid. *Trends Plant Sci.* **2017**, *22*, 830–841. [CrossRef]
71. Binenbaum, J.; Weinstain, R.; Shani, E. Gibberellin Localization and Transport in Plants. *Trends Plant Sci.* **2018**, *23*, 410–421. [CrossRef]
72. Achard, P.; Cheng, H.; De Grauwe, L.; Decat, J.; Schoutteten, H.; Moritz, T.; Van Der Straeten, D.; Peng, J.; Harberd, N.P. Integration of Plant Responses to Environmentally Activated Phytohormonal Signals. *Science* **2006**, *311*, 91–94. [CrossRef]
73. Iftikhar, A.; Ali, S.; Yasmeen, T.; Arif, M.S.; Zubair, M.; Rizwan, M.; Alhaithloul, H.A.S.; Alayafi, A.A.M.; Soliman, M.H. Effect of Gibberellic Acid on Growth, Photosynthesis and Antioxidant Defense System of Wheat under Zinc Oxide Nanoparticle Stress. *Environ. Pollut.* **2019**, *254*, 113109. [CrossRef]
74. Yang, F.; Fan, Y.; Wu, X.; Cheng, Y.; Liu, Q.; Feng, L.; Chen, J.; Wang, Z.; Wang, X.; Yong, T.; et al. Auxin-to-Gibberellin Ratio as a Signal for Light Intensity and Quality in Regulating Soybean Growth and Matter Partitioning. *Front. Plant Sci.* **2018**, *9*, 56. [CrossRef]
75. Adams, E.; Turner, J. COI1, a Jasmonate Receptor, Is Involved in Ethylene-Induced Inhibition of Arabidopsis Root Growth in the Light. *J. Exp. Bot.* **2010**, *61*, 4373–4386. [CrossRef]
76. Kaya, A.; Doganlar, Z.B. Exogenous Jasmonic Acid Induces Stress Tolerance in Tobacco (*Nicotiana tabacum*) Exposed to Imazapic. *Ecotoxicol. Environ. Saf.* **2016**, *124*, 470–479. [CrossRef]
77. Sirhindi, G.; Mushtaq, R.; Gill, S.S.; Sharma, P.; Abd_Allah, E.F.; Ahmad, P. Jasmonic Acid and Methyl Jasmonate Modulate Growth, Photosynthetic Activity and Expression of Photosystem II Subunit Genes in *Brassica oleracea* L. *Sci. Rep.* **2020**, *10*, 9322. [CrossRef]
78. Vanstraelen, M.; Benková, E. Hormonal Interactions in the Regulation of Plant Development. *Annu. Rev. Cell Dev. Biol.* **2012**, *28*, 463–487. [CrossRef]

agronomy

Article

Salt-Induced Autophagy and Programmed Cell Death in Wheat

Larisa I. Fedoreyeva [1,*], Elena M. Lazareva [1,2], Olga V. Shelepova [1,3], Ekaterina N. Baranova [1,3] and Neonila V. Kononenko [1]

1 All-Russia Research Institute of Agricultural Biotechnology, Timiryazevskaya 42, 127550 Moscow, Russia
2 Biological Department, M.V. Lomonosov Moscow State University, Leninskie Gory 1, 119991 Moscow, Russia
3 N.V. Tsitsin Main Botanical Garden of Russian Academy of Sciences, Botanicheskaya 4, 127276 Moscow, Russia
* Correspondence: fedlara@inbox.ru

Abstract: The high salinity of soil salts limits plant growth. Wheat is sensitive to toxic levels of mineral salts. Salinity leads to the accumulation of toxic ions in all organs of wheat. Depending on the level of ion accumulation, wheat is defined as salt stress-tolerant or -sensitive. The wheat variety Zolotaya accumulated Cl^- and Na^+ ions to a greater extent than the Orenburgskaya 22 variety. The accumulation of toxic ions was accompanied by an increase in ROS and an increase in damage to root tissues up to 80% in the Zolotaya variety. The formation of autophagosomes is considered a defense mechanism against abiotic stresses in plants. At a concentration of 150 mM NaCl, an increase in the expression level of *TOR*, which is a negative regulator of the formation of autophagosomes, occurred. The level of *TOR* expression in the Zolotaya variety was 2.8 times higher in the roots and 3.8 times higher in the leaves than in the Orenburgskaya 22 variety. Under the action of salinity, homeostasis was disturbed in the root cells and ROS production accumulated. In the unstable variety Zolotaya, ROS was found in the cap zone and the root meristem in contrast to the resistant variety Orenburgskaya 22 in which ROS production was found only in the cap zone. Accumulation of ROS production triggered autophagy and PCD. PCD markers revealed DNA breaks in the nuclei and metaphase chromosomes, cells with a surface location of phosphatidylserine, and the release of cytochrome c into the cytoplasm, which indicates a mitochondrial pathway for the death of part of the root cells during salinity. Based on electron microscopy data, mitophagy induction was revealed in wheat root and leaf cells under saline conditions.

Keywords: salt tolerance; *Triticum aestivum* L.; *Triticum durum* Desf.; autophagy; mitophagy; PCD; ROS

Citation: Fedoreyeva, L.I.; Lazareva, E.M.; Shelepova, O.V.; Baranova, E.N.; Kononenko, N.V. Salt-Induced Autophagy and Programmed Cell Death in Wheat. *Agronomy* **2022**, *12*, 1909. https://doi.org/10.3390/agronomy12081909

Academic Editors: Sara Álvarez and José Ramón Acosta-Motos

Received: 7 July 2022
Accepted: 8 August 2022
Published: 14 August 2022

1. Introduction

Wheat is one of the most important grain crops. Soil salinity is the most common abiotic stress that inhibits crop growth and yield. A high concentration of sodium chloride has a negative effect on all aspects of plant physiology and metabolism, mainly due to the disruption of the ionic and osmotic balance of cells. Plant resistance to salinity is due to the presence of specific and/or nonspecific mechanisms for ensuring stable metabolism, growth, and development in plant ontogenesis associated with sensitivity to one or more types of stress factors, namely, osmotic, oxidative, and toxic stress effects of NaCl [1]. Mechanisms of salt tolerance include the excretion of Na^+ and Cl^- ions from vacuoles, blocking the transport of Na^+ ions into the cell, exclusion of Na^+ from the transpiration flow, and some other mechanisms [2]. High concentrations of Na^+ ions are toxic for cell metabolism and can inhibit the activity of many important enzymes, cell division and reproduction, membrane disorganization, and osmotic imbalance, which can ultimately lead to growth inhibition and even plant death. In saline soils, high levels of sodium ions lead to inhibition of plant growth and even death.

The program of cell death is genetically inherent in multicellular organisms. Ontogenesis is impossible without the elimination of individual cells, tissue sections, and even entire organs. Programmed cell death (PCD) is necessary for the normal functioning of the organism in terms of removing diseases and terminating the life cycle of mutated dangerous cells and replacing them with new cells [3,4]. Cell death occurs constantly, not only as a result of pathologies in the body but also as a necessary process during normal life. The destruction and death of cells plays an important role during the period of embryonic development during organ laying and tissue differentiation. The aging process can also be considered to be a form of programmed cell death. However, the absence or excess of cell death is pathological and may be a main cause of disease. Thus, cell death responses must be tightly controlled and well-balanced with cell growth and division to ensure cellular and organismal survival. The regulation of life and death processes involves complex cellular pathways that include metabolic and hormonal signals, as well as abiotic and biotic stresses [5–7].

Programmed cell death comprises a series of events which occur in different tissue cells that are programmed to die but with a specific positive effect related to the function of the cell, the tissue itself, or full organism [8]. In land plants, this process can occur in many different highly specialized tissues in relation to their developmental stage; for instance, in the tapetum cells during their lysis, prior to pollen release [9,10], during the death of abnormal megaspores during megasporogenesis in angiosperms [11] by forming antipodal cells [12], or for nucellus dissolution during gametophyte formation [13]. Furthermore, we can see the formation of different tissue structures, such as absorbing trichomes [14] and nectary [15], involving some PCD and autophagic events. Processes associated with an increase in reactive oxygen species (ROS) and their neutralization by organelles have a special role in the processes of programmed cell death [16].

Autophagy is the main pathway for the degradation and processing of cytoplasmic material, including individual proteins, aggregates, and whole organelles [17]. In plants, the role of autophagy in the regulation of programmed cell death (PCD) is still a matter of controversy; however, recent evidence has led to a consensus that autophagy can both promote and limit various forms of PCD [18,19]. In the past few years, the study of autophagy in plants has intensified. It has expanded through studies of the model plant *Arabidopsis thaliana*, photosynthetic organisms, including aquatic photosynthetic eukaryotes [20], gymnosperms [21], and angiosperms, including monocots [22] and dicots [23]. In addition, new data have emerged on the role of autophagy in cell survival [24] and cell death [25].

Autophagy is involved in almost every aspect of plant life, including germination, seedling formation, development, reproduction, metabolism, and plant responses to biotic and abiotic stresses, including malnutrition, oxidation, osmosis, drought, and pathogenic infections [26,27].

Despite the fact that the study of autophagy in plants is lagging significantly behind the study of the process of autophagy in mammals and yeast, some mechanisms that function in plants have recently been revealed [28,29]. The autophagy process proceeds through the formation of autophagosomes [30]. Under the action of signals initiating macroautophagy, the so-called phagophore is formed, which consists of a lipid membrane and a number of autophagy-related proteins (ATG) encoded by *ATG* genes or *ATG* gene homologues. With the help of a complex system of regulation, multicomponent complexes are assembled, the membrane grows, and an open structure is formed. Then, the bilayer membrane closes, and inside the resulting vesicle called autophagosome are macromolecules and organelles (ribosomes, mitochondria, and fragments of the endoplasmic reticulum). Autophagosomes are two membrane vesicles separating part of the cytoplasm. It is believed that, ultimately, the outer membrane of the autophagosome fuses with the vacuole (in yeast and plants) or lysosome (in mammals) to release cytoplasmic material for hydrolytic degradation. More than 30 genes associated with autophagy (*ATG*) are involved in the processes of membrane remodeling and transport [31,32].

Autophagy-related ATG proteins play an essential role in the process of programmed cell death. The autophagy process is highly conserved, and ATG orthologues are present in fungi, plants, and mammals [33,34]. Autophagic proteins regulate the formation of a double membrane-coated phagophore assembly (PAS, also known as the pre-autophagosomal structure), which then expands to form autophagosomes 500–1000 nm in diameter, and the subsequent fusion of these vesicles with lysosomal or vacuolar compartments to degrade the components [35].

Proteins that are essential for the autophagy process have been grouped into four major functional groups [36,37]. The first group includes the ATG1/ATG13 kinase complex, which initiates the formation of autophagosomes in response to signals of the cell's need for nutrients, as well as to various stress factors. Further, ATG9 contributes to the expansion of the phagophore by moving membrane components from different sources, and the complex with ubiquitin-like proteins ATG8/ATG12 is involved in the expansion of the phagophore and its maturation [37]. The ATG5-ATG12/ATG16 complex functions as an E3 ligase that transfers phosphoethanoamine (PE) to *ATG*8 in vitro [38] to produce the autophagosome-localized lipid form of *ATG*8-PE [29,39].

Redox metabolism in plant cells inevitably includes the formation of ROS. There is evidence that ROS signals may be the primary targets of autophagy [16]. The ROS molecule can function as a signaling molecule to trigger autophagy as a survival mechanism [40]. Autophagy has been shown to be strongly induced by oxidative stress [41]. Oxidative stress in wheat roots caused by prooxidants paraquat and salicylic acid leads to intensive formation of autophagosomes [42]. Dysregulation of autophagy leads to increased oxidative stress, as shown by inhibitory and knockout studies [43,44].

Although autophagy appears to be involved in plant responses to high salinity and osmotic and drought stresses [45], its precise role has yet to be determined. Salt and osmotic stress can also increase ROS production and cause protein damage; one possibility is that autophagy may be responsible for the degradation of oxidized proteins under salt and osmotic stress [46].

The aim of this study was to diagnose tissue damage in durum and soft wheat as a result of the action of sodium chloride at the stage of seedlings and the pathway of cell death of damaged tissues using PCD markers.

2. Materials and Methods

2.1. Plant Material

Two varieties of spring wheat *Triticum durum* Desf. Zolotaya (2n = 28) and *Triticum aestivum* L. Orenburgskaya 22 (2n = 42), developed by the Orenburg Research Institute of Agriculture of the steppe ecological group (FGBNU, Federal Scientific Center of the Russian Academy of Sciences, Orenburg, Russia), were used in this study. The sensitivity of wheat seedlings to salinity was assessed using the roll culture method [47]. Seedlings were grown in the presence of 150 mM NaCl, which corresponds to an osmotic pressure of 6 atm, at 24 °C under a 10 h light/14 h dark photoperiod and fluorescent lamps (5000 lx). After 10 days of growth (or 4 days, depending on the task), the fresh plant biomass, root length, and shoot length were measured. Data are expressed as mean ± standard deviation (SD; $n = 30$), and significant differences were determined $p < 0.05$.

2.2. Ion Detection

The cell walls of 10-day-old wheat shoots and roots (100–300 mg in 25 mL of deionized water) were destroyed by ultrasonication in an apparatus (Sapfir, Moscow, Russia) at 35 kHz for 30 min at 40 °C. The resulting suspension was filtered on a 0.45 µm Millipore membrane. Samples were analyzed on an ITAN ionometer (TomskAnalit, Tomsk, Russia). The content of ions in mg/L in the samples was determined from the calibration graph. The content of electrolytes in the samples was determined by the electrical conductivity of the solution using an Expert-002 conductometer (Ekoniks, Tomsk, Russia).

2.3. Trypan Blue Staining

Coleoptiles of 10-day-old seedlings were stained with 0.5% trypan blue for 5 min and then washed three times. Samples were visualized by light microscopy (Olympus BX51 microscope; 10× lens) and photographed using a Color View digital camera (Germany).

2.4. Fluorescence Microscopy

Root tips (4–5 mm) of all 10-day-old seedlings were excised and placed on a glass slide in a drop of water (five root tips per glass slide). To determine ROS levels in cells, the root tips were incubated in 25–50 nM carboxy-H2DFFDA (Thermo Fisher Scientific, Waltham, MA, USA) for 30 min and then washed three times with distilled water. The root tip samples were analyzed under Olympus BX51 fluorescence microscope (Japan), fitted with 10× objective lens, at a wavelength of 490 nm. Images were obtained using Color View digital camera (Germany).

2.5. Transmission Electron Microscopy (TEM)

Root apex segments of 4-day-old seedlings (4 mm) were fixed for 24 h in 2.5% glutaraldehyde (Merck, Darmstadt, Germany) dissolved in 0.1 M Sorensen's phosphate buffer with 1.5% sucrose (pH 7.2). Then, the samples were washed and post-fixed in 1% OsO^4 (Sigma-Aldrich, St. Louis, MO, USA) and dehydrated in ethanol of increased concentrations (30, 50, 70, 96, and 100%) and in propylene oxide (Fluka, Nuremberg, Germany). The samples were embedded in a mixture of Epon-812 and Araldite (Merck, Darmstadt, Germany) according to the standard procedure. For TEM, the embedded samples were sectioned using an ultramicrotome LKB-III (LKB, Sweden), placed on formvar coated grids, and stained with uranyl acetate and lead citrate. The ultrathin sections were examined and photographed with an electron microscope H-300 (Hitachi, Tokyo, Japan). The ultrastructure of mitochondria at root parenchyma cells was studied.

2.6. Apoptosis Detection Assay

Root apex segments of 4-day-old seedlings (15 mm) were cut off and fixed in a solution of 4% paraformaldehyde (Sigma Aldrich, St. Louis, MO, USA) in PHEM buffer pH = 6.9 (60 mM PIPES (Sigma Aldrich, St. Louis, MO, USA), 25 mM HEPES (Sigma Aldrich, St. Louis, MO, USA), 10 mM EGTA (Sigma Aldrich, St. Louis, MO, USA), and 2 mM $MgCl_2$ (Sigma Aldrich, St. Louis, MO, USA) for 1.5–2 h at room temperature. The fixative was washed in PHEM buffer.

To prepare preparations of macerated cells (without a cell wall), fixed root tips were incubated for 10–15 min in 0.4 M mannitol containing 1% cellulase (Sigma Aldrich, St. Louis, MO, USA) and 5 mM EGTA, washed in PBS buffer (2 times for 10 min), transferred to a drop of buffer onto a coverslip, and divided into cells with metal needles. The finished preparations were dried in a refrigerator at +4 °C for 24 h.

To identify root tissue cells at the stages of programmed cell death (PCD) under salinity, phosphatidylserine was detected using Xpert Annexin V-FITC Apoptosis Detection Assay (Grisp, Spain). A total of (100 μL) of Annexin V-FITC solution was used at a dilution of 5 μL per 500 μL of reaction buffer, incubated for 30 min in the dark at 22 °C, washed 3 times for 5 min, transferred to drops (100 μL) of propidium iodide at a dilution of 10 μL at 500 μL, washed twice for 5 min, and mounted in Mowiol U-44 (Hoechst, Frankfurt, Germany) supplemented with DAPI (1 μL/1 mL) (4,6-diamidino-2-phenilindole) (Sigma Aldrich, St. Louis, MO, USA).

2.7. TUNEL Analysis

Nuclear DNA breaks were detected by the terminal deoxynucleotidyl transferase dUTP nick end labeling (TUNEL) method. Preparations with macerated cells were permeabilized in a 0.5% solution of Triton X100 in PBS for 30 min, then after washing twice with buffer, they were placed in cocodelate buffer (pH 7.4) containing terminal deoxynucleotidyltransferase 20 units/μL (Silex, Moscow, Russia), 3′ labeled probes with 10 mM dATP (Silex, Moscow, Russia), and 1 mM fluorescein (Silex, Moscow, Russia). The reaction was stopped by placing the preparations in a 2× SSC solution for 15 min. After washing twice with buffer, preparations were mounted in Mowiol U-44 (Hoechst, Frankfurt, Germany) supplemented with DAPI (1 μL/1 mL).

2.8. Cytochrome c Detection

For immunocytochemical detection of cytochrome c, the preparations were placed in PHEM buffer for 5 min and transferred for 30 min to a solution of 0.5% Triton X-100 in PHEM buffer containing 5% DMSO. Then, it was washed in PBS (pH 7.4) and incubated for 16–18 h at room temperature with rabbit polyclonal antibodies to cytochrome c at a dilution of 1:100 in PBS (pH 7.4) containing 0.1% BSA. Then, the preparations were washed and incubated with goat anti-rabbit Igl antibodies conjugated with Texas Red (Sigma, St. Louis, MO, USA) at a dilution of 1:25 for 45 min at 37 °C, washed, stained with DAPI, and enclosed in Mowiol U-88 (Hoechst, Frankfurt, Germany). All obtained preparations were analyzed using an Axiovert 200 M microscope (Zeiss, Oberkochen, Germany) with epifluorescent illumination and a Neofluar ×10 and ×20 objective. Images were obtained using an AxioCam HRm camera.

2.9. Total RNA Isolation and Gene Expression Analysis

Total RNA was isolated from individual shoots and roots using reagent kits for the isolation of RNA-Extran RNA Syntol (Russia), according to the instructions. Then, cDNA was synthesized by reverse transcription using a standard method (Synthol, Moscow, Russia). The concentration of cDNA was determined spectrophotometrically on a nanophotometer IMPLEN.

To analyze gene expression, the cDNA was amplified by a real-time polymerase chain reaction (RT-PCR) using SYBR Green I (Syntol) on CFX 96 Real-Time System thermal cycler (BioRad, Hercules, CA, USA). Information on the structure of *Triticum aestivum* genes was obtained from the National Center for Biotechnology Information (NCBI). Gene-specific primers were designed using NCBI Primer-BLAST and synthesized by Syntol. The RT-PCR was carried out under the following conditions: 95 °C for 5 min, then 45 cycles of 94 °C for 30 s, 59 °C for 30 s, and 72 °C for 30 s. Each RT-PCR reaction was performed in three replicates.

2.10. Statistical Methods

The calculation of the main statistical parameters was carried out according to standard methods, and Statistica 6.0 and STATAN programs for statistical data processing were used. Values are presented as means ± standard deviation of the three biological replicates. All of the treatment effects were statistically analyzed using the Student's *t*-test (DPS software). Different letters indicate significant differences at $p < 0.05$.

3. Results

3.1. Morphometric Parameters

The impact of salinity in the studied wheat varieties caused a decrease in the growth of both the root system and aboveground organs (Figure 1). Morphometric evaluation of seedlings made it possible to assess the changes more clearly. As can be seen from Figure 1C,D, the variety Orenburgskaya 22 was more stable in the height of the aboveground part in the presence of 150 mM NaCl. In contrast to the variety Orenburgskaya 22, where the decrease in the length of the shoot was about 40%, that in the variety Zolotaya was 65%. Thus, the difference in the reduction in the length of the shoot between wheat varieties was about 25% in the presence of 150 mM NaCl. When salinized with sodium chloride, the difference between varieties along the length of the root was about 20%. Chloride salinity causes various morphological changes in varieties associated with changes in the habitus of plants and the shape of their organs.

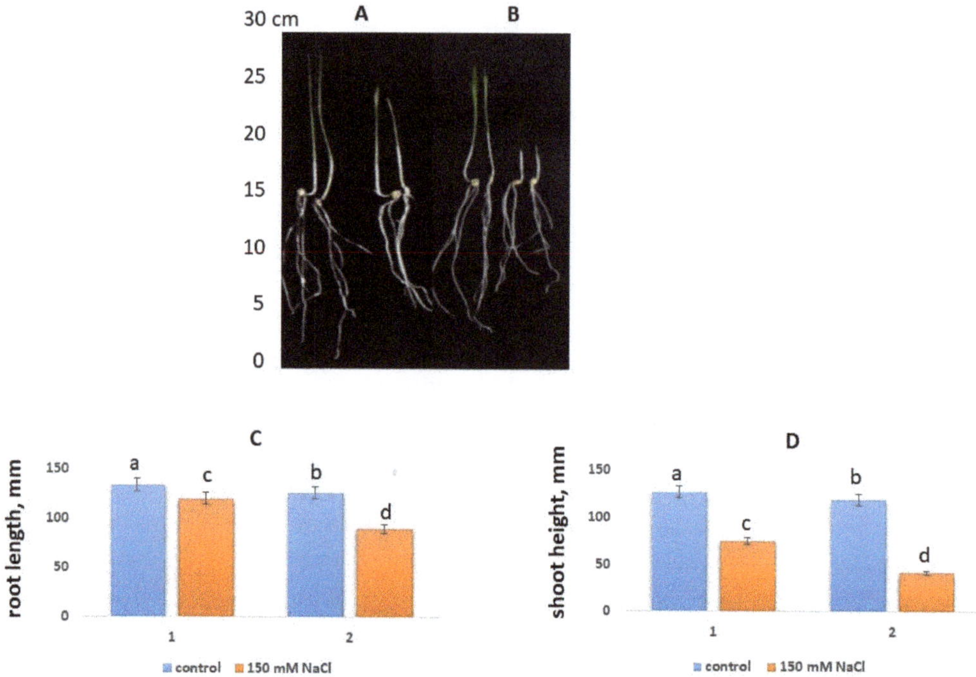

Figure 1. 10-day seedlings of wheat after germination under normal conditions (left) and chloride salinization (right) in roll culture. (**A**)—Variety Orenburgskaya 22; (**B**)—variety Zolotaya; (**C**,**D**)—biometric parameters; (**C**)—Root length: (**D**)—Shoot length: 1—variety Orenburgskaya 22; 2—variety Zolotaya. Blue—control, Orange—150 mM NaCl. The mean values (n = 30) and their standard deviations, $p < 0.05$. Different letters indicate significant differences at $p < 0.05$.

3.2. *Ionic Detection*

The genotypes of shoots and roots of wheat showed a different reaction in relation to the accumulation of Na^+ and K^+ cations and chloride ions. Since the roots are in direct contact with the soil and absorb nutrients, a higher accumulation of Na^+ was observed in the roots compared to the control. Orenburgskaya 22, according to morphometric parameters, is a salt-tolerant wheat variety compared to the Zolotaya variety. The tolerance of plant organisms to high concentrations of toxic ions is determined by various mechanisms of excretion and control over the content of ions in tissues [48,49].

From the results obtained (Table 1), it follows that Na^+ enters the roots of the resistant variety Orenburgskaya 22 to a limited extent, then the ions are transported to the shoots. In the unstable variety Zolotaya, accumulation of Na^+ ions was observed both in the roots and in the shoots, which may indicate that the mechanisms of excretion of ions from the xylem into the root vacuoles, as well as the outflow of excess Na^+ ions from the shoots, are disturbed in the variety Zolotaya. The results showed that the salt-tolerant variety retained K^+-to-Na^+ selectivity and maintained a lower Na^+/K^+ ratio in the shoots under stress, while the tolerant Zolotaya variety had an imbalance in the ion ratio.

Table 1. Ion content and electrical conductivity in root and shoot of Orenburgskaya 22 and Zolotaya wheat.

Wheat Variety		Ions				
		K^+, mkg/g	Na^+, mkg/g	Na^+/K^+	Cl^-, mkg/g	Electrical Conductivity, mkSm
Oren-22	control	0.43 ± 0.02 e	0.3 ± 0.01 e	0.70 ± 0.03 d	0.64 ± 0.03 e	95.81 ± 4.8 f
root	150 mM NaCl	0.64 ± 0.03 d	1.16 ± 0.06 c	1.81 ± 0.06 a	3.55 ± 0.18 b	234.73 ± 11.74 c
Oren-22	control	1.24 ± 0.06 c	0.74 ± 0.04 d	0.60 ± 0.03 e	0.53 ± 0.03 f	116.4 ± 5.82 e
shoot	150 mM NaCl	1.47 ± 0.07 b	1.22 ± 0.06 c	0.83 ± 0.04 c	1.65 ± 0.08 d	230.69 ± 11.53 c
Zolotaya	control	0.66 ± 0.03 d	0.35 ± 0.02 e	0.53 ± 0.02 f	0.58 ± 0.03 e	116.68 ± 5.83 e
root	150 mM NaCl	1.18 ± 0.06 c	1.39 ± 0.07 b	1.78 ± 0.05 a	4.63 ± 0.23 a	411.05 ± 20.55 a
Zolotaya	control	1.38 ± 0.07 b	0.80 ± 0.04 d	0.58 ± 0.02 e	0.54 ± 0.03 f	133.43 ± 6.67 d
shoot	150 mM NaCl	1.88 ± 0.09 a	2.25 ± 0.11 a	1.20 ± 0.04 b	2.42 ± 0.12 c	274.51 ± 13.72 b

The mean values (n = 3) and their standard deviations are shown according standard deviation, $p < 0.05$. Different letters indicate significant differences at $p < 0.05$.

Comparison of the distribution of Na^+ between shoot and root tissues under control conditions revealed the accumulation of Na^+ at a higher concentration in the shoot, indicating that the shoot is the main Na^+ accumulator. However, when growing wheat in the presence of 150 mM NaCl, an excess of Na^+ and Cl^- accumulated in the roots. It should be noted that while the content of Na^+ ions in the roots under stress increased in Orenburgskaya 22 and Zolotaya almost the same (by 3.9 and 4.0 times, respectively), the content of chloride ions in the Zolotaya variety turned out to be significantly higher than in the resistant variety Orenburgskaya 22 (by 8.0 and 5.5 times, respectively).

The electrical conductivity parameter is proportional to the concentration of electrolytes and reflects the degree of accumulation of the sum of ions ($K^+/Na^+/Cl^-$) in wheat tissues. The lowest electrical conductivity was in the roots of the Orenburgskaya 22 variety, and the highest was in the roots of the Zolotaya variety in the presence 150 mM NaCl. In the control in the roots, the electrical conductivity of both varieties was approximately the same. However, in the presence of an increased concentration of NaCl, the sum of ions accumulated in the roots of the unstable variety Zolotaya compared to Orenburgskaya 22 (by 3.5 and 2.4 times, respectively).

3.3. *Expression of Gene* HKT

The expression of HKT1;4 and HKT2;1 genes, class 1 and 2 ion transporters, was studied in the shoots and roots of wheat varieties Orenburgskaya 22 and Zolotaya (Figure 2).

The level of expression of HKT1;4 and HKT2;1 in variety Orenburgskaya 22 in the root was approximately equal to each other but almost two times higher than in the shoots. Although in the presence of 150 mM NaCl, the expression level of HKT2.*1* increased slightly (only by 1.1 times) in the root, we suggest that the role of the class 2 ion transporter, which removes both K^+ and Na^+ ions, in salt stress increases compared to the class 1 ion transporter, since its expression level decreased by 1.2 times. In the shoot, the activity of genes of both classes decreased in the presence of sodium chloride. It was noted that in the shoot, the level of expression of the HKT1;4 ion transporter under control conditions in the Zolotaya variety was higher than in the Orenburgskaya 22 variety.

3.4. *Expression of AFG Genes*

Autophagy proteins (ATGs) play an essential role in the formation of autophagosomes involved in the removal and processing of dying cells and their components. There are few data on the mechanisms and physiological functions of autophagy in cultivated plants, especially in wheat, which is one of the most important food crops in the world. It has been noted that soft wheat ATG4 and ATG8 proteins play the most important role in the process of autophagy in response to adverse environmental influences [50]. We also studied the expression of the *ATG4* and *ATG8* genes, as well as the *ATG1* and *TOR* genes, which are initiators of phagophore formation (Figure 3).

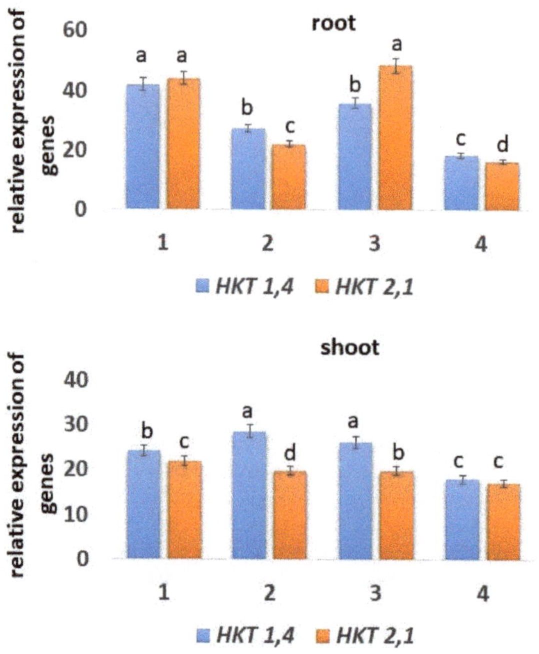

Figure 2. Expression of genes HKT in root and shoot of wheat Orenburgskaya 22 (1, 3) and Zolotaya (2, 4) grown in control condition (1, 2) and in presence of 150 mM NaCl. Bar represent standard deviation, $p < 0.05$. Different letters indicate significant differences at $p < 0.05$.

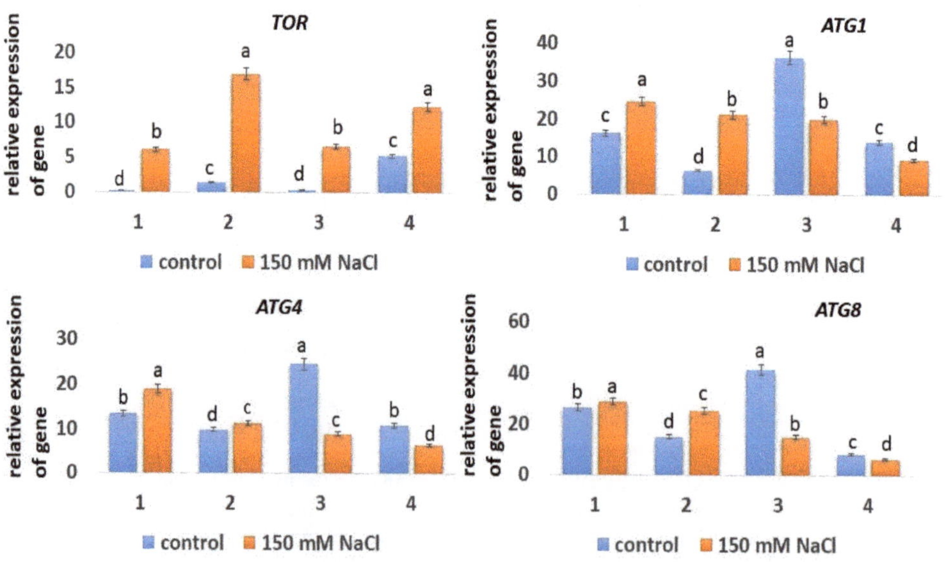

Figure 3. Expression of genes *TOR, ATG1, ATG4,* and *ATG8* in root (1, 2) and shoot (3, 4) of wheat Orenburgskaya 22 (1, 3) and Zolotaya (2, 4) grown in control condition (1, 2) and in presence of 150 mM NaCl. Bar represent standard deviation, $p < 0.05$. Different letters indicate significant differences at $p < 0.05$.

The relative level of *TOR* gene expression in the roots of both wheat varieties under control growing conditions was very low. However, in the shoots of the Zolotaya variety, the expression of the *TOR* gene was 3.8 times higher than in the roots. With an increase in the content of NaCl in solution, the expression of the *TOR* gene increased in all variants. The expression of the *TOR* gene in the variety Orenburgskaya 22 increased in the roots (by 30 times) and in the shoots (by 22 times) compared with the control, and in the variety Zolotaya, in the roots by 12 times and in the shoots by 2.3 times. However, the level of expression of the *TOR* gene in Orenburgskaya 22 in the roots was 2.8 times lower than in the Zolotaya variety and 3.8 times lower in the shoots. TOR is thought to be a negative regulator of autophagy in plants. *TOR* overexpression blocks autophagy activation under saline and osmotic stress [51]. Based on the obtained data, it follows that under the control conditions of wheat cultivation, the formation of autophagosomes was induced along the *TOR*-independent pathway. Under salt stress, *TOR* was activated, which led to the blocking of the autophagy process and an increase in the process of programmed cell death in wheat, especially the Zolotaya variety, thereby confirming that the Zolotaya variety is a salt stress-sensitive variety.

ATG1 protein binds and localizes in phagophores. The expansion of the phagophore depends on the degree of its phosphorylation. In the control variety Orenburgskaya 22, the relative expression of the *ATG1* gene both in roots and shoots was 2.6 times higher than in the Zolotaya variety. Under salinization, the expression of the *ATG1* gene in the roots of Orenburgskaya 22 increased by 1.5 times and in the Zolotaya variety by 3.3 times. Interestingly, when 150 mM NaCl was added to the solution, *ATG1* gene expression in shoots decreased in both wheat varieties. This fact can probably be explained by the fact that under salt stress, the autophagy process proceeds along the TOR-dependent pathway, and overexpression of the *TOR* gene plays a negative role in *ATG1* expression.

The ATG8 protein is one of the most important autophagy proteins. It binds to PE, localizes on the membrane, and participates in membrane closure and its fusion with the vacuole. In the variety Orenburgskaya 22 in the control shoots, the expression of the *ATG8* gene was 1.5 times higher. At the same time, in the Zolotaya variety, the expression of the *ATG8* gene in the control roots was 1.8 times higher than in the shoots. When sodium chloride was added to the solution, both varieties behaved similarly: there was an increase in the expression of the *ATG8* gene in the roots (Orenburgskaya 22 by about 1.1 times and Zolotaya by 1.7 times) and a decrease in the shoots (Orenburgskaya 22 by 2.7 times and Zolotaya by 1.3 times). Although the nature of changes in the *ATG8* gene expression was similar, the *ATG8* activity in the Orenburgskaya 22 variety was significantly higher than in the Zolotaya variety.

Activation of the ATG8 protein requires cleavage of its C-terminus to release the glycine residue responsible for its lipidation. The ATG4 protein is a cysteine protease. Under salinization, the expression of the *ATG4* gene in the roots of the Orenburgskaya 22 variety increased by 1.4 times and in the roots of the Zolotaya variety by 1.1 times. As follows from the data in the Figure 3, the expression profiles of the *ATG8* and *ATG4* genes in both wheat varieties were similar. This fact confirms that these two proteins are interconnected and necessary for each other.

Although the nature of changes in the expression profile of the *ATG8* and *ATG4* genes was similar in resistant- and unresistant-to-salinity wheat varieties, the level of the expression of these genes in the Orenburgskaya 22 (resistant) variety was higher than in the Zolotaya (unstable) variety. This fact is most likely associated with a more intensive formation of autophagosomes and thus confirms that the process of autophagy under salt stress is a protective system.

3.5. TEM Analysis

The cells of the initials of the cortex have a dense structure of the cytoplasm with rounded or oval organelles typical for the cells of the meristematic zone. These cells are characterized by the central position of the nucleus and the absence of vacuoles. Plastids and mitochondria have a structure characteristic of these cells: they do not have developed internal membranes and significant inclusions, and they contain a dense stroma and a matrix with randomly arranged ribosomes (Figure 4a). With further development, many small vacuoles are formed in these cells, which actively merge and push the nucleus and the organelle containing the cytoplasm to the cell periphery. The junctions of small and large vacuoles are shown by arrows. Vacuole inclusions do not have a pronounced structure and contain phenolic inclusions, and less often, proteins (dark spots in Figure 4c). Under the action of toxic salt concentrations, a part of the cytoplasm can be modified and localized as autophagosomes in vacuoles, which in this case perform a lytic function (Figure 4b). With the development of damage, most of the organelles are inside the vacuole, the turgor is disturbed, and the structure of the cytoplasm becomes uneven with areas of different density and further autophagosome degradation (Figure 4d).

The cells of the aboveground organs also transform from dense vacuoleless cells to cells with a large central vacuole and peripherally displaced organelles (Figure 4e). Under the action of salinity, the fusion of vacuoles is disrupted, the nucleus remains in the center of the cells, the organelles are located both near the nucleus and on the periphery, and the cytoplasm forms characteristic strands connecting the peripheral cytoplasm and perinuclear fragments (Figure 4f).

Normally, mitochondria of cells that have switched to specialization acquire developed cristae and have a developed matrix (Figure 4a,c,e insets). Salinity changes the structure of mitochondria, causing a change in density and impaired formation of cristae and their uneven location, which differs greatly depending on the location of the cells (Figure 4b,d,f inserts).

3.6. Fluorescence Analysis

Processes associated with an increase in reactive oxygen species (ROS) and their neutralization by organelles play a special role in the processes of programmed cell death. Single-stained cells are found on the surface of the roots; however, the intensity of staining differs between varieties in cells from different root zones. During salinization, the most intense coloration of ROS was observed in the zone of the cap and division, which was more intense in the Zolotaya variety. At the same time, compared with the control, the increase in the level of ROS was observed to the greatest extent in the cells of the epidermis and cortex and to a lesser extent in the zone of the central cylinder (Figure 5).

3.7. Trypan Blue Analysis

Trypan blue staining of coleoptiles of two wheat varieties was carried out to characterize viability under salt stress. Trypan blue penetrates the membrane of dead cells, and staining determines the degree of tissue damage during salinization and the number of dead cells (Figure 6). In the control, there were almost no visible changes in the coleoptile, while in the presence of sodium chloride, rather strong tissue damage was observed, depending on the wheat variety. Therefore, in the Zolotaya variety, more than 80% of the cells were damaged by salinity. At the same time, in the variety Orenburgskaya 22 in the variant with NaCl, there were up to 20% of dead cells.

3.8. Apoptosis Assay

On preparations macerated with the help of an enzyme that destroys the cell wall of the cells of the roots of seedlings of resistant- and tolerant-to-salinity Zolotaya varieties of wheat, we carried out the detection of phosphatidylserine using Annexin V-FITC (Figure 7a,c), while propidium iodide did not stain the DNA of these nuclei. Such localization of phosphatidylserine was found on the surface of 5% of root cells of 4-day-old seedlings of the Zolotaya variety.

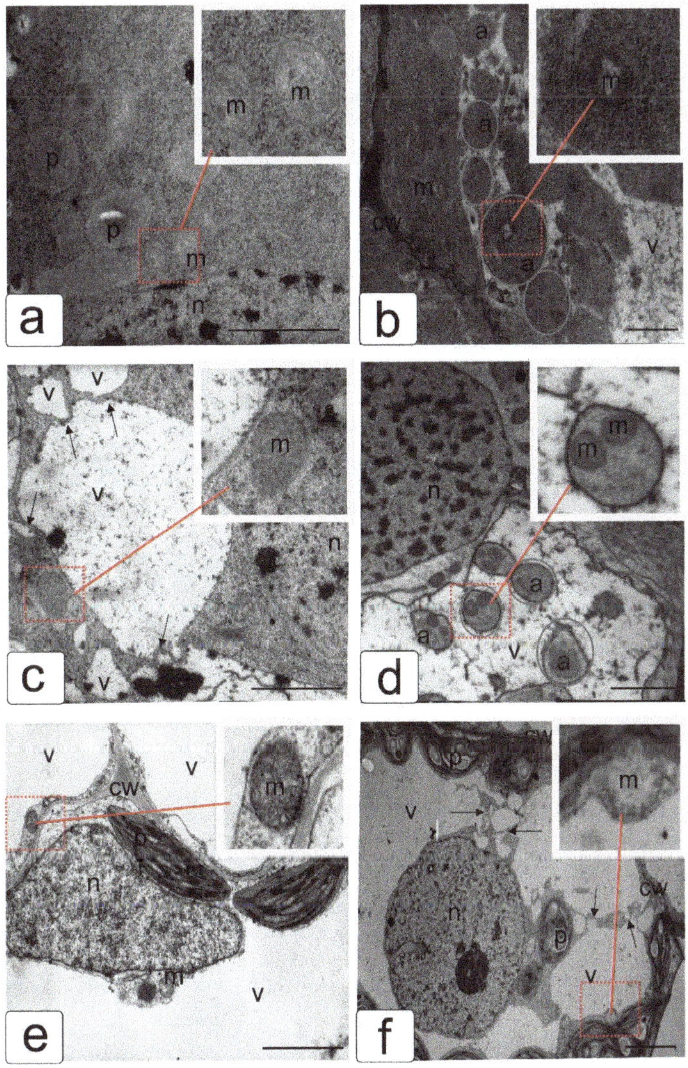

Figure 4. Formation and fusion of vacuoles and autophagosomes in cells of the root cortex (**a**–**d**) and aerial part (**e**,**f**) under control conditions and (**a**,**c**,**e**) under the influence of 150 mM NaCl. Designations: p—plastid; m—mitochondria; cw—cell wall; n—nucleus; v—vacuoles; a—autophagosomes; arrows point to the vacuole fusion sites; Red arrows indicate the position of the enlarged fragment with inserts with mitochondria. Bar 3 μm.

3.9. TUNEL Detection

Nuclear DNA breaks were detected by the TUNEL method (Figure 8). In the salinity-resistant variety Orenburgskaya 22, DNA breaks were observed in the nuclei of 0.4% of the control cells, and in the presence of NaCl, in 19% of the cells (Figure 8e,f). Breaks were observed in metaphase chromosomes and micronuclei (Figure 8f,i). In the salt-tolerant variety Zolotaya, DNA breaks were observed in the nuclei of 0.5% of control cells, and in the presence of NaCl, in 32% of cells.

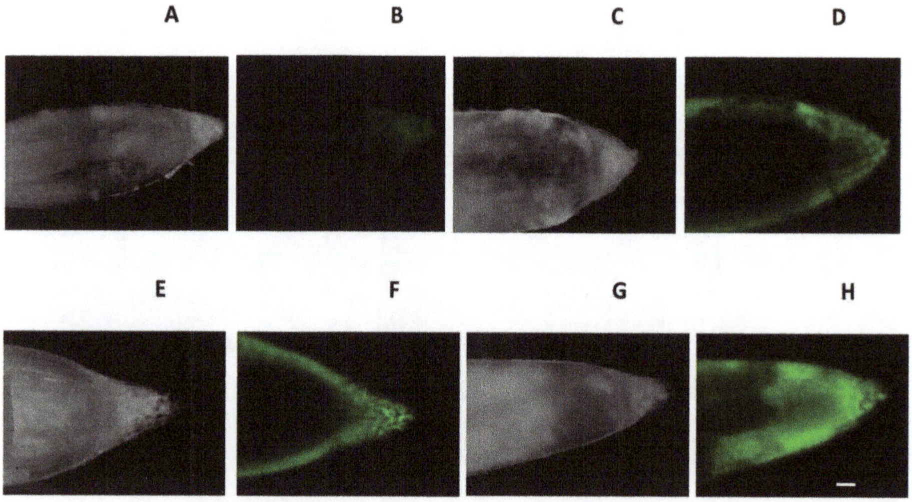

Figure 5. Distribution of ROS^+ and ROS^- cells in the zones of 10-day wheat roots. (**A–D**)—Orenburgskaya 22; (**E–H**)—Zolotaya; (**A,B,E,F**)—control; (**C,D,G,H**)—150 mM NaCl: (**A,C,E,G**)—light microscopy; (**B,D,F,H**)—fluorescence microscopy, respectively. Bar 400 µm.

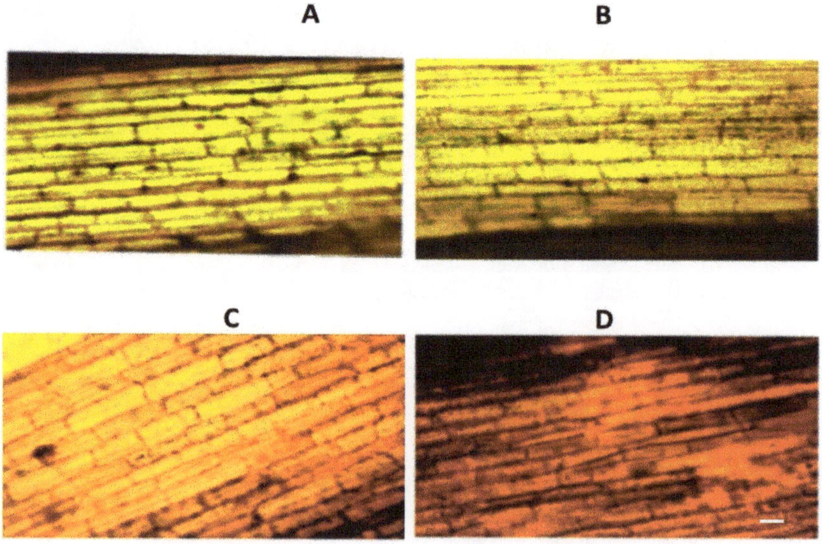

Figure 6. Trypan blue staining of coleoptile of 10-day seedlings differing in the number of dead cells. (**A,B**)—Orenburgskaya 22; (**C,D**)—Zolotaya; (**A,C**)—control; (**B,D**)—150 mM NaCl. Bar 400 µm.

3.10. Cytochrome c Assay

Cytochrome c immunodetection was carried out in the cytoplasm of root cells (Figure 9). In salinity-resistant variety Orenburgskaya 22, cytochrome c was detected both in mitochondria and in the cytoplasm of 9% of cells. In the salt-tolerant variety Zolotaya, cytochrome c was detected both in mitochondria and in the cytoplasm of 15% of cells.

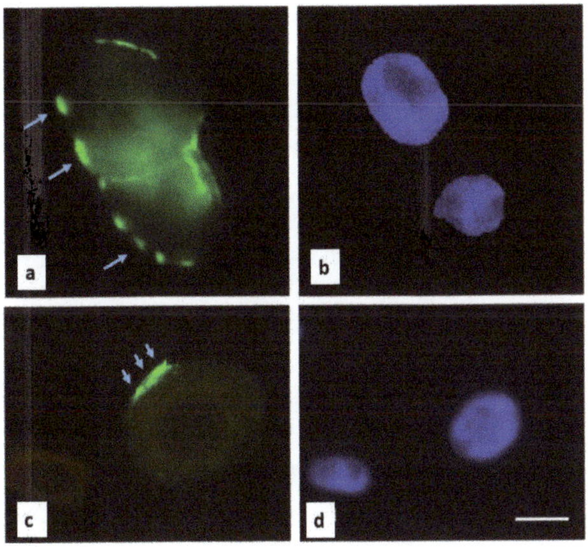

Figure 7. Localization of phosphatidylserine on the surface of plasma membranes of root cells of 4-day-old seedlings of Triticum durum variety Zolotaya in the presence of 150 mM NaCl. (**a**,**c**)—Clusters of phosphatidylserine on surface plasmatic membranes (Annexin V-FITC), blue arrows; (**b**,**d**)—nuclei of cells (DAPI). Bar 200 μm.

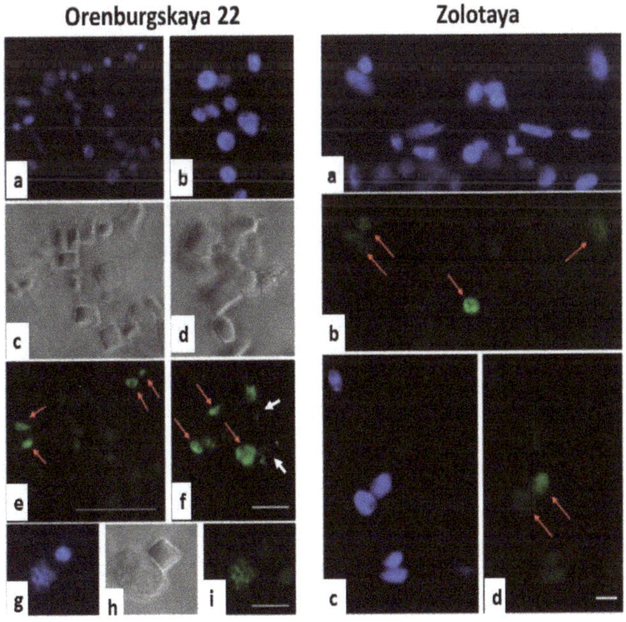

Figure 8. DNA breaks in the nuclei of cell tissues of the roots of 4-day-old seedlings of Triticum aestivum variety Orenburgskaya 22 and Triticum durum variety Zolotaya in the presence of 150 mM NaCl, detected by the TUNEL method. (**a**,**b**,**g**)—Root cell nuclei (DAPI); (**c**,**d**,**h**)—phase contrast of cells; (**e**,**f**,**i**)—nuclei, micronuclei (arrows—white), and chromosomes with DNA breaks (arrows—red). Bar 200 μm.

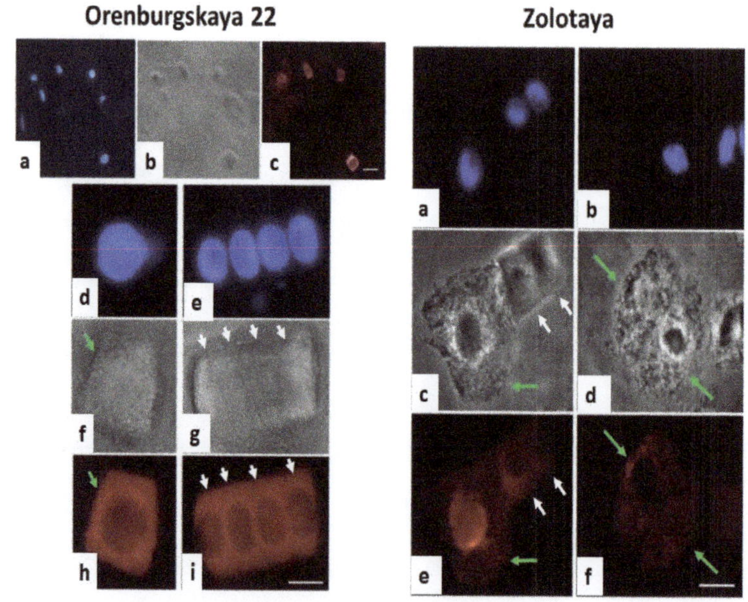

Figure 9. DNA breaks in the nuclei of cell tissues of the roots of 4-day-old seedlings in the presence of 150 mM NaCl, detected by the TUNEL method in Triticum aestivum variety Orenburgskaya 22: (**a**,**b**,**g**)—root cell nuclei (DAPI); (**c**,**d**,**h**)—phase contrast of cells; and (**e**,**f**,**i**)—nuclei, micronuclei (arrows—white), and chromosomes with DNA breaks (arrows—green); and in Triticum durum variety Zolotaya, (**c**)—nuclei of root tissue cells (DAPI); (**c**,**d**)—nuclei with DNA breaks (arrows—green), Bar 200 µm.

4. Discussion

For growth and development on saline soils, the plant needs to regulate the accumulation of toxic ions. In the course of evolution, plants have acquired several mechanisms of salt tolerance. It is known that more salt-tolerant plant species have high adaptive properties; they are able not to accumulate Na^+ and Cl^- in vacuoles and they maintain a low concentration in the cytoplasm [52].

Mechanisms of salt tolerance include the excretion of Na^+ and Cl^- ions from vacuoles, blocking the transport of Na^+ ions into the cell, exclusion of Na^+ from the transpiration flow, and some other mechanisms [53]. Susceptibility and tolerance to stress caused by the action of high concentrations of NaCl in plants is a coordinated action of many genes that respond to stress [54]. Generally, K^+ is preferred for uptake by roots from the soil, and most plants show a high degree of K^+/Na^+ discrimination in their uptake. High-affinity potassium transporters (*HKTs*) are active at the plasma membrane level and function as a Na^+/K^+ symporter as well as a Na^+ selective uniporter [55]. Plant gene *HKT* transporters of potassium and sodium ions are divided into two subfamilies [56]. The subfamily *HKT1* is found in all higher plants. Genes of this class encode selective ion transporters, while subfamily 2 genes encode transporters that are permeable to both K^+ and Na^+ ions. Violation of the expression of genes of the *HKT1* family leads to hypersensitivity to Na^+ ions and excessive accumulation of sodium in the shoots.

An excess of Na^+ and Cl^- accumulated in the roots of wheat grown in the presence of 150 mM NaCl (Table 1). The content of Na^+ and Cl^- ions in the Zolotaya variety in the roots turned out to be significantly higher than in the resistant variety Orenburgskaya 22 (Na^+ -1.39 ± 0.07 and 1.16 ± 0.06, respectively, and Cl^- -4.63 ± 0.23 and 3.55 ± 0.18, respectively). The level of expression of the genes of ion transporters of both classes in the root of Orenburgskaya 22 and Zolotaya was almost two times higher than in the shoots. In the presence of NaCl, the activity of genes of the *HKT* family increased in the roots of the Orenburgskaya 22 variety (Figure 2). In the root of the wheat variety Zolotaya, activity of the transporter genes was significantly lower than in the Orenburgskaya 22 variety (*HKT1;4* by 1.5 times and *HKT2;1* by 2 times). It should be noted that in the presence of sodium chloride, the expression level of the *HKT2;1* gene in the Orenburgskaya 22 variety was 1.3 times higher than that of the *HKT1;4* gene. At the same time, under these conditions, the expression level of *HKT2;1* in the root of the Zolotaya cultivar decreased by 1.1 times compared to the expression of *HKT1;4*. These indicators are important for characterizing wheat tolerance to salt stress. These data are consistent with morphometric characteristics, confirming that variety Zolotaya is less resistant to high concentrations of sodium chloride than variety Orenburgskaya 22.

TOR (serine/threonine protein kinase) is a regulator that controls ATG1/ATG13-mediated autophagy [57]. There are both TOR-dependent and independent pathways of autophagy regulation [58]. *TOR* overexpression blocks autophagy activation under saline and osmotic stress in addition to nutrient deficiency but does not affect autophagy activation under oxidative or endoplasmic reticulum stress. In Arabidopsis, activation of autophagy under various stress conditions requires a decrease in *TOR* activity [58]. Under control development conditions, *TOR* expression was almost completely inhibited in both wheat varieties, except in the shoots of the Zolotaya variety. However, under salt stress, a significant increase in *TOR* expression activity was observed, especially in the roots of the Zolotaya wheat variety. Overexpression blocked the activity of autophagy and led to an increase in instability and even death of the Zolotaya variety under conditions of high salt content.

The ubiquitin-like protein ATG8 is one of the active and essential proteins for autophagy [59]. The ATG8 protein is synthesized in an inactive form. The activation of the ATG8 protein occurs upon the release of C-terminal glycine as a result of cleavage by the cysteine protease ATG4. The released C-glycine residue binds to phosphatidylethanolamine to form an adduct [60]. Lipidation of ATG8 and its localization on the autophagosome membrane are critical for assembly, expansion, closure, and fusion of the autophagic membrane with the vacuole [61]. ATG4 can also release ATG8 from autophagic membranes, which promotes autophagosome maturation and fusion [62].

The expression of the *ATG4* and *ATG8* genes in the Orenburgskaya 22 wheat variety significantly exceeded the values in the Zolotaya variety. Under salt stress, the expression of these genes in the roots of both varieties increased; however, a decrease in expression activity was observed in the shoots. A similar picture occurred with the inhibition of *ATG1* expression in shoots under salt stress. Thus, it can be said that the shoots of both wheat varieties, which are more tolerant to the action of salts than the roots, had a lower formation of autophagosomes than the roots. However, under normal conditions, the formation of autophagosomes occurred more actively in shoots than in roots.

It can be assumed that under salt-free conditions of wheat growth, the autophagy process proceeds along a TOR-independent pathway. A complex of autophagosomal proteins, which includes the ATG1 protein, is responsible for the initiation of the formation of autophagosomes. With an increase in the content of sodium chloride, the TOR protein accumulates, resulting in the activation of the autophagy process along the TOR-dependent pathway. The more active the TOR protein, the more resistant the wheat variety to salt stress. Moreover, different wheat organs respond differently to abiotic stress. Thus, in the Zolotaya wheat variety in which the expression of the *TOR* gene was significantly activated under salt stress, the morphometric parameters of both roots and shoots were significantly reduced in contrast to the resistant variety Orenburgskaya 22 in which shoots, and to a lesser extent, roots, were subject to salt stress. This fact indicates that the processes of autophagosome formation in wheat can be carried out independently in different organs.

Two pathways of plant PCD are known: "apoptosis-like", the markers of which are DNA breaks, the release of cytochrome c from mitochondria, and the transfer of phosphatidylserine to the outer layer of the membrane, and "vacuolar death", characterized by the formation of large vacuoles and autophagosomes [63].

Autophagy is a vacuolar degradation pathway by which cells recycle their components, including macromolecules and organelles [64]. Macroautophagy, more commonly referred to simply as autophagy, is the most studied form of autophagy in plants and manifests itself under environmental stresses [29]. Selective autophagy involves the uptake of specific proteins or organelles into autophagosomes [65]. Selective autophagy promotes cellular homeostasis and quality control of proteins and organelles [66]. Selective autophagy includes chlorophagy, which is responsible for the uptake of whole chloroplasts [67], and mitophagy, which is an important mitochondrial control mechanism [68], as well as reticulophagy [69,70] and ribophagy [71,72].

Oxidative stress is the generation of reactive oxygen species, including superoxide anion, hydrogen peroxide, and hydroxyl radical [73]. ROS are a necessary component of normal cell metabolism and important signaling molecules involved in the regulation of many physiological processes associated with plant growth and development [74]. Abiotic and biotic stress factors induce a reaction in plants during which ROS production in cells increases and a series of cascade reactions is triggered to neutralize excess ROS. ROS in cells can participate in various biochemical and physiological reactions depending on the degree of damage to cellular structures, in which cell metabolism is rearranged and plants acclimatize to stress conditions or one or more variants of programmed cell death (PCD) are triggered [75–77]. Although reactive oxygen species are inevitable by-products of aerobic metabolism, they cause oxidation of plant lipids, leading to membrane damage, protein degradation, enzyme inactivation, base modification, and DNA breaks, thus generating mutations and ultimately leading to programmed cell death. [78,79].

In control wheat roots, the Carboxy-H2DFFDA marker detects ROS only in the apical part of the root cap. Under salt stress, Carboxy-H2DFFDA accumulates in cells of different root zones, which indicates an increase in the content of ROS in these cells or zones and the activation of oxidative stress and cellular damage. The most intense fluorescent coloration was observed in the root of the Zolotaya variety. Thus, the accumulation of the ROS fluorescent marker Carboxy-H2DFFDA in root cells under the action of salinity indicates that ROS homeostasis was disturbed in these cells and root tissues, which can trigger PCD.

We showed that germination and the subsequent 4-day acclimation of seedlings to salinity of resistant and non-resistant wheat varieties induced some of the root cells to be at the stages of programmed cell death. No significant signs of death were found in the root cells of the control seedlings. In cells of the resistant variety Orenburgskaya 22 compared with the non-resistant variety, no transfer of phosphatidylserine to the surface of the plasma membrane of cells was observed. However, DNA breaks were found in the nuclei and metaphase chromosomes, as well as the release of cytochrome c into the cytoplasm. The localization of cytochrome c in the cell cytoplasm indicates the mitochondrial pathway of root cell death under salinity. We observed similar markers of death only in a larger number of cells in the Zolotaya variety, which is unstable to salinity, in which cells with a superficial location of phosphatidylserine were detected.

The fact that the wheat variety Zolotaya is an unstable variety was also confirmed by the data on the staining of the coleoptile with Trypan blue. The wheat coleoptile is an ideal model for studying cell damage during salinity. Programmed for a relatively short period of development, the coleoptile functions and quickly dies during the growth of the seedling. Under the action of high concentrations of sodium chloride, the viability of wheat coleoptile cells was higher in the Orenburgskaya 22 variety than in the Zolotaya variety.

An important functional test for oxidative stress is in vivo staining of mitochondria using specific fluorescent markers of the mitotracker family. Using one of the mitotracker variants, which accumulates both in active and inactive mitochondria (in this case, the dye accumulated only in undamaged mitochondria), we previously showed that in the presence of NaCl, in which root cells produced an increased content of ROS, there was a change in the nature of mitochondrial staining [47]. When cultivated under normal conditions, all root cells had stained mitochondria, although the intensity of staining in different cells could vary. However, after incubation with salts, in certain areas of the root, there were cells in which mitochondrial staining was absent, indicating damage to them. The number of damaged mitochondria was much higher in sensitive wheat varieties.

Thus, on the basis of electron microscopy data, we revealed the induction of mitophagy in wheat root and leaf cells under saline conditions, which was confirmed by biochemical data and fluorescence microscopy data.

5. Conclusions

A high concentration of sodium chloride leads to the accumulation of toxic ions in all organs of wheat. The level of accumulation of ions in wheat can be an indicator of the resistance of a wheat variety to salt stress. The wheat variety Zolotaya accumulated Cl^- and Na^+ ions to a greater extent than the Orenburgskaya 22 variety. The accumulation of toxic ions was accompanied by an increase in ROS and an increase in damage to root tissues, especially in the Zolotaya variety. Under the action of salinity, ROS production accumulated in root cells, which led to the triggering of autophagy and PCD. At high salt concentrations, an increase in the expression level of *TOR*, which is a negative regulator of the formation of autophagosomes, occurred. The level of *TOR* expression in the Zolotaya variety was 2.8 times higher in the roots and 3.8 times higher in the leaves than in the Orenburgskaya 22 variety. With the help of PCD markers, in cells of the resistant variety Orenburgskaya 22 in comparison with the non-resistant variety Zolotaya, no transfer of phosphatidylserine to the cell surface was observed. However, DNA breaks in the nuclei and metaphase chromosomes were revealed, as well as the release of cytochrome c into the cytoplasm, which indicates a mitochondrial pathway for the death of part of the root cells during salinity. We observed similar markers of death only in a larger number of cells in the Zolotaya variety, which is non-resistant to salinity, where cells with a surface location of phosphatidylserine were also detected. Based on electron microscopy data, mitophagy induction was revealed in wheat root and leaf cells under saline conditions, which was confirmed by biochemical data.

Author Contributions: N.V.K. performed light and fluorescent microscopy, evaluated data, and wrote and finalized the manuscript the manuscript; E.N.B. performed electron microscopy and evaluated data; E.M.L. performed light and fluorescent microscopy, evaluated data, and wrote and finalized the manuscript the manuscript; O.V.S. performed ions detection and obtained and characterized plants; L.I.F. designed and performed the experiment and PCR, designed and prepared figures, evaluated data, and wrote and finalized the manuscript. All authors have read and agreed to the published version of the manuscript.

Funding: The reported study was supported by 0574-2019-002 of the Ministry of Science and Higher Education of the Russian Federation.

Institutional Review Board Statement: Not applicable.

Informed Consent Statement: Not applicable.

Data Availability Statement: Not applicable.

Acknowledgments: We are grateful to Ishen Besaliev for technical support.

Conflicts of Interest: The authors declare no conflict of interest.

References

1. Very, A.-A.; Sentenac, H. Molecular mechanisms and regulation of Na^+ transport in higher plants. *Annu. Rev. Plant Biol.* **2003**, *54*, 575–603. [CrossRef] [PubMed]
2. Munns, R.; Tester, M. Mechanisms of salinity tolerance. *Annu. Rev. Plant Biol.* **2008**, *59*, 651–681. [CrossRef] [PubMed]
3. Fuchs, Y.; Steller, H. Live to die another way: Modes of programmed cell death and the signals emanating from dying cells. *Nat. Rev. Mol. Cell Biol.* **2015**, *16*, 329–344. [CrossRef] [PubMed]
4. Huysmans, M.; Lema, A.S.; Coll, N.S.; Nowack, M.K. Dying two deaths—Programmed cell death regulation in development and disease. *Curr. Opin. Plant Biol.* **2017**, *35*, 37–44. [CrossRef] [PubMed]

5. He, C.; Klionsky, D.J. Regulation mechanisms and signaling pathways of autophagy. *Annu. Rev. Genet.* **2009**, *43*, 67–93. [CrossRef]
6. Farre, J.C.; Subramani, S. Mechanistic insights into selective autophagy pathways: Lessons from yeast. *Nat. Rev. Mol. Cell Biol.* **2016**, *17*, 537–552. [CrossRef]
7. Boya, P.; Reggiori, F.; Codogno, P. Emerging regulation and functions of autophagy. *Nat. Cell Biol.* **2013**, *15*, 1017. [CrossRef]
8. Papini, A. Investigation of morphological features of autophagy during plant programmed cell death. In *Plant Programmed Cell Death*; Humana Press: New York, NY, USA, 2018; pp. 9–19.
9. Papini, A.; Mosti, S.; van Doorn, W.G. Classical macroautophagy in *Lobivia rauschii* (Cactaceae) and possible plastidial autophagy in *Tillandsia albida* (Bromeliaceae) tapetum cells. *Protoplasma* **2014**, *251*, 719–725. [CrossRef]
10. Parish, R.W.; Li, S.F. Death of a tapetum: A programme of developmental altruism. *Plant Sci.* **2010**, *178*, 73–89. [CrossRef]
11. Papini, A.; Mosti, S.; Milocani, E.; Tani, G.; Di Falco, P.; Brighigna, L. Megasporogenesis and programmed cell death in *Tillandsia* (Bromeliaceae). *Protoplasma* **2011**, *248*, 651–662. [CrossRef]
12. Doronina, T.V.; Chaban, I.A.; Lazareva, E.M. Structural and Functional Features of the Wheat Embryo Sac's Antipodal Cells during Differentiation. *Russ. J. Dev. Biol.* **2019**, *50*, 194–208. [CrossRef]
13. Brighigna, L.; Milocani, E.; Papini, A.; Vesprini, J.L. Programmed cell death in the nucellus of *Tillandsia* (Bromeliaceae). *Caryologia* **2006**, *59*, 334–339. [CrossRef]
14. Papini, A.; Tani, G.; Di Falco, P.; Brighigna, L. The ultrastructure of the development of *Tillandsia* (Bromeliaceae) trichome. *Flora* **2010**, *205*, 94–100. [CrossRef]
15. Mosti, S.; Ross Friedman, C.; Pacini, E.; Brighigna, L.; Papini, A. Nectary ultrastructure and secretory modes in three species of *Tillandsia* L. (Bromeliaceae) that have different pollinators. *Botany* **2013**, *91*, 786–798. [CrossRef]
16. Broda, M.; Millar, A.H.; Van Aken, O. Mitophagy: A mechanism for plant growth and survival. *Trends Plant Sci.* **2018**, *23*, 434–450. [CrossRef]
17. Klionsky, D.J. Guidelines for the use and interpretation of assays for monitoring autophagy. *Autopagy* **2012**, *8*, 445–544. [CrossRef]
18. Anding, A.L.; Baehrecke, E.H. Autophagy in cell life and cell death. *Curr. Top. Dev. Biol.* **2015**, *114*, 67–91.
19. Üstün, S.; Hafrén, A.; Hofius, D. Autophagy as a mediator of life and death in plants. *Curr. Opin. Plant Biol.* **2017**, *40*, 122–130. [CrossRef]
20. Shemi, A.; Ben-Dor, S.; Vardi, A. Elucidating the composition and conservation of the autophagy pathway in photosynthetic eukaryotes. *Autophagy* **2015**, *11*, 701–715. [CrossRef]
21. Minina, E.A.; Filonova, L.H.; Fukada, K.; Savenkov, E.I.; Gogvadze, V.; Clapham, D.; Sanchez-Vera, V.; Suarez, M.F.; Zhivotovsky, B.; Daniel, G.; et al. Autophagy and metacaspase determine the mode of cell death in plants. *J. Cell Biol.* **2013**, *203*, 917–927. [CrossRef]
22. Wada, S.; Hayashida, Y.; Izumi, M.; Kurusu, T.; Hanamata, S.; Kanno, K.; Kojima, S.; Yamaya, T.; Kuchitsu, K.; Makino, A.; et al. Autophagy supports biomass production and nitrogen use efficiency at the vegetative stage in rice. *Plant Physiol.* **2015**, *168*, 60–73. [CrossRef] [PubMed]
23. Zhou, J.; Wang, J.; Yu, J.Q.; Chen, Z. Role and regulation of autophagy in heat stress responses of tomato plants. *Front. Plant Sci.* **2014**, *5*, 174–186. [CrossRef] [PubMed]
24. Henry, E.; Fung, N.; Liu, J.; Drakakaki, G.; Coaker, G. Beyond glycolysis: GAPDHs are multifunctional enzymes involved in regulation of ROS, autophagy, and plant immune responses. *PLoS Genet.* **2015**, *11*, e1005199. [CrossRef] [PubMed]
25. Minina, E.A.; Bozhkov, P.V.; Hofius, D. Autophagy as initiator or executioner of cell death. *Trends Plant Sci.* **2014**, *19*, 692–697. [CrossRef]
26. Bassham, D.C.; Laporte, M.; Marty, F.; Moriyasu, Y.; Ohsumi, Y.; Olsen, L.J.; Yoshimoto, K. Autophagy in development and stress responses of plants. *Autophagy* **2006**, *2*, 2–11. [CrossRef]
27. Hayward, A.P.; Dinesh-Kumar, S.P. What can plant autophagy do for an innate immune response? *Annu. Rev. Phytopathol.* **2011**, *49*, 557–576. [CrossRef]
28. Michaeli, S.; Galili, G.; Genschik, P.; Fernie, A.R.; Avin-Wittenberg, T. Autophagy in plants—What's new on the menu? *Trends Plant Sci.* **2016**, *21*, 34–144. [CrossRef] [PubMed]
29. Yang, X.; Bassham, D.C. New insight into the mechanism and function of autophagy in plant cells. *Int. Rev. Cell Mol. Biol.* **2015**, *320*, 1–40.
30. Lamb, C.A.; Yoshimori, T.; Tooze, S.A. The autophagosome: Origins unknown, biogenesis complex. *Nat. Rev. Mol. Cell Biol.* **2013**, *14*, 759–774. [CrossRef]
31. Mizushima, N.; Yoshimori, T.; Ohsumi, Y. The role of Atg proteins in autophagosome formation. *Annu. Rev. Cell Dev. Biol.* **2011**, *27*, 107–132. [CrossRef]
32. Reggiori, F.; Klionsky, D.J. Autophagic processes in yeast: Mechanism, machinery and regulation. *Genetics* **2013**, *194*, 341–361. [CrossRef] [PubMed]
33. Bassham, D.C. Function and regulation of macroautophagy in plants. *Biochim. Biophys. Acta* **2009**, *1793*, 1397–1403. [CrossRef] [PubMed]
34. Liu, Y.; Bassham, D.C. Autophagy: Pathways for self-eating in plant cells. *Annu. Rev. Plant Biol.* **2012**, *63*, 215–237. [CrossRef] [PubMed]
35. Klionsky, D.J.; Ohsumi, Y. Vacuolar import of proteins and organelles from the cytoplasm. *Annu. Rev. Cell Dev. Biol.* **1999**, *15*, 1. [CrossRef]

36. Behrends, C.; Sowa, M.E.; Gygi, S.P.; Harper, J.W. Network organization of the human autophagy system. *Nature* **2010**, *466*, 68–76. [CrossRef]
37. Suzuki, K.; Ohsumi, Y. Current knowledge of the pre-autophagosomal structure (PAS). *FEBS Lett.* **2010**, *584*, 1280–1286. [CrossRef]
38. Chung, T.; Phillips, A.R.; Vierstra, R.D. ATG8 lipidation and ATG8-mediated autophagy in *Arabidopsis* require ATG12 expressed from the differentially controlled ATG12A and ATG12 B loci. *Plant J.* **2010**, *62*, 483–493. [CrossRef]
39. Lv, X.; Pu, X.; Qin, G.; Zhu, T.; Lin, H. The roles of autophagy in development and stress responses in *Arabidopsis thaliana*. *Apoptosis* **2014**, *19*, 905–921. [CrossRef]
40. Scherz-Shouval, R.; Shvets, E.; Fass, E.; Shorer, H.; Gil, L.; Elazar, Z. Reactive oxygen species are essential for autophagy and specifically regulate the activity of Atg4. *EMBO J.* **2007**, *26*, 1749–1760. [CrossRef]
41. Bassham, D.C. Plant autophagy—More than a starvation response. *Curr. Opin. Plant Biol.* **2007**, *10*, 587–593. [CrossRef]
42. Sarkar, S.; Floto, R.A.; Berger, Z.; Imarisio, S.; Cordenier, A.; Pasco, M.; Cook, L.J.; Rubinsztein, D.C. Lithium. Lithium induces autophagy by inhibiting inositol monophosphatase. *J. Cell Biol.* **2005**, *170*, 1101–1111. [CrossRef] [PubMed]
43. Kim, I.; Rodriguez-Enriquez, S.; Lemasters, J.J. Selective degradation of mitochondria by mitophagy. *Arch. Biochem. Biophys.* **2007**, *462*, 245–253. [CrossRef] [PubMed]
44. Lee, J.; Giordano, S.; Zhang, J. Autophagy, mitochondria and oxidative stress: Cross-talk and redox signaling. *Biochem. J.* **2012**, *41*, 523–540. [CrossRef] [PubMed]
45. Liu, Y.; Xiong, Y.; Bassham, D.C. Autophagy is required for tolerance of drought and salt stress in plants. *Autophagy* **2009**, *5*, 954–963. [CrossRef] [PubMed]
46. Toyooka, K.; Moriyasu, Y.; Goto, Y.; Takeuchi, M.; Fukuda, H.; Matsuoka, K. Protein aggregates are transported to vacuoles by a macroautophagic mechanism in nutrient-starved plant cells. *Autophagy* **2006**, *2*, 96–106. [CrossRef] [PubMed]
47. Kononenko, N.V.; Sharova, A.A.; Fedoreyeva, L.I. Tissue damage to wheat seedlings (*Triticum aestivum*) under salt exposure. *AIMS Agric. Food* **2020**, *5*, 395–407. [CrossRef]
48. Munns, R.; James, R.A.; Lauchli, A. Approaches to increasing the salt tolerance of wheat and other cereals. *J. Exp. Bot.* **2006**, *57*, 1025–1043. [CrossRef]
49. Møller, I.S.; Gilliham, M.; Jha, D.; Mayo, G.M.; Roy, S.J.; Coates, J.C. Shoot Na^+ exclusion and increased salinity tolerance engineered by cell type-specific alteration of Na^+ transport in *Arabidopsis*. *Plant Cell* **2009**, *21*, 2163–2178. [CrossRef]
50. Pei, D.; Zhang, W.; Sun, H.; Wei, X.; Yue, J.; Wang, H. Identification of autophagy-related genes *ATG4* and *ATG8* from wheat (*Triticum aestivum* L.) and profiling of their expression patterns responding to biotic and abiotic stresses. *Plant Cell Rep.* **2014**, *33*, 1697–1710. [CrossRef]
51. Wanga, W.; Mugumca, Y.; Bacchama, D.C. New advances in autophagy in plants: Regulation, selectivity and function. *Semin. Cell Dev. Biol.* **2018**, *80*, 113–122. [CrossRef]
52. Shi, H.; Lee, B.H.; Wu, S.J.; Zhu, J.K. Overexpression of a plasma membrane Na+/H+ antiporter gene improves salt tolerance in *Arabidopsis thaliana*. *Nat. Biotechnol.* **2003**, *21*, 81–85. [CrossRef] [PubMed]
53. Deinlein, U.; Stephan, A.B.; Horie, T.; Luo, W.; Xu, G.; Schroeder, J.I. Plant salt-tolerance mechanisms. *Trends Plant Sci.* **2014**, *19*, 371–379. [CrossRef] [PubMed]
54. Yamaguchi, T.; Hamamoto, S.; Uozumi, N. Sodium transport system in plant cells. *Front. Plant Sci.* **2013**, *4*, 410–417. [CrossRef]
55. Horie, T.; Hauser, F.; Schroeder, J.I. *HKT* transporter mediated salinity resistance mechanisms in *Arabidopsis* and monocot crop plants. *Trends Plant Sci.* **2009**, *14*, 660–668. [CrossRef] [PubMed]
56. Platten, J.D.; Cotsaftis, O.; Berthomieu, P.; Bohnert, H.; Davenport, R.J.; Fairbairn, D.J. Nomenclature for *HKT* transporters, key determinants of plant salinity tolerance. *Trends Plant Sci.* **2006**, *11*, 372–374. [CrossRef] [PubMed]
57. Noda, T.; Ohsumi, Y. Tor, a phosphatidylinositol kinase homologue, controls autophagy in yeast. *J. Biol. Chem.* **1998**, *273*, 3963–3966. [CrossRef]
58. Pu, Y.; Luo, X.; Bassham, D.C. TOR-dependent and -independent pathways regulate autophagy in *Arabidopsis thaliana*. *Front. Plant Sci.* **2017**, *8*, 1204–1217. [CrossRef] [PubMed]
59. Ryabovol, V.V.; Minibayeva, F.V. Autophagic Proteins ATG4 and ATG8 in Wheat: Structural Characteristics and Their Role under Stress Conditions Doklady. *Biochem. Biophys.* **2014**, *458*, 179–181. [CrossRef]
60. Romanov, J.; Walczak, M.; Ibiricu, I.; Schüchner, S.; Ogris, E.; Kraft, C.; Martens, S. Mechanism and functions of membrane binding by the Atg5-Atg12/Atg16 complex during autophagosome formation. *EMBO J.* **2012**, *31*, 4304–4317. [CrossRef]
61. Xie, Z.; Nair, U.; Klionsky, D.J. Atg8 controls phagophore expansion during autophagosome formation. *Mol. Biol. Cell* **2008**, *19*, 3290–3298. [CrossRef]
62. Nair, U.; Yen, W.L.; Mari, M.; Cao, Y.; Xie, Z.; Baba, M. A role for Atg8-PE deconjugation in autophagosome biogenesis. *Autophagy* **2012**, *8*, 780–793. [CrossRef] [PubMed]
63. Van Doorn, W.G.; Beers, E.P.; Dangl, J.L.; Franklin-Tong, V.E.; Gallois, P.; Hara-Nishimura, I.; Jones, A.M.; Kawai-Yamada, M.; Lam, E.; Mundy, J.; et al. Morphological classification of plant cell deaths. *Cell Death Differ.* **2011**, *18*, 1241–1246. [CrossRef] [PubMed]
64. Inoue, Y.; Suzuki, T.; Hattori, M.; Yoshimoto, K.; Ohsumi, Y.; Moriyasu, Y. AtATG genes, homologs of yeast autophagy genes, are involved in constitutive autophagy in Arabidopsis root tip cells. *Plant Cell Physiol.* **2006**, *47*, 1641–1652. [CrossRef]
65. Birgisdottir, A.B.; Lamark, T.; Johansen, T. The LIR motif—Crucial for selective autophagy. *J. Cell Sci.* **2013**, *126*, 3237–3247. [CrossRef]

66. Zaffagnini, G.; Martens, S. Mechanisms of selective autophagy. *J. Mol. Biol.* **2016**, *428*, 1714–1724. [CrossRef] [PubMed]
67. Izumi, M.; Ishida, H.; Nakamura, S.; Hidema, J. Entire photodamaged chloroplasts are transported to the central vacuole by autophagy. *Plant Cell* **2017**, *29*, 377–394. [CrossRef]
68. Ashrafi, G.; Schwarz, T.L. The pathways of mitophagy for quality control and clearance of mitochondria. *Cell Death Differ.* **2013**, *20*, 31–42. [CrossRef]
69. Bernales, S.; McDonald, K.L.; Walter, P. Autophagy counterbalances endoplasmic reticulum expansion during the unfolded protein response. *PLoS Biol.* **2006**, *4*, 2311–2324. [CrossRef]
70. Yorimitsu, T.; Nair, U.; Yang, Z.; Klionsky, D.J. Endoplasmic reticulum stress triggers autophagy. *J. Biol. Chem.* **2006**, *281*, 30299–30304. [CrossRef]
71. Kraft, C.; Deplazes, A.; Sohrmann, M.; Peter, M. Mature ribosomes are selectively degraded upon starvation by an autophagy pathway requiring the Ubp3p/Bre5p ubiquitin protease. *Nat. Cell Biol.* **2008**, *10*, 602–610. [CrossRef]
72. Waliullah, T.M.; Yeasmin, A.M.; Kaneko, A.; Koike, N.; Terasawa, M.; Totsuka, T.; Ushimaru, T. Rim15 and Sch9 kinases are involved in induction of autophagic degradation of ribosomes in budding yeast. *Biosci. Biotechnol. Biochem.* **2017**, *81*, 307–310. [CrossRef] [PubMed]
73. Inze, I.; Van Montagu, M. Oxidative stress in plants. *Curr. Opin. Biotechnol.* **1995**, *6*, 153–158. [CrossRef]
74. Apel, K.; Hirt, H. Reactive oxygen species: Metabolism, oxidative stress and signal transduction. *Annu. Rev. Plant Biol.* **2004**, *55*, 373–399. [CrossRef] [PubMed]
75. Pasternak, T.; Rudas, V.; Potter, G.; Jansen, K.M.A. Morphogenic effects of abiotic stress: Reorientation of growth in *Arabidopsis thaliana* seedlings. *Environ. Exp. Bot.* **2005**, *53*, 299–314. [CrossRef]
76. Potters, G.; Pasternak, T.; Guisez, Y.; Palme, K.J.; Jansen, K.M.A. Stress-induced morphogenic responses: Growing out of trouble? *Trends Plant Sci.* **2007**, *12*, 98–105. [CrossRef]
77. Potters, G.; Pasternak, T.; Guisez, Y.; Jansen, M.A.K.; Potters, G. Different stresses, similar morphogenic responses: Integrating a plethora of pathways. *Plant Cell Environ.* **2009**, *32*, 158. [CrossRef]
78. Wang, L.Y.; Wang, G. Salt stress-induced programmed cell death in tobacco protoplastsis mediated by reactive oxygen species and mitochondrial permeability transition pore status. *J. Plant Physiol.* **2006**, *63*, 731–739.
79. Mancini, A.; Buschini, A.; Maria Restivo, F.M.; Rossi, C.; Poli, P. Oxidative stress as DNA damage in different transgenic tobacco plants. *Plant Sci.* **2006**, *170*, 845–852. [CrossRef]

agronomy

MDPI

Article

Health Status of Oilseed Rape Plants Grown under Potential Future Climatic Conditions Assessed by Invasive and Non-Invasive Techniques

Mónica Pineda * and Matilde Barón

Department of Biochemistry and Molecular and Cell Biology of Plants, Estación Experimental del Zaidín, Consejo Superior de Investigaciones Científicas (CSIC), 18008 Granada, Spain
* Correspondence: monica.pineda@eez.csic.es; Tel.: +34-958-526-622

Abstract: Environmental conditions affect many plant traits such as biochemistry, physiology, morphology, and even their distribution around the world. Human activities have increased greenhouse gas emissions, which will promote a global rise in temperatures. The impact of climate change on natural vegetation and crops is difficult to predict, making it necessary to conduct experiments that mimic potential future climate conditions. Here, oilseed rape has been grown under environmental conditions that reproduce severe and intermediate climate change, setting the current climatic conditions as a control, with the main objective of evaluating the impact of climate change on the health status of this plant of agronomic interest. For such a purpose, two approaches (invasive and non-invasive) have been applied. Invasive quantitative measurements are based on the absorbance of biochemical compounds. Non-invasive methods such as thermal, multicolor fluorescence, and hyperspectral reflectance imaging sensors rely on the spectral properties of the plants. The results revealed that climate change induced lipid peroxidation, as well as alterations in pigment composition, transpiration, photosynthesis, and secondary plant metabolism. Those changes were more drastic the more severe the climatic condition imposed. Novel vegetation indices obtained from hyperspectral reflectance and specifically tailored to detect stress in brassicas correlated with physiological traits such as lipid peroxidation and secondary plant metabolism.

Keywords: brassicas; climate change; hyperspectral reflectance; imaging sensors; plant phenotyping; multicolor fluorescence imaging; thermography; vegetation index

Citation: Pineda, M.; Barón, M. Health Status of Oilseed Rape Plants Grown under Potential Future Climatic Conditions Assessed by Invasive and Non-Invasive Techniques. *Agronomy* **2022**, *12*, 1845. https://doi.org/10.3390/agronomy12081845

Academic Editors: José Ramón Acosta-Motos and Sara Álvarez

Received: 7 July 2022
Accepted: 1 August 2022
Published: 4 August 2022

1. Introduction

Oilseed rape (*Brassica napus* L.) is the main oilseed crop in Europe. It is used as a source of feed for livestock and as oil for human consumption, as well as a source for bio-products such as biodiesel [1]. In addition, recent research points to the possible use of rapeseed in the food industry to produce preparations classified as healthy food, rich in minerals, proteins, fiber, and compounds with anti-neoplastic properties [2]. In the European Union (EU), with 5.17 million hectares of oilseed rape planted in 2020, the most productive countries are Germany and France, with half of the European production, followed by Poland (according to data from DG AGRI EU for the Spanish Ministry of Agriculture, Fishing and Food; www.mapa.gob.es, accessed on 23 June 2022). In Spain, oilseed rape production has increased exponentially in recent years, and it is expected to continue rising in the future. Almost 60% of Spanish oilseed rape is intended for animal feed, whereas there is little tradition to destine the oil for human consumption, in contrast to central Europe. The largest oilseed rape producer in Spain is Castilla y León, with a third of the total harvest. In 2021, 40,568 hectares were planted, yielding 108,428 tons (Consejería de Agricultura, Ganadería y Desarrollo Rural, Junta de Castilla y León; https://gobierno.jcyl.es/web/es/consejerias/consejeria-agricultura-ganaderia-desarrollo-rural.html, accessed on 23 June 2022).

Greenhouse gases responsible for global warming are leading to changes in climate worldwide that would have a major impact on both the ecology of wildlife and agriculture. The direct effects of climate change (increasing CO_2 levels; changes in temperature, ultraviolet-B radiation, rainfall, and humidity; extreme events), as well as the indirect ones (shifts in crop suitability, changes in plant nutrition, degradation of resources, and an increase in the incidence of pests), could compromise the yield and quality of food production with serious consequences on our society [3]. The negative effects of climate change were analyzed exhaustively by the Intergovernmental Panel on Climate Change (IPCC) in the Assessment Report (AR5) [4]. This assessment evaluated several potential scenarios for greenhouse gases and air pollutants emissions and their atmospheric concentrations, as well as for land use by 2100. Based on those estimates, future climate conditions were projected on representative concentration pathways (RCPs). According to AR5, a stringent mitigation scenario that keeps global warming below 2 °C is unlikely to occur. In an extreme scenario, RCP 8.5 would represent the effects of not restricting greenhouse gas emissions. However, the climate change scenario most probable to occur under the current governments' climate change policies is the intermediate scenario RCP 4.5.

Based on the modeling for the RCP 4.5 and RCP 8.5 climate change scenarios, northern Europe would offer the most favorable conditions for oilseed rape growth in the future [5]. Nonetheless, several studies have highlighted the negative effects of abiotic stress factors brought about by climate change on rapeseed crops. Elevated atmospheric CO_2 concentrations, high temperatures, drought, and high ultraviolet-B radiations damage photosynthetic structures, thus reducing photosynthesis and affecting crop production and quality of yields [6–8]. The combination of elevated CO_2, O_3, and temperature in oilseed rape simulating several climate change scenarios by 2075 in southern Scandinavia (including RCP 8.5), caused a decrease in plant growth, as well as reductions in oil production and quality, and negative effects on the saturated to unsaturated fatty acids ratio [9,10]. High temperatures of different climate change projections (RCP 4.5 and RCP 8.5 among them) for Poland also caused reductions in linolenic acid contents, one of the fatty acids playing a significant role in the quality of oilseed rape oil [11]. However, more research is required to find out the consequences of future climate conditions on oilseed rape plants grown in southern Europe.

To analyze plant physiology, traditional techniques can be used that provide quantitative data on a parameter of interest and imaging techniques that perform robust estimates of these parameters. Traditional sampling to measure physiological traits based on the absorbance of biochemical compounds, although a powerful and reliable method, is invasive and laborious [12]. That means that it requires a large amount of plant material and time to process. On the other hand, stress causes alterations in tissue color, leaf structure, the transpiration rate, as well as in the way in which plants interact with solar radiation. Thus, stress alters the optical properties of the leaves within different regions of the electromagnetic spectrum, which can be measured by imaging techniques [13,14]. As the relative intensities of optical signals are highly sensitive to intrinsic plant properties, numerous pieces of evidence exist that unequivocally link image measurements with different physiological traits. Blue (F440) and green fluorescence (F520) induced by ultraviolet light and measured by multicolor fluorescence imaging (MCFI) are emitted by phenolic compounds covalently bound to cell walls and involved in plant defense and structure; whereas red fluorescence (F680) and far red fluorescence (F740) are emitted by chlorophylls (Chl). The ratio F680/F740 is inversely correlated to Chl content [15,16]. Thermal imaging visualizes leaf temperature, which is inversely correlated to leaf transpiration [17,18]. The availability of reflectance spectra recorded by hyperspectral cameras allows the calculation of numerous vegetation indexes (VIs) that provide information about many traits of the leaf, such as photosynthesis performance [19], pigment composition [20,21], and fitness [22,23]. Moreover, the acquisition of whole reflectance spectra let researchers design novel VIs to sense the effects of stress in plants [24].

Imaging sensors provide information on plant metabolism in a rapid, non-destructive way, which means that the same leaf or plant could be imaged using different sensors in subsequent days along an experiment [12,25]. Moreover, one of the main advantages of imaging techniques is that they have the ability to scan large populations, enabling high-throughput screening [13,26]. In addition, sensors often provide a presymptomatic diagnosis of stress, i.e., they can detect alterations in plant physiology before symptoms are visible [14]. Nonetheless, invasive and non-invasive methods are not mutually exclusive options, and their combined use often improves the results obtained in studies on plant physiology under stress. Indeed, the body of knowledge obtained through different approaches such as "-omics", microscopy, biochemistry, metabolic engineering, or computer vision is being successfully incorporated into machine learning algorithms coming from Artificial Intelligence to predict plant growth and development under specific environmental constraints [27–29].

The main aim of this work was to assess the consequences of global climate change on the health status of oilseed rape plants cultivated in southern Europe. For this purpose, three climatic conditions were selected: (i) current climate conditions (CCC), (ii) an intermediate climate change scenario (RCP 4.5), and (iii) a projection of extreme climate change (RCP 8.5). The ambient temperatures and the atmospheric CO_2 concentrations were selected according to regionalized data for Castilla y León (Spain) by the AEMet (Spanish State Meteorology Agency) for the years 2081–2100. The health status of oilseed rape plants grown under these three climatic conditions was evaluated using invasive colorimetric assays and non-invasive methods (MCFI, thermography, and hyperspectral reflectance imaging). All of these techniques provided information on antioxidant activity, lipid peroxidation, phenolic compounds and pigment contents, leaf temperature, photosynthesis performance, and vigor. In addition, a novel VI specifically designed to detect the stress caused by climate change in oilseed rape plants was developed. Finally, bivariate correlation analyses were performed to elucidate the physiological traits underlying several VIs.

2. Materials and Methods

2.1. Plant Growth

Oilseed rape plants (*Brassica napus* L. var. *napobrassica*; Franchi Sementi, Grassobbio, Italy) were grown in pots located in a growth chamber in a 16/8 h day/night regime with 60% relative humidity and 200 μmol photon m^{-2} s^{-1} of PAR light. Three climatic conditions were selected as treatments (Table 1): CCC, RCP 4.5, and RCP 8.5. CCC was considered the control treatment. For each experiment, the plants were maintained from the moment of sowing in the corresponding treatment. The ambient temperature and CO_2 concentration shown in Table 1 were selected according to the data regionalized for Castilla y León by the AEMet for CCC and those matching to RCP 4.5 and RCP 8.5 in years 2081–2100. Day and night temperatures correspond to the average values during the sowing and first stages of the growth season (September). The plants were watered every day and they took water ad libitum.

Table 1. Climatic conditions (treatments) used for oilseed rape growth: current climate conditions (CCC) and Representative Concentration Pathways 4.5 and 8.5 (RCP 4.5 and RCP 8.5, respectively) regionalized for September (season of sowing and first stages of oilseed rape growth) in Castilla y León for years 2081–2100.

Treatment	Average Day Temperature (°C)	Average Night Temperature (°C)	CO_2 Concentration (ppm)
CCC	26	12	408
RCP 4.5	29	15	650
RCP 8.5	32	18	1000

2.2. *Invasive Methods for Determining Plant Health Status*

The sampling for conventional techniques based on colorimetric reactions was conducted using 4.15 cm^2 leaf disks practiced on leaf number 4 of oilseed rape plants at 28 and 32 days after sowing (das). Six different leaves were sampled for every measurement, das, and treatment. Once collected, the samples were immediately frozen in liquid nitrogen and kept at −80 °C until the moment they were processed. All of the spectrophotometric measurements were carried out using the Shimadzu UV1800 spectrophotometer (Shimadzu Corporation, Tokyo, Japan). All of the chemicals products used were reagent grade.

The method for measuring the total antioxidant activity (TAA) is based on that published by Miller et al. [30] with some modifications. The concentration of the antioxidant substances present in a leaf sample is estimated as the capacity to scavenge the radical cation of 2,2′-azinobis(3-ethylbenzothiazoline-6-sulfonate) ($ABTS^+$) in the aqueous phase, as compared to standard amounts of the antioxidant ascorbic acid. Leaf disks were homogenized in liquid nitrogen, and the antioxidants were extracted in 0.45 mL of 50 mM potassium phosphate pH 7.8 (Merck, Darmstadt, Germany). The $ABTS^+$ radicals were obtained in the spectrophotometer cuvette using 0.995 mL 50 mM glycine (Panreac Quimica, Barcelona, Spain)-hydrochloric acid (VWR International, Radnor, PA, USA) buffer, 10 µL of 10 µM horseradish peroxidase (Sigma-Aldrich, Saint Louis, MO, USA), 10 µL of 2 mM hydrogen peroxide (Sigma-Aldrich) and 25 µL of 10 mM ABTS (Sigma-Aldrich). Once the absorbance at 730 nm reached stable valor (20 s), the samples (or standard ascorbic acid, Sigma-Aldrich) were added to the cuvette, and the reaction of $ABTS^+$ scavenging was followed during 120 s at 730 nm. The results are expressed as TAA by ascorbic acid (Sigma-Aldrich) equivalent and referred to the sampled area (nmol·cm^{-2}).

The determination of the total soluble phenolic compounds by the Folin–Ciocalteu method was performed on the leaf extracts prepared by the homogenization of leaf disks in liquid nitrogen and extraction in 0.5 mL of chloroform (Scharlau, Barcelona, Spain), 0.5 mL of methanol (Panreac), and 0.25 mL of 1% (*w*/*v*) sodium chloride (Merck). The extracts were added to a combination of 2% (*w*/*v*) sodium carbonate (Merck) in 0.1 N sodium hydroxide (Panreac) and 50% (*v*/*v*) Folin–Ciocalteu reagent (Merck). The absorbance was measured at 735 nm using a caffeic acid (Sigma-Aldrich) standard curve. The results are expressed as the content of total phenolics by caffeic acid equivalent and referred to the sampled area (µg·cm^{-2}) [31,32].

Lipid peroxidation was estimated by analyzing malondialdehyde (MDA) content obtained using the thiobarbituric acid reaction according to the method described by Rodríguez-Serrano et al. [33]. Leaf disks were homogenized in liquid nitrogen and extracted in 0.65 mL of 50 mM Tris (Scharlau)-HCl (VWR) pH 7 with 0.2% (*v*/*v*) Triton X-100 (Boehringer Mannheim, Mannheim, Germany) and 0.1 mM EDTA (Panreac) buffer. Then, 0.5 mL of the leaf extract were combined with 0.1 mL of a solution of 15% (*w*/*v*) trichloroacetic acid (Panreac), 0.375% (*w*/*v*) thiobarbituric acid (Sigma-Aldrich), 0.1% (*w*/*v*) butylated hydroxytoluene (Sigma-Aldrich), and 0.2 M hydrochloric acid (VWR) and heated at 95 °C for 15 min. The supernatant of the subsequent centrifugation was measured at 535 nm. The MDA concentration was determined using a standard curve with commercial MDA (Sigma-Aldrich), and the results are referred to the sampled area (nmol MDA·cm^{-2}).

The measurements of Chl, carotenoids (Car), and xanthophylls (Xanth) were conducted according to Lichtenthaler and Buschmann [34]. Briefly, the pigments were extracted in 1 mL of 80% (*v*/*v*) acetone (Panreac) by homogenization in liquid nitrogen. Absorbance was measured at 470 nm, 647 nm, and 663 nm. Then, the equations provided by the authors were applied to the measured absorbance to calculate the total amount of pigments and referred to the sampled area (µg·cm^{-2}). On the other hand, anthocyanins (Anth) quantification was conducted using the protocol described by Solfanelli et al. [35]. Leaf disks were homogenized in liquid nitrogen, and anthocyanins were extracted in 0.6 mL of methanol (Panreac) in 1% (*v*/*v*) hydrochloric acid (VWR) and 0.4 mL of distilled water. Then, the samples were purified in the same volume of chloroform (Scharlau). The aqueous phase was measured at 535 nm.

2.3. Non-Invasive Imaging Techniques for Stress Detection

All of the measurements using non-destructive imaging sensors were acquired over leaf number 4 of oilseed rape plants at 25, 28, and 32 das. The leaf remained attached to the plant until the end of the experiments at 32 das. Twelve leaves per treatment were sequentially imaged using three different sensors.

MCFI was recorded using a customized Open FluorCam FC 800-O (Photon Systems Instruments, Brno, Czechia) vertically positioned 50 cm above the leaves. Autofluorescence leaf emission was excited using a 355 nm ultraviolet lamp. Then, F440, F520, F680, and F740 were each acquired during 30 s and using the appropriate cutoff filters, as described before [31,36]. The entire measurement lasted 2 min. Nine images per fluorescence region were averaged to extract fluorescence values for whole leaves using the FluorCam v. 7.1.0.3 software (Arlington, VA, USA) as well as the fluorescence ratio F680/F740. For a better understanding of the pictures, the software also allows the application of a false color scale to the images captured in black and white. The images presented in the figures correspond to the most representative data.

An FLIR A305sc camera (FLIR Systems, Wilsonville, OR, USA) was used to capture thermal images of the whole leaves as described before [37,38]. The camera was vertically positioned 50 cm above the leaves in the plant growth chamber. For each measurement, 10 thermal images were collected at a rate of one image per second. Then, all of the images were averaged to extract the temperature values for entire leaves using the FLIR ResearchIR v. 3.4 software (Arlington, VA, USA). The parameter $T_L - T_A$ (leaf temperature corrected by ambient temperature) was calculated for a better comparison between the treatments. To interpret the images captured in black and white, the software also allows the application of a false color scale. The images presented in the figures correspond to the most representative data.

The hyperspectral reflectance of attached leaves was recorded using a Pika L hyperspectral imaging camera (Resonon, Bozeman, MT, USA) in the visible (400–700 nm) to near-infrared spectral range (700–1000 nm), according to Pineda et al. [24]. Prior to the leaf measurements, dark and light corrections were carried out. The dark correction was performed in darkness; then, the light correction was performed by illuminating a white homogenous calibration tile provided by Resonon with four calibrated xenon lamps with a homogeneous light intensity between 400 and 1000 nm. Finally, the leaves were placed on a translation stage while they were homogeneously illuminated, and the camera took the images vertically positioned 50 cm over the sample. Spectronon v. 2.134 (Resonon) software was used for dark and light corrections, build-up of 281 images-datacubes, and measurement analysis. The reflectance spectra averaged for whole leaves were obtained and used to calculate VIs summarized in Table 2, as well as to design novel VIs specific for abiotic stress detection.

Table 2. Vegetation indices (VIs) obtained from hyperspectral reflectance imaging assays.

Vegetation Index	Related to	Equation	References
Anthocyanins reflectance index	Anthocyanins content	$ARI = R800 \cdot \left(\frac{1}{R550} + \frac{1}{R700}\right)$	[20]
Carotenoids reflectance index	Carotenoids content	$CRI = \frac{1}{R510} - \frac{1}{R700}$	[21]
Diseased broccoli index 1	Biotic stress	$DBI_1 = \frac{R400-R690}{R850}$	[24]
Diseased broccoli index 2	Biotic stress	$DBI_2 = \frac{R400}{R850}$	[24]
Diseased broccoli index 3	Biotic stress	$DBI_3 = \frac{R578}{R529}$	[24]
Normalized difference vegetation index	Vigor	$NDVI = \frac{R800-R670}{R800+R670}$	[22]
Photochemical reflectance index	Photosynthesis	$PRI = \frac{R531-R570}{R531+R570}$	[19]

2.4. Statistics

Microsoft Office Excel 2016 (Microsoft Corporation, Redmond, WA, USA) was used to create the databases and plot the figure graphs (averages ± standard errors). Boxplots were performed to discard outliers in the sample distributions. Later, SigmaPlot v. 14.5 (Systat Software Inc., San José, CA, USA) was chosen for statistical analysis. The sample distributions passed both the Shapiro–Wilk normality test and Brown–Forsythe equal variance test. Then, a two-tailed Student *t*-test could be used to compare treatments at every das assayed. Differences were considered significant at $p < 0.05$ and were indicated by different lowercase letters in the bar graphs. When comparing reflectance spectra, differences were considered significant at $p < 0.001$. Bivariate correlations were developed using SPSS v. 28.0 software (IBM Corporation, Armonk, NY, USA); the Pearson correlation coefficient (R) was considered significant at $p < 0.01$.

3. Results

3.1. Visual Evolution of the Fourth Leaf under the Three Climatic Conditions Assayed

The fourth leaf of oilseed rape plants was selected for this study. The leaf was fully developed at 25 das and did not undergo any drastic visual changes throughout the experiment under CCC (Figure 1). However, the plants grown under the analyzed RCPs were visually different from the control treatment, with the most dramatic changes occurring the more intense the condition imposed. The leaves of plants grown under RCP 4.5 and RPC 8.5 appeared progressively less green from 28 das onwards, whereas yellow and purple colors were clearly evident at 32 das, suggesting considerable changes in leaf metabolism and leaf health status.

Figure 1. Visual evolution of the fourth leaf of oilseed rape plants grown under current climate conditions (CCC) and under two representative concentration pathways (RCP): an intermediate and an extreme scenario of climate change (RCP 4.5 and RCP 8.5, respectively). das: days after sowing.

3.2. Climate Change Induces Alterations in Plant Health Status

To evaluate the alterations caused by climate change on the general health status of the fourth leaf of oilseed rape plants, the TAA, total phenolics content, and lipid peroxidation were quantified using invasive methods based on the absorbance of biochemical compounds (Figure 2).

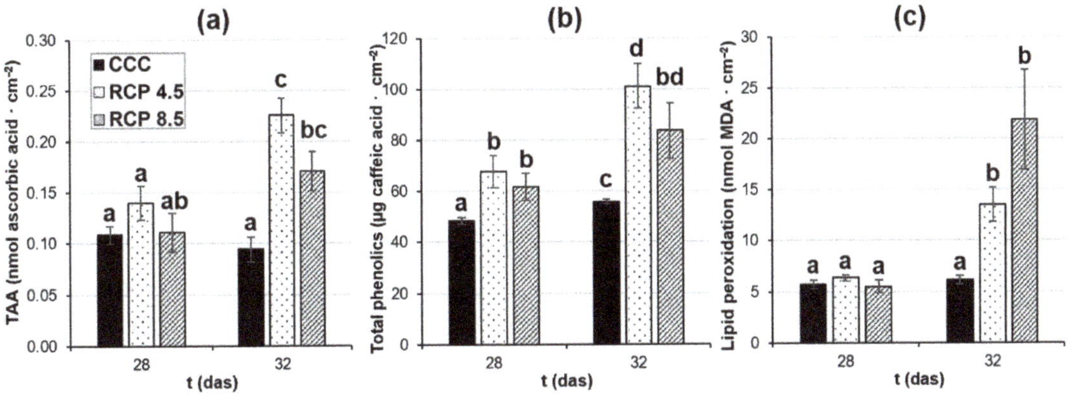

Figure 2. Measurements of (**a**) total antioxidant activity (TAA), (**b**) total phenolic content, and (**c**) lipid peroxidation in the fourth leaf of oilseed rape plants at 28 and 32 days after sowing (das). Graphs display averages ± standard errors of six leaf disks collected from six plants. Different lowercase letters indicate the significance of the measurements at $p < 0.05$. CCC: current climate conditions; RCP 4.5 and RCP 8.5: representative concentration pathways 4.5 and 8.5, respectively.

TAA (Figure 2a) measures the capacity of the antioxidant substances present in the leaf samples to scavenge free oxidant radicals. Moreover, several phenolic compounds (Figure 2b) possess antioxidant activity, among other functions. Lastly, the level of lipid peroxidation (Figure 2c) is a widely used marker of oxidative damage. Neither TAA nor lipid peroxidation increased in the leaves of plants grown at CCC during the measured period; however, a slight increment of soluble phenolic content could be registered during the last das assayed. At 28 das, no differences in TAA or lipid peroxidation were detected in plants grown under RCP 4.5 and RCP 8.5 relative to the controls, whereas soluble phenolic compounds increased in both treatments. At 32 das, increments in TAA and phenolic compounds could be registered in plants cultured under RCP 4.5 and RCP 8.5. However, in spite of this increment of antioxidant substances, oxidative damage was registered as lipid peroxidation in the leaves of plants grown under the studied RCPs with respect to the controls.

Since the visual changes in the plants grown under the analyzed RCPs suggested an impairment of pigment composition respecting to those grown at CCC, the content of Chl, Car, Xanth, and Anth were measured using invasive techniques at 28 and 32 das (Figure 3).

The total Chl content of the controls plants slightly decreased from 28 to 32 das (Figure 3a). However, the Chl content in the leaves of plants grown under the analyzed RCPs was considerably lower with respect to the controls at both 28 and 32 das. The Chl *a*/Chl *b* ratio (Figure 3b) remained constant in the leaves of plants grown under CCC during the period measured, whereas the leaves of plants grown under the studied RCPs showed ratio decreases relative to the controls at both 28 and 32 das. The content of (Xanth + Car) remained constant in the leaves of the control plants from 28 to 32 das (Figure 3c). The plants grown at RCP 4.5 did not show any significant differences with respect to the controls at 28 das, whereas the content of (Xanth + Car) decreased at 32 das. However, plants grown under RCP 8.5 displayed significant differences relative to the control condition at both 28 and 32 das. Chl/(Xanth + Car) ratio decreased in leaves of plants grown under CCC from 28 to 32 das (Figure 3d). The plants grown under the analyzed RCPs showed lower ratio values than the controls at both 28 and 32 das. Finally, the plants grown at CCC did not significantly synthetize Anth from 28 to 32 das (Figure 3e). Only the leaves of the plants grown at RCP 8.5 displayed significant increments of Anth with respect to the controls at both 28 and 32 das, while the leaves of the plants grown at RCP 4.5 only showed increments at 32 das.

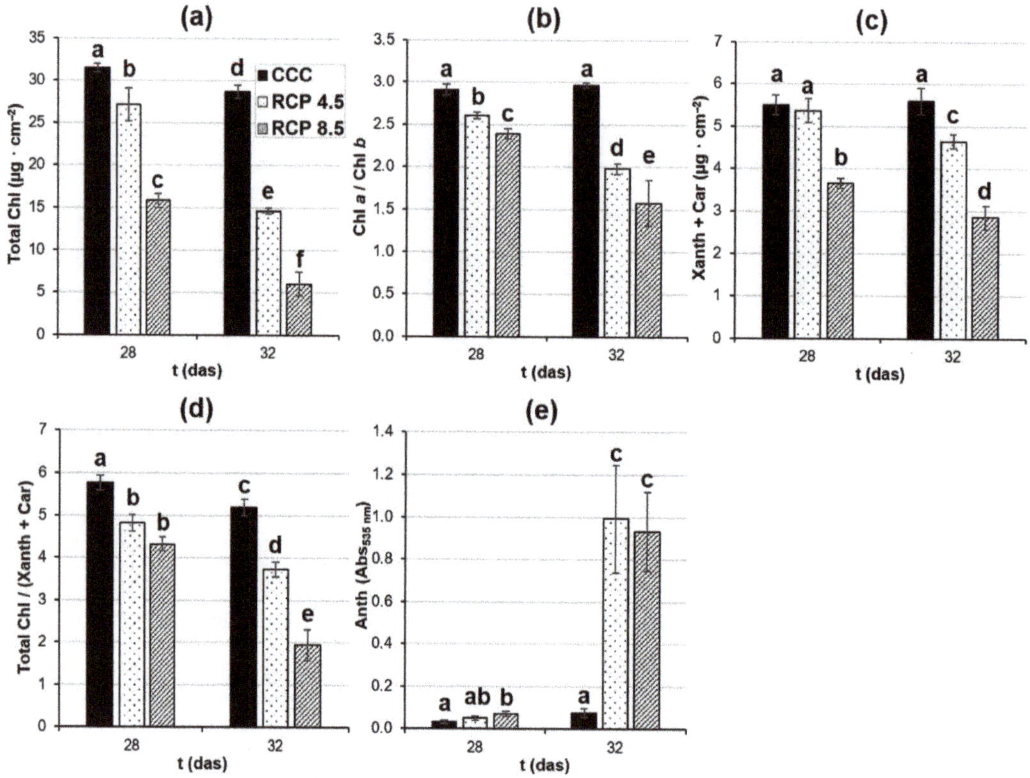

Figure 3. Pigment composition of the fourth leaf of oilseed rape plants at 28 and 32 days after sowing (das). (**a**) Total chlorophyll (Chl) content obtained as Chl *a* plus Chl *b*; (**b**) Chl *a* over Chl *b* ratio; (**c**) xanthophylls (Xanth) plus carotenoids (Car) content; (**d**) total Chl over (Xanth + Car) ratio; (**e**) anthocyanins (Anth). Graphs display averages ± standard errors of six leaf disks collected from six plants. Different lowercase letters indicate the significance of the measurements at $p < 0.05$. CCC: current climate conditions; RCP 4.5 and RCP 8.5: representative concentration pathways 4.5 and 8.5, respectively.

In general terms, the more drastic the climate condition imposed on the plants, the greater the change in pigment composition, except for Anth. In this case, there were no significant differences between the plants grown under RCP 4.5 and those grown under RCP 8.5.

3.3. Non-Invasive Techniques to Assess Plant Health Status

The health status of the leaves of oilseed rape plants grown under CCC, RCP 4.5, and RCP 8.5 was also evaluated by several imaging techniques: MCFI, thermography, and hyperspectral reflectance imaging. The sensors provide results in short periods of time and do not consume plant material. Therefore, one additional das could be included in the study: 25, 28, and 32 das; i.e., before stress symptoms were apparent, with mild and clearly evident symptoms, respectively.

The F440 (Figures 4a and 5a) and F520 (Figures 4b and 5b) that were excited with ultraviolet light and measured by MCFI correlate with the insoluble phenolic content of the leaves. The leaves of the plants grown at CCC displayed slightly increasing F440 values from 25 to 32 das, whereas F520 increased only at 32 das. The F440 and F520 increases were mainly related to the primary and secondary veins. The F440 and F520 leaf values of the plants grown at the studied RCPs were higher than the control plants from 25 das onwards, except for those plants grown at RCP 4.5 at 32 das when no significant differences related to the controls could be detected in F520. The increments registered mostly affected the main veins (primary and secondary) at 25 das, but they also could be measured in interveinal tissues on successive days. As in the previous measurements, the greater changes with respect to the controls could be detected in the most dramatic climate change projection (RCP 8.5).

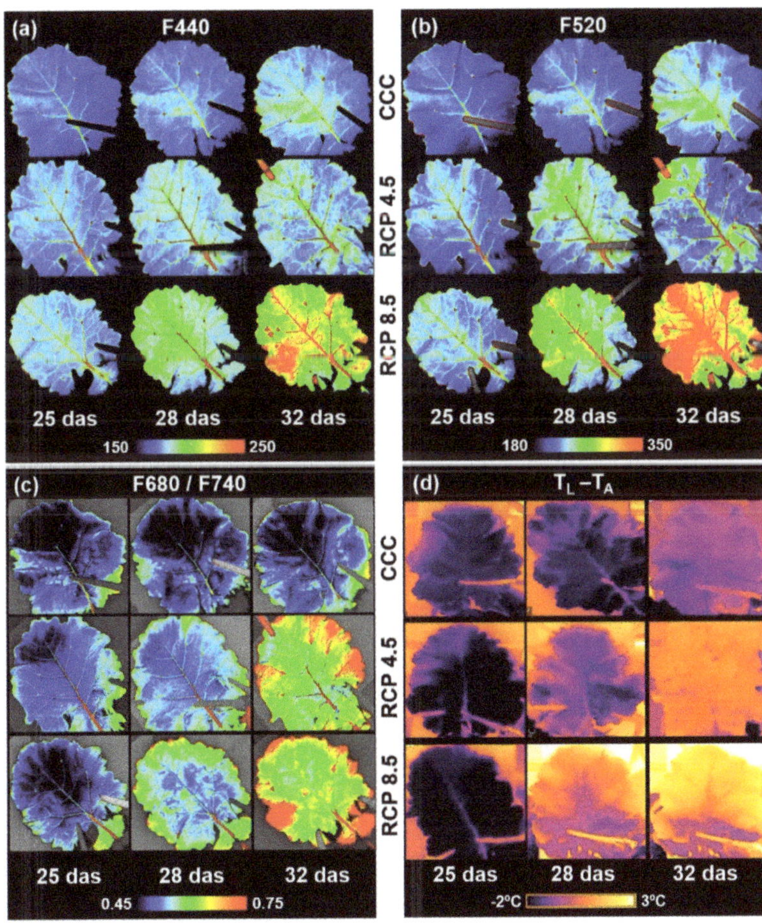

Figure 4. Images of the fourth leaf of oilseed rape plants captured with two different imaging sensors: a multicolor fluorescence camera (**a**–**c**) and a thermal camera (**d**). (**a**,**b**) blue (F440) and green (F520) fluorescence, respectively; (**c**) red over far red fluorescence ratio (F680/F740); (**d**) leaf temperature corrected

by ambient temperature ($T_L - T_A$), showing transpiration patterns of the leaves. Experiments were made using twelve leaves per day after sowing (das) and climate conditions. Images correspond to the most representative data. False color scale is provided by the software of the corresponding cameras for a better interpretation of the images. CCC: current climate conditions; RCP 4.5 and RCP 8.5: representative concentration pathways 4.5 and 8.5, respectively.

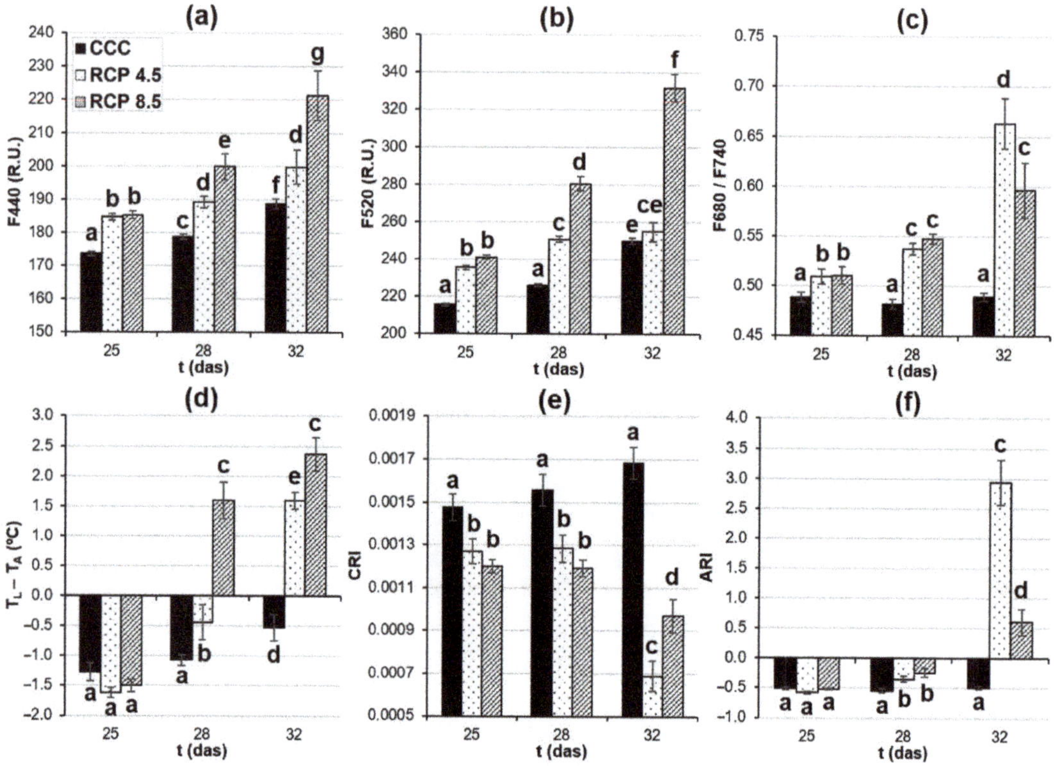

Figure 5. Numerical data obtained from images of whole leaves of oilseed rape plants. (**a**) Blue (F440) and (**b**) green fluorescence (F520), and (**c**) red over far-red fluorescence ratio (F680/F740). (**a**–**c**) were recorded using multicolor fluorescence imaging. (**d**) Leaf temperature corrected by ambient temperature ($T_L - T_A$) was obtained by a thermal camera; (**e**) carotenoids reflectance index (CRI) and (**f**) anthocyanins reflectance index (ARI) were measured by hyperspectral reflectance imaging. Graphs display averages ± standard errors of twelve leaves. Different lowercase letters indicate the significance of the measurements at $p < 0.05$. das: days after sowing. CCC: current climate conditions; RCP 4.5 and RCP 8.5: representative concentration pathways 4.5 and 8.5, respectively.

The F680/F740 ratio obtained from MCFI is inversely correlated with the Chl content of the leaves (Figures 4c and 5c). No changes in this ratio could be measured in the leaves of the plants grown under the control conditions within the experiment. On the other hand, the leaves of those plants grown at RCP 4.5 and RCP 8.5 displayed significantly higher F680/F740 values with respect to the controls from 25 to 32 das. Those increments increased as the experiment progressed, especially in the case of the plants grown under RCP 4.5. However, no significant differences could be found between the plants grown under RCP 4.5 and RCP 8.5 until 32 das.

$T_L - T_A$ is a parameter measured by thermography and is inversely correlated to the transpiration rates of the leaves (Figures 4d and 5d). The leaves of the plants grown under CCC showed an increase in $T_L - T_A$ and, thus, a decrease in leaf transpiration from 28 to

32 das. The leaves of the plants grown under the assessed RCPs displayed an increment in $T_L - T_A$ relative to the controls from 28 das onwards; the more drastic the climate conditions imposed on the plants, the higher the temperature registered in the leaves.

Hyperspectral reflectance sensors allow for the calculation of many VIs related to different vegetation traits such as vigor, fitness, and pigment composition. CRI is a VI related to Car content (Table 2; Figure 5e). No differences in CRI could be detected in the plants grown under CCC within the experiment. However, the plants grown under the studied RCPs showed lower CRI values than the controls at every das assayed. However, no significant differences could be detected between the leaves of plants grown at RCP 4.5 and RCP 8.5 until 32 das. ARI is a VI related to the Anth content (Table 2; Figure 5f). There were no significant increases in the ARI values in the leaves of plants grown under the control conditions within the experiment. On the contrary, significant increments in the ARI values of the leaves of the plants grown under the analyzed RCPs could be measured at both 28 and 32 das. Significant differences between the ARI values of the leaves of the plants grown at RCP 4.5 and RCP 8.5 were only evident at 32 das.

3.4. Common and Novel Vegetation Indices to Evaluate Plant Fitness

There is a collection of VIs related to plant fitness that can be calculated from hyperspectral reflectance experiments (Figure 6). The most widely used are NDVI, related to plant vigor and greenness, and PRI, strongly correlated with plant photosynthesis (Table 2). The NDVI values of the leaves of the plants grown under the control conditions did not show changes during the experiment (Figure 6a). However, the leaves of the plants grown under the studied RCPs displayed lower NDVI values than the controls from 25 das onwards; the strongest decay of leaf vigor was registered at 32 das. No significant differences between the NDVI values of the leaves of the plants grown at RCP 4.5 and those grown at RCP 8.5 could be found. Regarding PRI (Figure 6b), the leaves of the plants grown at CCC registered a progressive decrease in photosynthesis from 25 das onwards. Moreover, the leaves of the plants grown at RCP 4.5 and RCP 8.5 showed lower PRI values with respect to those grown at CCC during the entire experiment. In the case of the plants grown at RCP 8.5, the extent of the PRI decline was greater than in the case of those grown at RCP 4.5 only at 25 das; there were no significant differences between these two treatments for the rest of the das analyzed.

The great advantage of having complete reflectance spectra is that it allows researchers to create novel VIs tailored to their specific needs. That was the case of the VIs DBI_{1-3}, specifically designed to detect biotic stress challenged by *Xanthomonas campestris* pv. *campestris* on broccoli plants, another plant of the *Brassica* genus (Table 2). These three indices were applied to this piece of work to investigate their capacity to detect the abiotic stress caused by climate change on oilseed rape plants with a different extent of success. DBI_1 was the only DBI displaying differences between the plants grown under RCP 4.5 and RCP 8.5 from 28 das onwards (Figure 6c). However, DBI_1 was the least sensitive of them to presymptomatically detect significant changes between the controls and the plants growing under the analyzed RCPs: at 28 and 32 das for RCP 4.5 and RCP 8.5 treatments, respectively. Indeed, DBI_2 and DBI_3 showed those differences presymptomatically from 25 das onwards (Figure 6d,e, respectively). Nevertheless, the DBI_2 values did not develop any changes during the experiment for any of the treatments assayed, while the DBI_3 values showed increments compared to previous days in the case of the plants growing under the studied RCPs.

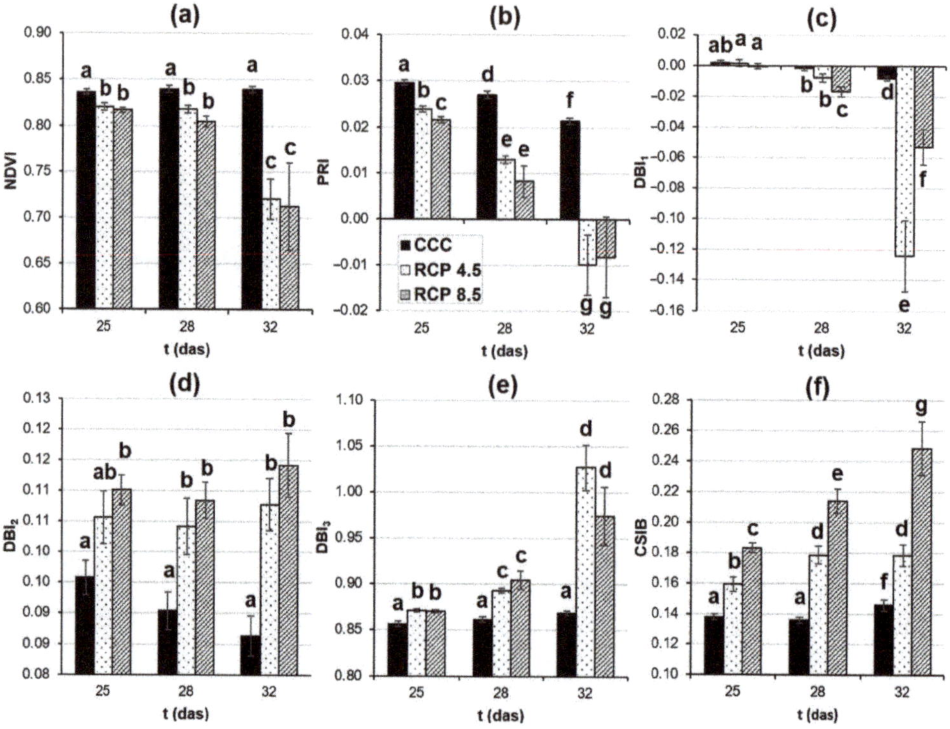

Figure 6. Vegetation indices obtained from hyperspectral reflectance measurements of the fourth leaf of oilseed rape plants. (**a**) Normalized difference vegetation index (NDVI). (**b**) Photochemical reflectance index (PRI). (**c–e**) Diseased broccoli index 1, 2, and 3 (DBI_{1-3}). (**f**) Climatic stress index for brassicas (CSIB) was specifically developed to monitor the stress caused by the analyzed climate change conditions. Graphs display averages ± standard errors of twelve leaves. Different lowercase letters indicate the significance of the measurements at $p < 0.05$. das: days after sowing; CCC: current climate conditions; RCP 4.5 and RCP 8.5: representative concentration pathways 4.5 and 8.5, respectively.

In this piece of work, two requirements have been established to be indispensable when considering VI as an optimal reporter of the abiotic stress caused by climate change in oilseed rape plants: (i) to provide information before the symptoms become visible; and (ii) to distinguish between the plants grown at RCP 4.5 and RCP 8.5 treatments. None of the analyzed VIs meet both criteria at the same time. Therefore, novel VIs have been designed for this particular interest, following the process described by Pineda et al. [24]. Briefly, the whole leaf reflectance spectra were averaged for each treatment and das assayed (Figure S1) and then compared by a two-tailed Student *t*-test. Then, two wavelength ranges were selected: the first one (510–630 nm) for showing the maximal differences between the three treatments at every das assayed ($p < 0.001$); and the second one (790–815 nm) for displaying little or no differences between them. The purpose of selecting the most stable range is for the "normalization" of the designed parameters. Finally, selected wavelengths were combined by different mathematical calculations (subtractions, additions, divisions, and combinations thereof) to construct novel VIs showing statistical differences ($p < 0.001$ according to a Student *t*-test) between the three climatic conditions along the entire experiment. Thus, one VI was found that met requirements (i) and (ii) and was therefore named the climatic stress index for brassicas (CSIB):

$$CSIB = R525/R800, \quad (1)$$

CSIB showed significant differences between the plants growing under the three climatic conditions as early as 25 das (Figure 6f) when the oilseed rape leaves were visually similar to each other. Moreover, these differences increased in the subsequent days and were greater for the more severe climatic conditions imposed. In regard to the control plants, there was a small increase in CSIB at 32 das with respect to previous days.

3.5. Assessing the Relationship between Novel VIs and Underlying Physiological Traits of Plants

The novel VIs DBI_{1-3} developed by Pineda et al. [24] and the mainly CSIB first proposed here have been shown to be indicative of the abiotic stress induced by climatic conditions in oilseed rape plants to different extents. However, the physiological processes underlying these VIs remains unknown. To try to establish a relationship between those indices and the physiological traits analyzed in this piece of work, bivariate correlations have been developed. The parameters available for 25, 28, and 32 das (i.e., those measured by imaging sensors: F440, F520, F680/F740, CRI, ARI, $T_L - T_A$, NDVI, and PRI) were selected for this analysis. Only those variables with a Pearson correlation coefficient of $R \geq 0.85$ at $p < 0.01$ were considered to be true correlations (Figure 7). Thus, DBI_1 and DBI_2 did not correlate with any of the selected parameters representing the physiological traits. On the other hand, DBI_3 correlated well with F680/F740 (Figure 7a), NDVI (Figure 7b), and PRI (Figure 7c), with R = 0.946, 0.914, and 0.958, respectively. F680/F740 is an MCFI parameter linked to chlorophyll content, as described above, whereas NDVI and PRI are measured by hyperspectral reflectance and are associated with the plant's vitality and photosynthesis, respectively. Finally, CSIB only correlated with F520 (Figure 7d), although to a lesser extent (R = 0.870). F520 is linked to phenolic compounds synthetized by plant secondary metabolism.

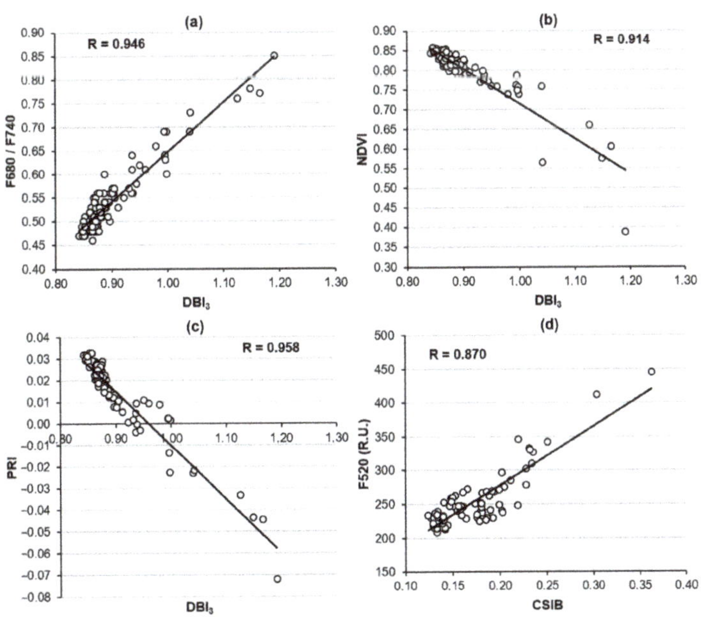

Figure 7. Correlation of the diseased broccoli index 3 (DBI_3) with F680/F740 (**a**), normalized difference vegetation index NDVI (**b**), and photochemical reflectance index PRI (**c**); as well as the correlation between

climatic stress index for brassicas CSIB with green fluorescence F520 (**d**). Pearson correlation coefficient ($p < 0.01$) is shown for every couple of variables.

4. Discussion

Climate change is one of the greatest challenges we face, especially in terms of its effects on global ecosystems, natural resources, and agriculture, on which humanity depends [29]. However, studies on the alterations that climate change causes in the metabolism of plants of agronomic interest (such as oilseed rape) grown under combined elevated temperatures and CO_2 remain scarce, despite being clearly needed. Several potential scenarios should be tested to assay the magnitude of the alterations that climate change would cause in oilseed rape metabolism. In this piece of work, the intermediate projection RCP 4.5 has been selected as it would represent the most probable scenario of climate change by the year 2100; whereas the extreme projection RCP 8.5 would take place if governments do not take any action against climate change, according to IPCC [4]. A combination of both invasive and non-invasive methods was chosen to assess the plant health status of oilseed rape plants to gain a broad insight into the physiological processes affected by climate change.

Some plant traits related to oxidative stress, such as TAA and lipid peroxidation, are of particular interest for the assessment of the plant health status, as the exposure to extreme environmental factors leads to an imbalance in the production of reactive oxygen species, which in turn results in oxidative stress that could damage cellular components, including lipids [39]. The oilseed rape plants grown under CCC showed no changes in TAA or lipid peroxidation, suggesting that these leaves did not undergo oxidative stress during the period analyzed. On the contrary, TAA increased at 32 das in the plants grown under the studied RCPs with respect to 28 das and compared to the controls. This was in accordance with the previous results from the plants grown under extreme temperatures [40–42] and elevated CO_2 concentrations [43]. Maintaining a high level of TAA might be a good strategy to scavenge the excessive reactive oxygen species produced by environmental stresses [44]. However, the increase in the antioxidant substances registered on the oilseed rape leaves grown under the analyzed RCPs was not sufficient to protect the plants from oxidative damage, as measured as lipid peroxidation. Heat stress also induces lipid peroxidation in plants such as lentils [45] or wheat [42]. Indeed, it is known that leaf senescence is related to the membrane deterioration induced by the increasing levels of lipid peroxidation [46,47]. Thus, climate change may be accelerating the oilseed rape leaf aging process.

Among plant secondary metabolites, phenolic compounds have the highest antioxidant activity both in vivo and in vitro [48–50], although they also have other important functions such as providing structural support [51] or protection from pathogens and pests [52]. In the case of the plants grown under CCC, a slight increment in the total soluble phenolic content and both F440 and F520 could be detected from the first das assayed, while TAA remained stable. This fluorescence increase was mainly associated with principal veins; thus, the progressive accumulation of compounds with structural function could be responsible for the increase in phenolics in the control plants [14]. On the other hand, the oilseed rape plants grown under the assessed RCPs presented a higher soluble phenolic content than those grown at CCC from 28 das as measured by colorimetric measurements. It has been previously described that other plants growin under climate change conditions also accumulated phenolic compounds with a high antioxidant capacity [40,41,53], as discussed above. Furthermore, MCFI registered higher F440 and, particularly, higher F520 from 25 das onwards across the whole leaves. The content of insoluble phenolics increased as a more drastic climatic condition was imposed. This could be due to the more pronounced decrease in Chl and Car contents measured in the leaves of the plants grown under RCP 8.5 since these pigments are responsible for the reabsorption of part of F440 and F520 emitted by phenolic compounds [54]. In any case, it is known that the antioxidant activity of plant phenolics benefits human health [48], but more research is needed to identify the nature of those compounds accumulated in oilseed rape leaves grown under the analyzed RCPs to elucidate how they will affect the quality of the products yielded from this plant in the future.

Lipid peroxidation causes ruptures in cell membranes, including the thylakoids where pigment–protein complexes are located, thus leading to leaf yellowing [47]. Yellowing is an indicator of leaf senescence, a programmed physiological process mainly associated with aging but which can also be induced by environmental stress [55,56]. Indeed, exposure to elevated temperatures or high atmospheric CO_2 concentrations accelerates plant senescence [57–59]. During leaf senescence, Chl catabolism is the main process, while the degradation of Xanth and Car occurs to a lesser extent [60–62]. The ratio between Chl *a* and Chl *b* is fine-tuned in leaves; however, chlorophyll *a* is more easily degraded by oxidative stress than chlorophyll *b*, causing decreases in Chl a/Chl *b* ratio [63]. The leaves of the oilseed rape plants grown under CCC did not display any symptoms of leaf yellowing and thus of senescence, as the total Chl content and total Chl/(Xanth + Car) ratio only decreased slightly in the last das assayed. However, the content of (Xanth + Car) and the ratio Chl *a*/Chl *b* remained stable. On the other hand, the damage to membranes due to lipid peroxidation can account for the decrease in photosynthetic pigments registered on leaves of plants grown under the studied RCPs, as well as for the impairment of the Chl *a*/Chl *b* ratio. Therefore, the oilseed rape plants grown under climate change conditions showed a clear early onset of leaf senescence, with the earlier the response, the more dramatic the climate change projection imposed.

The red-purple pigments Anth are phenylpropanoids, thus products of secondary metabolism in contrast to Chl, Xanth, and Car. Anth accumulation is considered to be a stress response to reduce photo oxidative damage [64,65]. In addition, Anth is synthetized during leaf senescence, and the degradation of Chl causes them to be even more noticeable [66]. Thus, the oilseed rape control plants maintained low levels of Anth throughout the experiment, whereas increasing amounts of this pigment were found in those plants grown under the studied RCPs. These results are in agreement with previous works showing alterations in the Anth composition of grape berries [67] and flowers [68] of plants growing under climate change conditions. Concerning green tissues, Anth accumulation under cold temperatures is well documented [69], but more research is needed to refer high temperatures, although some clues suggest that high CO_2 concentrations would induce up-regulation of genes responsible for Anth biosynthesis to prolong leaf longevity [70]. Despite that, the parameters related to pigment content calculated from imaging techniques (F680/F740 ratio, CRI, and ARI) could not distinguish between RCP 4.5 and RCP 8.5 treatments or do so late at 32 das, the tendency of their measurements agreed with those obtained by invasive methods.

Carbon and water balances are two of the most important physiological traits for plant growth and development. Indeed, stomata are able to sense environmental changes to ensure adequate CO_2 and water fluxes. As a consequence, the stomatal conductance and transpiration are parameters affected by environmental stress [71,72]. Cultivation under the studied RCPs caused stomatal closure in oilseed rape plants, measured by thermography as higher leaf temperatures with respect to the controls from 28 das. The plants grown under RCP 8.5 showed the largest increases in leaf temperature. This is in agreement with previous works that found that stomatal conductance generally increases with air temperature up to an optimum value and then decreases progressively with increasing ambient temperatures [73]; moreover, long exposures to elevated CO_2 concentrations reduce stomatal conductance, particularly in plants grown in pots [7,59].

The photosynthesis rate correlates with stomatal conductance since the stomatal aperture regulates the CO_2 available for photosynthesis [74]. That could be the case for the leaves of oilseed rape plants grown under warming conditions and high atmospheric CO_2, as the PRI decreased throughout the experiment together with the decrease in leaf transpiration. Indeed, high night temperatures reduce the photosynthetic efficiency of C_3 plants such as oilseed rape [6], as well as long exposures to high CO_2 concentrations [7]. Moreover, lipid peroxidation may lead to disorganization of the thylakoid membrane, resulting in loss of photosynthetic pigments (as discussed above) and preventing the proper performance of photosynthesis. However, further microscopic research is required to prove this hypothesis. Together with PRI decreases, leaf vitality measured as NDVI also diminished in plants grown under climate change conditions. Several works have compared the relationship between climate change and NDVI in several locations worldwide, showing that this parameter is inversely proportional to environmental temperature [75–77]. In the case of control plants, photosynthesis only decayed slightly, possibly as a consequence of normal leaf aging, without affecting NDVI.

The ability of VIs DBI_{1-3} to estimate the health status of the oilseed rape plants grown under the studied RCPs was also assessed, as they were developed to sense stress in other brassica, such as broccoli [24]. However, among all of the VIs analyzed here, DBIs were the least adequate indices to distinguish between the two RCPs analyzed in a presymptomatic way. This apparent lack of sensitivity could be due to the fact that DBI_{1-3} were designed to detect the biotic stress caused by *X. campestris* pv. *campestris* [24]. For this reason, a novel VI sensitive to the abiotic stress induced by climate change was specifically designed in this piece of work and thus named the climatic stress index for brassicas (CSIB). CSIB was demonstrated to be very sensitive to abiotic stress since it could presymptomatically distinguish between the three treatments, being directly proportional to the degree of abiotic stress. Defining novel VIs is a good strategy to maximize differences between treatments when standard reflectance parameters are not sensitive enough. Thus, the copper stress vegetation index was developed to detect the stress caused by this heavy metal in several plants [78], whereas the mangrove forest index was good at revealing the damage triggered by periodical submersions caused by tidal floods [79].

One step further, an attempt was made to elucidate what physiological traits underlie DBI_{1-3} and CSIB by correlation analysis. Only DBI_3 and CSIB were significantly correlated with other parameters assayed in this work. Whereas an increase in CSIB could be related to an enhancement of plant secondary metabolism, DBI_3 was linked to the primary metabolism and may reflect the damages caused by lipid peroxidation to thylakoid membranes: loss of photosynthetic pigments, as well as a decrease in photosynthesis and plant vigor, as it correlated to F680/F740, PRI, and NDVI, respectively. Nonetheless, more research is required to establish a direct relationship between the DBI_3 parameter and lipid peroxidation.

In conclusion, climate change would induce premature senescence in the leaves of oilseed rape plants, probably as a consequence of lipid peroxidation triggered by the production of reactive oxygen species that exceeds the antioxidant capacity of the plant. The extent of induced premature senescence is proportional to the climatic condition imposed. The use of invasive and non-invasive methods to determine the health status of plants proved to be complementary, although imaging techniques alone provide a reliable, sustainable, fast, and efficient estimate of it. Moreover, the novel parameter DBI_3 could reflect the degree of lipid peroxidation of leaves, whereas CSIB revealed presymptomatically the stress caused by climate change in oilseed rape plants.

Supplementary Materials: The following supporting information can be downloaded at: https://www.mdpi.com/article/10.3390/agronomy12081845/s1, Figure S1: Spectral profiles of whole leaves of oilseed rape grown under three different treatments: current climate conditions (blue; CCC), RCP 4.5 (orange; Representative Concentration Pathway 4.5) and RCP 8.5 (green; Representative Concentration Pathway 8.5) at 25 (a), 28 (b) and 32 (c) days after sowing (das), respectively. RCP 4.5 and RCP 8.5 are intermediate and extreme conditions of climate change according to the Intergovernmental Panel on Climate Change [4] and regionalized for Castilla y León for years 2081–2100. Graphs represent average values (n = 9–12) for every treatment and das assayed.

Author Contributions: Conceptualization and writing, M.P. and M.B.; experimental design, data acquisition and analyzing, figures author, M.P., resources and funding acquisition, M.B. All authors have read and agreed to the published version of the manuscript.

Funding: This work was supported by grant number RTI2018-094652-B-I00 funded by Ministerio de Ciencia e Innovación and Agencia Estatal de Investigación: MCIN/AEI/10.13039/501100011033, and by "European Regional Development Fund, ERDF: A way of making Europe". The free open access publication was partially funded by Consejo Superior de Investigaciones Científicas (CSIC) through the Unidad de Recursos de Información Científica para la Investigación (URICI).

Institutional Review Board Statement: Not applicable.

Informed Consent Statement: Not applicable.

Data Availability Statement: The original contributions presented in the study are included in the article/Supplementary Material. Further inquiries can be directed to the corresponding authors.

Acknowledgments: Authors would like to thank E. Cartea for guidelines for brassicas cultivation; T. Moreno-Martín for the care of the plants used in the experiments; as well as M.L. Pérez-Bueno for providing useful ideas for the experimental design and data management.

Conflicts of Interest: The authors declare no conflict of interest. The funders had no role in the design of the study; in the collection, analyses, or interpretation of data; in the writing of the manuscript, or in the decision to publish the results.

References

1. Nesi, N.; Delourme, R.; Brégeon, M.; Falentin, C.; Renard, M. Genetic and molecular approaches to improve nutritional value of *Brassica napus* L. seed. *C. R. Biol.* **2008**, *331*, 763–771. [CrossRef] [PubMed]
2. Kowalska, G.; Kowalski, R.; Hawlena, J.; Rowinski, R. Seeds of oilseed rape as an alternative source of protein and minerals. *J. Elem.* **2020**, *25*, 513–522. [CrossRef]
3. Rial-Lovera, K.; Davies, W.P.; Cannon, N.D. Implications of climate change predictions for UK cropping and prospects for possible mitigation: A review of challenges and potential responses. *J. Sci. Food Agric.* **2017**, *97*, 17–32. [CrossRef] [PubMed]
4. IPCC. *Climate Change 2014: Synthesis Report. Contribution of Working Groups I, II and III to the Fifth Assessment Report of the Intergovernmental Panel on Climate Change*; IPCC: Geneva, Switzerland, 2015.
5. Pullens, J.W.M.; Sharif, B.; Trnka, M.; Balek, J.; Semenov, M.A.; Olesen, J.E. Risk factors for European winter oilseed rape production under climate change. *Agric. For. Meteorol.* **2019**, *272*, 30–39. [CrossRef]
6. Alemu, S.T. Photosynthesis limiting stresses under climate change scenarios and role of chlorophyll fluorescence: A review article. *Cogent Food Agric.* **2020**, *6*, 1785136. [CrossRef]
7. Dusenge, M.E.; Duarte, A.G.; Way, D.A. Plant carbon metabolism and climate change: Elevated CO_2 and temperature impacts on photosynthesis, photorespiration and respiration. *New Phytol.* **2019**, *221*, 32–49. [CrossRef]
8. Bakhshi, B. Heat and drought stress response and related management strategies in oilseed rape. *Agrotech. Ind. Crops* **2021**, *1*, 170–181. [CrossRef]
9. Namazkar, S.; Stockmarr, A.; Frenck, G.; Egsgaard, H.; Terkelsen, T.; Mikkelsen, T.; Ingvordsen, C.H.; Jørgensen, R.B. Concurrent elevation of CO_2, O_3 and temperature severely affects oil quality and quantity in rapeseed. *J. Exp. Bot.* **2016**, *67*, 4117–4125. [CrossRef] [PubMed]
10. Namazkar, S.; Egsgaard, H.; Frenck, G.; Terkelsen, T.; Jørgensen, R.B. Significant reductions in oil quality and lipid content of oilseed rape (*Brassica napus* L.) under climate change. *Procedia Environ. Sci.* **2015**, *29*, 121–122. [CrossRef]
11. Wójtowicz, M.; Wójtowicz, A. The effect of climate change on linolenic fatty acid in oilseed rape. *Agronomy* **2020**, *10*, 2003. [CrossRef]
12. Chaerle, L.; Van der Straeten, D. Seeing is believing: Imaging techniques to monitor plant health. *Biochim. Biophys. Acta* **2001**, *1519*, 153–166. [CrossRef]
13. Mahlein, A.-K. Plant disease detection by imaging sensors—Parallels and specific demands for precision agriculture and plant phenotyping. *Plant Dis.* **2016**, *100*, 241–251. [CrossRef] [PubMed]
14. Barón, M.; Pineda, M.; Pérez-Bueno, M.L. Picturing pathogen infection in plants. *Z. Naturforsch. C* **2016**, *71*, 355–368. [CrossRef]
15. Buschmann, C.; Lichtenthaler, H.K. Principles and characteristics of multi-colour fluorescence imaging of plants. *J. Plant Physiol.* **1998**, *152*, 297–314. [CrossRef]
16. Cerovic, Z.G.; Samson, G.; Morales, F.; Tremblay, N.; Moya, I. Ultraviolet-induced fluorescence for plant monitoring: Present state and prospects. *Agronomie* **1999**, *19*, 543–578. [CrossRef]
17. Jones, H.G. Application of thermal imaging and infrared sensing in plant physiology and ecophysiology. *Adv. Bot. Res.* **2004**, *41*, 107–163.
18. Pineda, M.; Barón, M.; Pérez-Bueno, M.-L. Thermal imaging for plant stress detection and phenotyping. *Remote Sens.* **2021**, *13*, 68. [CrossRef]
19. Gamon, J.A.; Peñuelas, J.; Field, C.B. A narrow-waveband spectral index that tracks diurnal changes in photosynthetic efficiency. *Remote Sens. Environ.* **1992**, *41*, 35–44. [CrossRef]

20. Gitelson, A.A.; Merzlyak, M.N.; Chivkunova, O.B. Optical properties and nondestructive estimation of anthocyanin content in plant leaves. *Photochem. Photobiol.* **2001**, *74*, 38–45. [CrossRef]
21. Gitelson, A.A.; Zur, Y.; Chivkunova, O.B.; Merzlyak, M.N. Assessing carotenoid content in plant leaves with reflectance spectroscopy. *Photochem. Photobiol.* **2002**, *75*, 272–281. [CrossRef]
22. Tucker, C.J. Red and photographic infrared linear combinations for monitoring vegetation. *Remote Sens. Environ.* **1979**, *8*, 127–150. [CrossRef]
23. Pettorelli, N. *The Normalized Difference Vegetation Index*; Oxford University Press: Oxford, UK, 2013.
24. Pineda, M.; Pérez-Bueno, M.L.; Barón, M. Novel vegetation indices to identify broccoli plants infected with *Xanthomonas campestris* pv. *campestris*. *Front. Plant Sci.* **2022**, *13*, 790268. [CrossRef] [PubMed]
25. Chaerle, L.; Lenk, S.; Leinonen, I.; Jones, H.G.; Van Der Straeten, D.; Buschmann, C. Multi-sensor plant imaging: Towards the development of a stress-catalogue. *Biotechnol. J.* **2009**, *4*, 1152–1167. [CrossRef] [PubMed]
26. Ninomiya, S. High-throughput field crop phenotyping: Current status and challenges. *Breed. Sci.* **2022**, *72*, 3–18. [CrossRef]
27. Baslam, M.; Mitsui, T.; Hodges, M.; Priesack, E.; Herritt, M.T.; Aranjuelo, I.; Sanz-Sáez, Á. Photosynthesis in a changing global climate: Scaling up and scaling down in crops. *Front. Plant Sci.* **2020**, *11*, 882. [CrossRef]
28. Arya, S.; Sandhu, K.S.; Singh, J.; Kumar, S. Deep learning: As the new frontier in high-throughput plant phenotyping. *Euphytica* **2022**, *218*, 47. [CrossRef]
29. Rolnick, D.; Donti, P.L.; Kaack, L.H.; Kochanski, K.; Lacoste, A.; Sankaran, K.; Ross, A.S.; Milojevic-Dupont, N.; Jaques, N.; Waldman-Brown, A.; et al. Tackling climate change with machine learning. *ACM Comput. Surv.* **2022**, *55*, 42. [CrossRef]
30. Miller, N.J.; Diplock, A.T.; Rice-Evans, C.A. Evaluation of the total antioxidant activity as a marker of the deterioration of apple juice on storage. *J. Agric. Food Chem.* **1995**, *43*, 1794–1801. [CrossRef]
31. Pérez-Bueno, M.L.; Pineda, M.; Díaz-Casado, E.; Barón, M. Spatial and temporal dynamics of primary and secondary metabolism in *Phaseolus vulgaris* challenged by *Pseudomonas syringae*. *Physiol. Plant.* **2015**, *153*, 161–174. [CrossRef]
32. Chun, O.K.; Kim, D.O. Consideration on equivalent chemicals in total phenolic assay of chlorogenic acid-rich plums. *Food Res. Int.* **2004**, *37*, 337–342. [CrossRef]
33. Rodríguez-Serrano, M.; Romero-Puertas, M.C.; Sanz-Fernández, M.; Hu, J.; Sandalio, L.M. Peroxisomes extend peroxules in a fast response to stress via a reactive oxygen species-mediated induction of the peroxin PEX11a. *Plant Physiol.* **2016**, *171*, 1665–1674. [CrossRef] [PubMed]
34. Lichtenthaler, H.K.; Buschmann, C. Chlorophylls and carotenoids: Measurement and characterization by UV-VIS spectroscopy. In *Current Protocols in Food Analytical Chemistry*; Wrolstad, R.E., Acree, T.E., An, H., Decker, E.A., Penner, M.H., Reid, D.S., Schwartz, S.J., Shoemaker, C.F., Sporns, P., Eds.; John Wiley and Sons: New York, NY, USA, 2001; pp. F4.3.1–F4.3.8.
35. Solfanelli, C.; Poggi, A.; Loreti, E.; Alpi, A.; Perata, P. Sucrose-specific induction of the anthocyanin biosynthetic pathway in Arabidopsis. *Plant Physiol.* **2005**, *140*, 637–646. [CrossRef] [PubMed]
36. Pérez-Bueno, M.L.; Granum, E.; Pineda, M.; Flors, V.; Rodríguez-Palenzuela, P.; López-Solanilla, E.; Barón, M. Temporal and spatial resolution of activated plant defense responses in leaves of *Nicotiana benthamiana* infected with *Dickeya dadantii*. *Front. Plant Sci.* **2016**, *6*, 1209. [CrossRef] [PubMed]
37. Pineda, M.; Luisa Perez-Bueno, M.; Paredes, V.; Baron, M. Use of multicolour fluorescence imaging for diagnosis of bacterial and fungal infection on zucchini by implementing machine learning. *Funct. Plant Biol.* **2017**, *44*, 563–572. [CrossRef] [PubMed]
38. Pérez-Bueno, M.L.; Pineda, M.; Cabeza, F.; Barón Ayala, M. Multicolor fluorescence imaging as a candidate for disease detection in plant phenotyping. *Front. Plant Sci.* **2016**, *7*, 1790. [CrossRef] [PubMed]
39. Xie, X.; He, Z.; Chen, N.; Tang, Z.; Wang, Q.; Cai, Y. The roles of environmental factors in regulation of oxidative stress in plant. *Biomed. Res. Int.* **2019**, *2019*, 9732325. [CrossRef] [PubMed]
40. Sárosi, S.; Bernáth, J.; Burchi, G.; Antonetti, M.; Bertoli, A.; Pistelli, L.; Benvenuti, S. Effect of different plant origins and climatic conditions on the total phenolic content and total antioxidant capacity of self-heal (*Prunella vulgaris* L.). *Acta Hortic.* **2011**, *925*, 49–55. [CrossRef]
41. Ben Mansour-Gueddes, S.; Saidana-Naija, D.; Bchir, A.; Braham, M. Climate change effects on phytochemical compounds and antioxidant activity of *Olea europaea* L. *Not. Bot. Horti Agrobot. Cluj-Napoca* **2020**, *48*, 436–455. [CrossRef]
42. Sairam, R.K.; Srivastava, G.C.; Saxena, D.C. Increased antioxidant activity under elevated temperatures: A mechanism of heat stress tolerance in wheat genotypes. *Biol. Plant.* **2000**, *43*, 245–251. [CrossRef]
43. Naudts, K.; Van den Berge, J.; Farfan, E.; Rose, P.; AbdElgawad, H.; Ceulemans, R.; Janssens, I.A.; Asard, H.; Nijs, I. Future climate alleviates stress impact on grassland productivity through altered antioxidant capacity. *Environ. Exp. Bot.* **2014**, *99*, 150–158. [CrossRef]
44. Sharma, P.; Dubey, R.S. Drought induces oxidative stress and enhances the activities of antioxidant enzymes in growing rice seedlings. *Plant Growth Regul.* **2005**, *46*, 209–221. [CrossRef]
45. Chakraborty, U.; Pradhan, D. High temperature-induced oxidative stress in *Lens culinaris*, role of antioxidants and amelioration of stress by chemical pre-treatments. *J. Plant Interact.* **2011**, *6*, 43–52. [CrossRef]
46. Strother, S. The role of free radicals in leaf senescence. *Gerontology* **1988**, *34*, 151–156. [CrossRef]
47. Dhindsa, R.S.; Plumb-Dhindsa, P.; Thorpe, T.A. Leaf senescence: Correlated with increased levels of membrane permeability and lipid peroxidation, and decreased levels of superoxide dismutase and catalase. *J. Exp. Bot.* **1981**, *32*, 93–101. [CrossRef]

48. Kasote, D.M.; Katyare, S.S.; Hegde, M.V.; Bae, H. Significance of antioxidant potential of plants and its relevance to therapeutic applications. *Int. J. Biol. Sci.* **2015**, *11*, 982–991. [CrossRef] [PubMed]
49. Vuolo, M.M.; Lima, V.S.; Maróstica Junior, M.R. Chapter 2—Phenolic compounds: Structure, classification, and antioxidant power. In *Bioactive Compounds*; Campos, M.R.S., Ed.; Woodhead Publishing: Duxford, UK, 2019; pp. 33–50. [CrossRef]
50. Sharma, A.; Shahzad, B.; Rehman, A.; Bhardwaj, R.; Landi, M.; Zheng, B. Response of phenylpropanoid pathway and the role of polyphenols in plants under abiotic stress. *Molecules* **2019**, *24*, 2452. [CrossRef] [PubMed]
51. de la Rosa, L.A.; Moreno-Escamilla, J.O.; Rodrigo-García, J.; Alvarez-Parrilla, E. Chapter 12—Phenolic compounds. In *Postharvest Physiology and Biochemistry of Fruits and Vegetables*; Yahia, E.M., Ed.; Woodhead Publishing: Duxford, UK, 2019; pp. 253–271. [CrossRef]
52. Dixon, R.A.; Achnine, L.; Kota, P.; Liu, C.J.; Srinivasa Reddy, M.S.; Wang, L. The phenylpropanoid pathway and plant defence —A genomics perspective. *Mol. Plant Pathol.* **2002**, *3*, 371–390. [CrossRef]
53. Kumar, S.; Yadav, A.; Yadav, M.; Yadav, J.P. Effect of climate change on phytochemical diversity, total phenolic content and in vitro antioxidant activity of *Aloe vera* (L.) Burm.f. *BMC Res. Notes* **2017**, *10*, 60. [CrossRef]
54. Buschmann, C.; Langsdorf, G.; Lichtenthaler, H.K. Imaging of the blue, green, and red fluorescence emission of plants: An overview. *Photosynthetica* **2000**, *38*, 483–491. [CrossRef]
55. Mayta, M.L.; Hajirezaei, M.-R.; Carrillo, N.; Lodeyro, A.F. Leaf senescence: The chloroplast connection comes of age. *Plants* **2019**, *8*, 495. [CrossRef]
56. Thakur, N.; Sharma, V.; Kishore, K. Leaf senescence: An overview. *Indian J. Plant Physiol.* **2016**, *21*, 225–238. [CrossRef]
57. Rossi, S.; Burgess, P.; Jespersen, D.; Huang, B. Heat-induced leaf senescence associated with chlorophyll metabolism in bentgrass lines differing in heat tolerance. *Crop Sci.* **2017**, *57*, S-169. [CrossRef]
58. Jochum, G.M.; Mudge, K.W.; Thomas, R.B. Elevated temperatures increase leaf senescence and root secondary metabolite concentrations in the understory herb *Panax quinquefolius* (Araliaceae). *Am. J. Bot.* **2007**, *94*, 819–826. [CrossRef] [PubMed]
59. Agüera, E.; De la Haba, P. Leaf senescence in response to elevated atmospheric CO_2 concentration and low nitrogen supply. *Biol. Plant.* **2018**, *62*, 401–408. [CrossRef]
60. Kusaba, M.; Tanaka, A.; Tanaka, R. Stay-green plants: What do they tell us about the molecular mechanism of leaf senescence. *Photosynth. Res.* **2013**, *117*, 221–234. [CrossRef]
61. Ougham, H.; Hörtensteiner, S.; Armstead, I.; Donnison, I.; King, I.; Thomas, H.; Mur, L. The control of chlorophyll catabolism and the status of yellowing as a biomarker of leaf senescence. *Plant Biol.* **2008**, *10*, 4–14. [CrossRef] [PubMed]
62. Biswal, B. Carotenoid catabolism during leaf senescence and its control by light. *J. Photochem. Photobiol. B* **1995**, *30*, 3–13. [CrossRef]
63. Kasajima, I. Difference in oxidative stress tolerance between rice cultivars estimated with chlorophyll fluorescence analysis. *BMC Res. Notes* **2017**, *10*, 168. [CrossRef] [PubMed]
64. Diaz, C.; Saliba-Colombani, V.; Loudet, O.; Belluomo, P.; Moreau, L.; Daniel-Vedele, F.; Morot-Gaudry, J.-F.; Masclaux-Daubresse, C. Leaf yellowing and anthocyanin accumulation are two genetically independent strategies in response to nitrogen limitation in *Arabidopsis thaliana*. *Plant Cell Physiol.* **2006**, *47*, 74–83. [CrossRef]
65. Steyn, W.J.; Wand, S.J.E.; Holcroft, D.M.; Jacobs, G. Anthocyanins in vegetative tissues: A proposed unified function in photoprotection. *New Phytol.* **2002**, *155*, 349–361. [CrossRef]
66. Lee, D.W. Anthocyanins in autumn leaf senescence. *Adv. Bot. Res.* **2002**, *37*, 147–165. [CrossRef]
67. de Rosas, I.; Deis, L.; Baldo, Y.; Cavagnaro, J.B.; Cavagnaro, P.F. High temperature alters anthocyanin concentration and composition in grape berries of malbec, merlot, and pinot noir in a cultivar-dependent manner. *Plants* **2022**, *11*, 926. [CrossRef]
68. Sullivan, C.N.; Koski, M.H. The effects of climate change on floral anthocyanin polymorphisms. *Proc. R. Soc. B Biol. Sci.* **2021**, *288*, 20202693. [CrossRef]
69. Chalker-Scott, L. Environmental significance of anthocyanins in plant stress responses. *Photochem. Photobiol.* **1999**, *70*, 1–9. [CrossRef]
70. Tallis, M.J.; Lin, Y.; Rogers, A.; Zhang, J.; Street, N.R.; Miglietta, F.; Karnosky, D.F.; De Angelis, P.; Calfapietra, C.; Taylor, G. The transcriptome of *Populus* in elevated CO_2 reveals increased anthocyanin biosynthesis during delayed autumnal senescence. *New Phytol.* **2010**, *186*, 415–428. [CrossRef] [PubMed]
71. Jones, H. *Plants and Microclimate: A Quantitative Approach to Environmental Plant Physiology*, 3rd ed.; Cambridge University Press: Cambrigde, UK, 2014; Volume 56.
72. Prashar, A.; Yildiz, J.; McNicol, J.W.; Bryan, G.J.; Jones, H.G. Infra-red thermography for high throughput field phenotyping in *Solanum tuberosum*. *PLoS ONE* **2013**, *8*, e65816. [CrossRef]
73. Turner, N.C. Measurement and influence of environmental and plant factors on stomatal conductance in the field. *Agric. For. Meteorol.* **1991**, *54*, 137–154. [CrossRef]
74. Lawson, T.; Blatt, M.R. Stomatal size, speed, and responsiveness impact on photosynthesis and water use efficiency. *Plant Physiol.* **2014**, *164*, 1556–1570. [CrossRef]
75. Ghebrezgabher, M.G.; Yang, T.; Yang, X.; Eyassu Sereke, T. Assessment of NDVI variations in responses to climate change in the Horn of Africa. *Egypt. J. Remote Sens. Space Sci.* **2020**, *23*, 249–261. [CrossRef]
76. Zhao, W.; Yu, X.; Jiao, C.; Xu, C.; Liu, Y.; Wu, G. Increased association between climate change and vegetation index variation promotes the coupling of dominant factors and vegetation growth. *Sci. Total Environ.* **2021**, *767*, 144669. [CrossRef]

77. Bagherzadeh, A.; Hoseini, A.V.; Totmaj, L.H. The effects of climate change on normalized difference vegetation index (NDVI) in the Northeast of Iran. *Model. Earth Syst. Environ.* **2020**, *6*, 671–683. [CrossRef]
78. Zhang, C.; Ren, H.; Qin, Q.; Ersoy, O.K. A new narrow band vegetation index for characterizing the degree of vegetation stress due to copper: The copper stress vegetation index (CSVI). *Remote Sens. Lett.* **2017**, *8*, 576–585. [CrossRef]
79. Jia, M.; Wang, Z.; Wang, C.; Mao, D.; Zhang, Y. A new vegetation index to detect periodically submerged mangrove forest using single-tide Sentinel-2 imagery. *Remote Sens.* **2019**, *11*, 2043. [CrossRef]

agronomy

Article

Root Reserves Ascertain Postharvest Sensitivity to Water Deficit of Nectarine Trees

María R. Conesa, Wenceslao Conejero, Juan Vera and Mª Carmen Ruiz-Sánchez *

Irrigation Department, Centro de Edafología y Biología Aplicada del Segura (CEBAS-CSIC), P.O. Box 164, 30100 Murcia, Spain; mrconesa@cebas.csic.es (M.R.C.); wenceslao@cebas.csic.es (W.C.); jvera@cebas.csic.es (J.V.)
* Correspondence: mcruiz@cebas.csic.es; Tel.: +34-968-396-200

Abstract: This work studied the sensitivity of the postharvest period of early maturing nectarine trees (*Prunus persica* L. Batsch, cv. Flariba) to water stresses. Along with a well-irrigated treatment (T-0), three water deficit treatments (by withholding irrigation) were applied: T-1: early postharvest (June–July), T-2: late postharvest (August–September), and T-3: the whole postharvest period (June–September). Soil water content (θ_v) and midday stem water potential (Ψ_{stem}) were measured throughout the study. During winter dormancy, L-arginine, starch, and phosphorus content in the roots were analyzed. Yield, fruit quality, and metabolites were determined at harvest. Ψ_{stem} reached −1.7 and −2.3 MPa at the end of the early and late postharvest periods, respectively. Total yield and number of fruits per tree were significantly reduced in all deficit treatments with respect to T-0, while no significant differences were observed in physicochemical fruit quality. The T-2 treatment showed the highest percentage of cracked fruits. Significantly, lower values of L-arginine and phosphorus were observed in the roots of T-2 trees, with respect to T-0, while they were similar in T-1 trees. Although the early postharvest stage is key for the application of RDI strategies, our results indicated that the late postharvest period was also a sensitive period to severe drought, as the accumulation of winter root reserves (L-arginine and phosphorus) was reduced, which limited yield.

Keywords: L-arginine; phenological periods; regulated deficit irrigation (RDI); root reserves; water relations; yield

Citation: Conesa, M.R.; Conejero, W.; Vera, J.; Ruiz-Sánchez, M.C. Root Reserves Ascertain Postharvest Sensitivity to Water Deficit of Nectarine Trees. *Agronomy* **2022**, *12*, 1805. https://doi.org/10.3390/agronomy12081805

Academic Editor: Alberto San Bautista

Received: 11 July 2022
Accepted: 28 July 2022
Published: 30 July 2022

1. Introduction

The Mediterranean region is one of the most water scarce regions of the world and is considered a climate change 'hotspot' due to the aridity of the climate and the increasing persistent shortage of water resources [1,2]. Spain is one of the driest Mediterranean countries in the European Union. In fact, February 2022 was the second driest February of the 21st century, after 2020, and the third driest since the series began in 1961, being the warmest in 30 years in Europe [3]. Management of water resources is a difficult challenge in these water-scarce regions that are highly dependent on irrigation development. In Spain, irrigated agriculture is responsible for more than 70% of total water withdrawals [2,4].

Peach and nectarine trees (*Prunus persica* L. Batsch) represent two of the main fruit species in the Mediterranean region, with an average production of 1158 Mt per year in the period 2015–2020 [5]. The establishment of early maturing varieties in this area is associated with a high export value of the fruits, the first ones being marketed in early spring [6–9]. These tree crops have high irrigation requirements, especially during dry and hot seasons, when irrigated orchards are often subjected to a sharp decrease in water supply [10,11]. Moreover, this situation is aggravated in early maturing cultivars, as their higher water requirements during the postharvest season coincide with the summer months [12].

The scarcity of water for irrigation drives the application of deficit irrigation strategies, which consists of applying less irrigation water than the full crop evapotranspiration requirements, but increasing water productivity [2,4,13].

Since 1970, a large amount of evidence has demonstrated the utility of regulated deficit irrigation (RDI) in fruit tree cultivation. RDI is a water-saving strategy, in which trees are deficit irrigated during certain phenological periods less sensitive to water deficit (named as non-critical periods), while irrigation is applied to meet full water requirements during the most sensitive stages of the crop [10,11,14–17].

As indicated in Torrecillas et al. [18], in stone fruits (to which nectarines belongs), two critical periods are defined: (i) the second rapid fruit growth period (stage III), in which drought stress induces a reduction in yield due to the smaller fruit size at the ongoing harvest, and (ii) the early postharvest period, due to its influence on the floral induction and differentiation processes, which leads to a lower pollen germination potential in the next bloom, encouraging young fruit to drop in the following season's harvest [19,20].

A key factor on RDI is the precise management of deficit and full irrigation episodes, since RDI can reduce yield if the recovery of tree water status is delayed, mainly when the water deficit extends to stage III of fruit development [21,22]. This particularly applies to early maturing cultivars, with a very short period from fruit set to harvest, and a very long postharvest period, and for this reason, deficit irrigation should be applied only during the postharvest season to avoid any adverse effect on yield components [23]. In this sense, recent research in early maturing peach and nectarine trees has shown that the summer postharvest season is the most advisable phenological period to apply deficit irrigation in RDI scheduling under Mediterranean conditions [6–9,23–27]. However, as the annual yield also depends on previous postharvest conditions, it is necessary to assess the sensitivity of this phenological period in a context of climate change (severe drought conditions) that defines early and late postharvest periods in order to accurately program such RDI strategies.

Root nutritional reserves are essential for fruit development. L-arginine is one of the 20 amino-acids synthesized in all plant species, considered as the main source of nitrogen reserves in stone fruit trees, and could be considered a general metabolic crop indicator [28]. Its use by the tree depends on its starch reserves. The balance between both nutritional root reserves is fundamental for the good progress of the sprouting and fruit set processes. In addition, adequate absorption of phosphorus storage by the root system determines tree development [29].

It is well known that the processes of shoot/flower development, root growth, nutrient uptake, and reserve storage might be affected by water stress [30]. Ebel et al. [31] reported significant reductions in apple productivity in years enduring severe water deficit, because of the reduced root reserves accumulation. Other studies describe crop yield reduction because of reduced root reserves due to water deficits that decrease the availability of carbohydrates to support flowering and fruit set the following season [32,33]. In mid to late-maturing peach trees, Lopez et al. [34] indicated that root starch reserves are a key factor in supporting early reproductive development, influencing peach productivity. However, the information reported in early maturing cultivars is still incomplete, particularly when postharvest RDI strategies are applied.

To simulate a climate change scenario with severe restrictions in irrigation water availability, the present study investigated the effects of severe drought stress (by withholding irrigation) during the postharvest period in an early maturing nectarine orchard, grown under Mediterranean conditions, on yield and fruit quality components, with special attention paid to the role of winter nutritional reserves in the roots (arginine, starch, and phosphorus). More importantly, the study highlights the contribution of early and late postharvest phases in determining the feasibility of the long postharvest period in the management of RDI strategies in early maturing cultivars.

2. Materials and Methods

2.1. Experimental Conditions

The experiment was conducted during the 2020–2021 growing season in a 0.5 ha plot of adult early maturing nectarine trees (*Prunus persica* (L.) Batsch *cv.* Flariba, on GxN-15 rootstock), spaced 6.5 m × 3.5 m and trained to an open-center canopy. The plot was located at the CEBAS-CSIC experimental field station in Santomera, Murcia, Spain (38°06′31″ N, 1°02′14″ W, 110 m altitude). The soil in the 0–0.5 m layer was highly calcareous (45% calcium carbonate), with a clay loam texture (clay fraction: 41% illite, 17% smectite, and 30% palygorskite), low organic matter content (1.3%), and a cationic exchange capacity of 97.9 mmol kg^{-1}. The average bulk density was 1.43 g cm^{-3}. The volumetric soil water content (θ_v) at field capacity and permanent wilting point were 0.29 and 0.14 m^3 m^{-3}, respectively. The irrigation system consisted of a single emitter-holder line, with 4 emitters of 4 L h^{-1} arranged at 0.5 and 1.3 m on both sides of the tree trunks. The irrigation water came from the Tagus-Segura transfer system with an average electrical conductivity ($EC_{25\,°C}$) of 1.3 dS m^{-1} and pH ≈ 8. The trees were fertilized with 83-56-109 UF $year^{-1}$ of N, P_2O_5, and K_2O, independently, applied through the drip irrigation system [35]. Standard cultural practices (i.e., weed control, phytosanitary treatments, thinning, and pruning) were the usual practices for stone fruit trees grown in the studied area.

The climate of the area was typically Mediterranean, with an average annual rainfall of 206 mm, irregularly distributed, and with annual reference crop evapotranspiration (ET_0) of 1097 mm. Agrometeorological data were recorded by an automatic weather station located at the CEBAS-CSIC experimental station, less than 100 m from the nectarine orchard (http://www.cebas.csic.es/general_spain/est_meteo.html) (accessed on 2 May 2021).

2.2. Irrigation Periods and Treatments

Three different irrigation periods were established in the experiment:

(i) **Pre-conditioning period (May):** All trees were fully irrigated to ensure non-limiting soil water conditions (T-0 treatment, hereafter) at 100% of crop evapotranspiration (ETc) from 1 May to 28 May 2020. ETc was estimated as the product of crop reference evapotranspiration (ET_0) from the Penman–Monteith equation [36] and local crop coefficient (Kc) obtained in the same experimental orchard for *Prunus* sp. [37].

(ii) **Stress period (June–September):** Along with T-0, three soil water deficit treatments were imposed by withholding irrigation water during the postharvest period (Figure 1B):

- Stress 1 (T-1 treatment): during early postharvest (from 28 May to 30 July 2020).
- Stress 2 (T-2 treatment): during late postharvest (from 30 July to 24 September 2020).
- Stress 3 (T-3 treatment): during the whole postharvest (from 28 May to 24 September 2020).

(iii) **Recovery period (October):** All trees were irrigated as T-0 (100% ETc) from 24 September to 5 October 2020.

Treatments were distributed according to a completely randomized blocks design with four replicates. Each one consisted of four trees, taking the two central trees for measurements, and the remaining ones served as guard trees, with a total of 16 trees per irrigation treatment. No active roots were observed more than 1.5 m from the drip line, as revealed by root distribution studies [38].

2.3. Soil Water Status

Soil water status was estimated by measuring soil water content (θ_v) at the end of each irrigation period (Pre-conditioning, Stress 1, Stress 2, and Recovery) using a Acclima-TDR-315H probe (Time Domain Reflectometry, Acclima Inc., Meridian, ID, USA). Measurements were performed near the emitter located at 0.5 m from the tree trunk, in the first 0.15 m of soil depth, in eight representative trees per treatment (two trees per replicate), following the recommendations described in Vera et al. [39].

2.4. Tree Water Status

Tree water status was estimated by measuring predawn leaf water potential (Ψ_{pd}) (at 05:30 h solar time) and midday stem water potential (Ψ_{stem}) (≈12:00 h solar time). Measurements were made weekly and coincided with the end of each irrigation period (Pre-conditioning, Stress 1, Stress 2, and Recovery) using a pressure chamber (Soil Moisture Equipment Corp. Model 3000, Santa Bárbara, CA, USA) on one leaf per tree and one tree per replicate of each treatment (n = 4). For Ψ_{stem}, leaves were selected from the northern part of the tree and close to the tree trunk and placed in plastic bags covered with aluminum foil for at least 2 h prior to the measurements following the recommendations indicated in Hsiao [40].

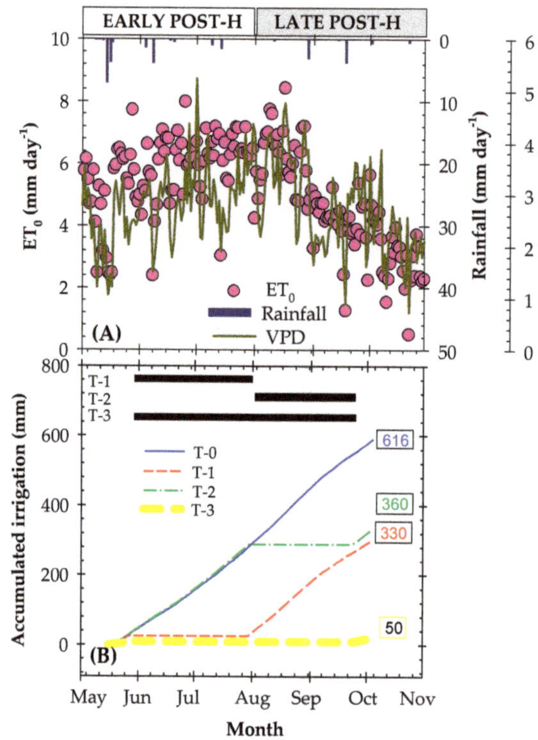

Figure 1. Seasonal evolution of (**A**) daily rainfall (mm), reference crop evapotranspiration (ET_0, mm day^{-1}), and mean vapor pressure deficit (VPD, kPa), and (**B**) accumulated irrigation (mm) in early maturing nectarine trees in the different treatments (T-0: non-stressed, T-1: stress during early postharvest; T-2: stress during late postharvest, and T-3: stress during the whole postharvest). Framed figures in (**B**) indicate the total of the irrigation volume applied to each treatment. Black horizontal bars in (**B**) indicate the water withholding period in T-1, T-2, and T-3 treatments.

The intensity of water stress endured by each treatment was calculated by the water stress integral ($S\Psi_{stem}$), using the equation defined by Myers [41]:

$$S\Psi_{stem}(\text{MPa day}) = \sum(\Psi_{i,i+1} - \Psi_c)\text{n} \quad (1)$$

where $\sum$ is the sum of the Ψ_{stem} measurements; $\Psi_{i,i+1}$ is the mean Ψ_{stem} for any measurement i and i + 1; Ψ_c is the maximum Ψ_{stem} value measured during the experiment; and n is the number of days in the interval.

2.5. *Leaf Gas Exchange*

Net photosynthesis (P_n, µmol m^{-2} s^{-1}), stomatal conductance (g_s, mmol m^{-2} s^{-1}), and transpiration rate (E, mmol m^{-2} s^{-1}) were measured weekly and at the end of each irrigation period (Pre-conditioning, Stress 1, Stress 2, and Recovery), on one mature sunny leaf per replicate (n = 4), at about 10:00 h solar time at mean values of ambient photosynthetic photon flux density (PPFD) ≈ 1200 µmol m^{-2} s^{-1} and near constant ambient CO_2 concentration (Ca ≈ 400 µmol mol^{-1}), using a portable gas exchange system (LI-COR, LI-

6400 Li-Cor, Lincoln, NE, USA). Water use efficiency transpiration (WUE_T) was calculated as the ratio of P_n and E (μmol $mmol^{-1}$).

2.6. Leaf Mineral Content

Leaf mineral content was determined on leaf samples taken at the end of September (coinciding with the end of stress period of S2) in the four replicates of each per treatment (n = 100 leaves). Leaves, taken from the middle third of trees and from extension shoots of the current year's growth, were washed, oven-dried, and grounded. Analyses were performed by the Ionomic Service of CEBAS-CSIC (http://www.cebas.csic.es/general_english/ionomics.html) (accessed on 10 December 2020). Cations were analyzed by inductively coupled plasma (ICP-ICAP 6500 DUO Thermo, Manchester, UK), and anions by ion chromatography with a liquid chromatograph (Metrohm, Herisau, Switzerland). The total nitrogen (N_{TOTAL}) and total carbon (C_{TOTAL}) were measured with an elemental analyzer Flash EA 1112 Series- Leco Truspec.

2.7. Winter Root Reserves

During winter dormancy (10 December 2020), nutritional root reserves, including arginine, starch, and phosphorus content, were determined in coarse suberized root samples collected from soil samples in one tree in three out of the four replicates of the different treatments (n = 3). Trenches were excavated with a backhoe (Optimal Model OP03) at 0.1 m from the tree row, between both emitters, up to 0.5 m depth. All roots (4 to 8 mm diameter) were taken at 0.3 m depth and carefully separated from the soil in the field, since no rainfall or irrigation occurred in the previous days and the ground was dry. Measurements were performed in an accredited laboratory (AGQ-Labs: http://www.agq.com.es/) (accessed on 2 February 2021). Arginine was extracted in an ice bath by grounding samples in a chilled mortar at a ratio of 0.1 g fresh weight per mL of 10% (w/v) cold trichloroacetic acid. The extract was centrifuged at 13,000 × g for 10 min at 4 °C, and the supernatant was used directly for arginine analysis as described by Sakaguchi [42]. The starch and phosphorous concentration were measured using the methods described in Clegg [43] and Teixeria et al. [44], respectively.

2.8. Phenological Stages

The reproductive phenological stages, identified according to the Baggiolini code [45], were studied as described in Mounzer et al. [46] for early maturing *Prunus* cultivars. In brief, full bloom (50% of open flowers at stage F) for *'Flariba'* nectarine trees occurred at the end of January, and the fruit set (50% of visible fruits at stage H) occurred at the beginning of March, before commercial hand-thinning (performed on 10 March 2021). Harvest occurred during the first week of May in the orchard.

The number of flowers and fruits were counted on 8 February and 2 March 2021, respectively, on four marked shoots (one on each cardinal direction, in one tree per replicate of the different treatments). The percentage of flower and fruit drops was evaluated.

2.9. Yield Measurements

At harvest, nectarine fruits were harvested at commercial maturity on 1 May 2021 (as corresponds for early maturing cultivars). Total yield was weighed from all the experimental trees (n = 24 trees per treatment) with an electronic scale (0–6000 ± 2 g, Scaltec, Model SSH 92, Dania Beach, FL, USA), and the number of fruits per tree was counted. Fruit mass was calculated from total mass and number of fruits per tree. Nectarine fruits classified as 'not extra' or 'not marketable' (equatorial diameter <51 mm) were eliminated and not considered in the study. Fruit cracking was calculated as the ratio of the cracked nectarine fruits to the total number of fruits. Irrigation water use efficiency (IWUE) was defined as the ratio of total marketable yield and the total amount of irrigation applied the previous year postharvest plus the current pre-harvest period (kg m^{-3}).

2.10. Fruit Quality Measurements

At harvest, a sample of 15 fruits from each replicate and treatment (n = 60) was taken to the lab in isolated boxes with individual alveolus to determine the fruit quality measurements. Fruit diameter was measured using a digital caliper (0–150 ± 0.01 mm; Mitutoyo, CD-15D, Japan). External fruit color was measured using a Konica Minolta Chroma Meter CR-10 (Osaka, Japan), and the results were expressed in CIEL* a^* b^* chromatic coordinates: L^* (lightness), a^* (red–green component), b^* (blue–yellow component). From these values, the color parameters chrome or chromaticity (C*) and hue angle or tone ($h°$) were calculated as

$$C^* = \sqrt{(a^*)^2 + (b^*)^2} \tag{2}$$

$$h^{\circ} = \tan^{-1}(b^*/a^*) \tag{3}$$

Total soluble solids (TSS) and titratable acidity (TA) were evaluated using a digital pocket Brix/Acidity meter (Atago PAL-BX/ACID F5 Master Kit- Multifruits, Tokyo, Japan) in the juice of five fruits of each replicate and treatments (n = 20 fruits), which was squeezed with a juice machine (Orbegozo LI-5000, Murcia, Spain). Values were expressed as oBrix and g of citric acid, respectively. The maturity index (MI) was calculated as the soluble solids/acidity ratio.

2.11. NMR-Based Metabolite Analysis

The concentration of primary metabolites: sugars, free amino acids, and organic acids were analyzed. A sample of 100 g of fresh sliced nectarine fruit, including the peel, per replicate and treatment (n = 4) was lyophilized, grounded, and dried, and then sent to the Metabolomics Platform of CEBAS-CSIC (http://www.cebas.csic.es/general_english/metabolomics.html) (accessed on 4 July 2021). Metabolites were quantified following Choi et al. [47] by 1H NMR spectra, recorded at 298 K on a Bruker AVIII HD 500 NMR spectrometer (500.13 MHz for 1H) equipped with a 5 mm CPP BBO cryogenic probe (Bruker Biospin, Ettlingen, Germany). The 1H spectra were referenced to the TSP signal (δ = 0.00 ppm), whereas ^{13}C spectra were referenced to CH-1 resonance of α-D-glucose (δ = 93.10 ppm). A standard one-dimensional pulse sequence "noesypr1d" (recycle delay-90°-t1-90°-tm -90°-acquisition) was used to obtain metabolic profiles of nectarine fruits with the 90° pulse length of about 11 µs and t1 of 3 µs.

2.12. Statistical Analysis

All data were analyzed using SPSS v. 20 (IBM, Armonk, NY, USA) by using a one-way analysis of variance (ANOVA) to determine significant differences between irrigation treatments. Previously, normal distribution and variance homogeneity of data were verified, complying with the ANOVA requirements. Means were compared with the least significant difference test at a confidential level of 95% ($LSD_{0.05}$).

3. Results

3.1. Meteorological Conditions and Water Applied

The seasonal evolution of agro-meteorological data: ET_0, VPD, and rainfall, together with the cumulative amount of water applied to each irrigation treatment during the postharvest period (from May to October), is represented in Figure 1. The climate was typically Mediterranean, with a total annual rainfall of 206 mm, mainly concentrated during the autumn–winter season, and a total annual crop reference evapotranspiration (ET_0) of 1097 mm. The postharvest period coincided with the maximum water requirements for early maturing nectarine trees that accounted for ≈85% of the water needs required for the complete growing season. The pattern of VPD and ET_0 followed a similar pattern, although a greater day-to-day variability was observed in VPD. The highest VPD values were obtained at the beginning of postharvest with a daily maximum of 5.24 kPa (July).

The control treatment (T-0) received a total water volume of 616 mm, including 27 mm (which were equally applied during the fruit growth period to all treatments). The stressed treatments (by withholding irrigation) accounted for a water reduction of 41, 46, and 92% in T-1 (early postharvest), T-2 (late postharvest), and T-3 (the whole postharvest), respectively, compared to the T-0 treatment.

3.2. Soil and Plant Water Relations

Soil water content (θ_v) values at the end of the preconditioning and recovery periods were similar in all treatments since all trees were fully irrigated. The θ_v ranged around 29–34%, close to field capacity value. At the end of early (S1) and late (S2) postharvest periods, a reduction of around 75% with respect to the T-0 treatment was observed, with θ_v values of 7–9%, similar to those obtained in the T-3 treatment during the summer season (data not shown).

Figure 2 shows plant water status, evaluated as Ψ_{pd} and Ψ_{stem} values at the end of each irrigation period. At the end of the preconditioning and recovery periods (corresponding to non-stressed water conditions), Ψ_{stem} ranged from −0.59 to −0.88 MPa in all irrigation treatments. At the end of S1 (water-withholding during early postharvest), the T-1 and T-3 treatments showed significant differences with respect to T-0 treatment, with mean Ψ_{stem} values of −1.86 and −2.27 MPa, respectively. At the end of the S2 period (water-withholding during late postharvest), the T-2 and T-3 treatments showed a reduction of ≈60% in Ψ_{stem}, with respect to the T-0 treatment, with mean values of −1.72 and −2.00 MPa, in T-2 and T-3, respectively. The seasonal trend of Ψ_{pd} was similar to that of Ψ_{stem}, as verified by the close relationship found between both plant water status indicators ($\Psi_{stem} = -0.90 + 1.21\ \Psi_{pd}$, $r^2 = 0.67$, $p \leq 0.001$).

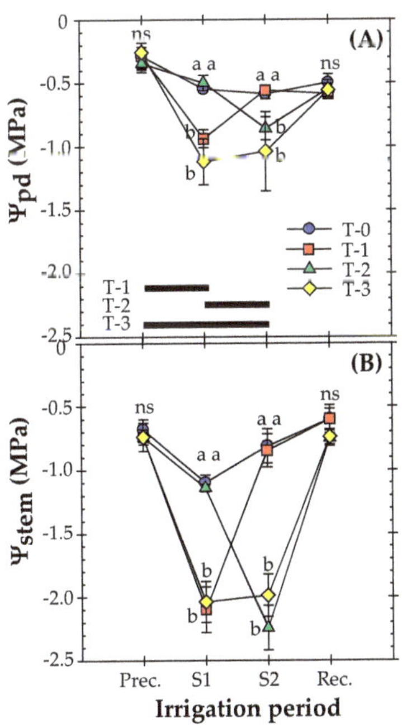

Figure 2. Values of (**A**) predawn leaf water potential (Ψ_{pd}) and (**B**) midday stem water potential (Ψ_{stem}) in early maturing nectarine trees at the end of each irrigation period: pre-conditioning (Prec.), stress 1 (S1), stress 2 (S2), and recovery (Rec.), in the different treatments (T-0: non-stressed, T-1: stress during early postharvest; T-2: stress during late postharvest, and T-3: stress during the whole postharvest). Each point is the mean ± SE (n = 4). Different letters indicate significant differences among treatments by $LSD_{0.05}$ test, ns: not significant. The black horizontal bars in (**A**) indicate the water withholding period in T-1, T-2, and T-3 treatments.

Considering the whole postharvest, the maximum values of the accumulated water stress integral ($S\Psi_{stem}$) were registered in T-3, followed by T-2 and T-1 stressed treatments (Figure 3). As expected, no differences in $S\Psi_{stem}$ values were noted between T-2 and T-3 treatments at the end of S1 stress period, and similarly between T-0 and T-1 at the end of S2 stress period, respectively. Mean $S\Psi_{stem}$ values varied in the range of 90 and 170 MPa day during the postharvest season.

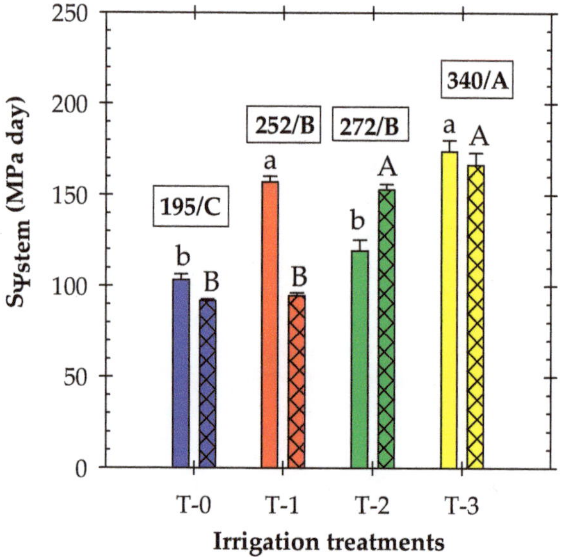

Figure 3. Values of accumulated water stress integral ($S\Psi_{stem}$) during the early postharvest (filled bars) and late postharvest (stripped bars) periods in early maturing nectarine trees in the different treatments (T-0: non-stressed; T-1: stress during early postharvest; T-2: stress during late postharvest; and T-3: stress during the whole postharvest). Bars are means + SE (n = 4). Different upper- and lower-case letters indicate significant differences between treatments by $LSD_{0.05}$ test, for early and late postharvest periods, respectively. Framed figures indicate the accumulated $S\Psi_{stem}$ during the entire postharvest period, followed by its significance by $LSD_{0.05}$ test.

Similar leaf gas exchange values were found in all irrigation treatments during the preconditioning period due to the absence of deficit irrigation conditions. However, water stress significantly decreased P_n and g_s as the postharvest season progressed, with a more marked decrease at the end of late postharvest (Figure 4A,B). In this sense, at the end of S1, P_n was reduced by ≈17 and ≈27% relative to T-0 values, in T-1 and T-3 treatments, respectively, whereas the corresponding reductions in g_s was ≈46 and ≈50% lower in T-1 and T-3 treatments than in the T-0 treatment, respectively. At the end of S2, P_n was ≈52 and ≈65% lower in T-2 and T-3 treatments than in the T-0 treatment, respectively, whereas the respective reductions in g_s were ≈69% and 78% lower in T-2 and T-3 treatments than in the T-0 treatment, respectively. The significant reductions in the P_n and g_s values observed at the end of S2 remained during the recovery period. Meanwhile, the WUE_T was only significantly increased at the S2 period in the T-2 and T-3 treatments, with average values of ≈4 mmol mol^{-1}, while it was ≈7 mmol mol^{-1} in well-irrigated control plants (Figure 4C).

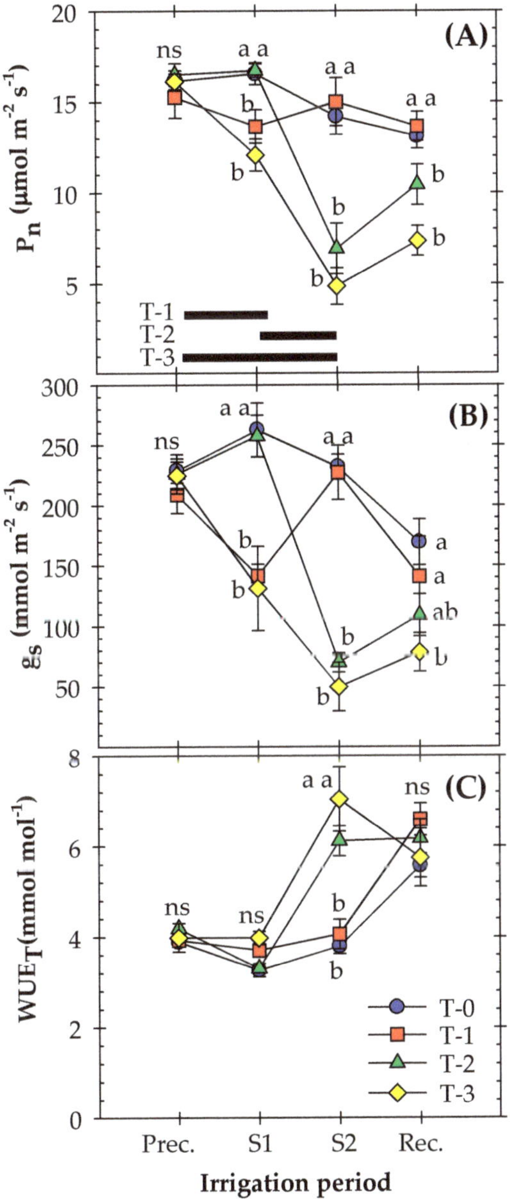

Figure 4. Values of (**A**) net photosynthesis (P_n); (**B**) stomatal conductance (g_s); and (**C**) water use efficiency transpiration (WUE_T) in early maturing nectarine trees at the end of each irrigation period: preconditioning (Prec.), stress 1 (S1), stress 2 (S2), and recovery (Rec.), in the different treatments (T-0: non-stressed; T-1: stress during early postharvest; T-2: stress during late postharvest; and T-3: stress during the whole postharvest). Each point is the mean ± SE (*n* = 4). Different letters indicate significant differences among treatments by $LSD_{0.05}$ test, ns: not significant. The black horizontal bars in Figure 4A. indicate the water-withholding period in T-1, T-2, and T-3 treatments.

3.3. *Leaf Mineral Content*

Severe water stress had a negative influence on leaf mineral nutrient composition (Table 1). The T-3 treatment registered the lowest values of N and C, followed by the T-2 treatment, in agreement with the intensity of water stress patterns (Figure 3). Moreover, the T-3 treatment had the highest values of NO_3^-, SO_4^{2-}, and PO_4^{3-} (although without statistical significance). On the other hand, F^- and Cl^- values registered a significant increase in the T-3 and T-2 treatments at the end of the experiment, respectively.

Table 1. Leaf mineral content (g 100 g^{-1}) in early maturing nectarine trees in the different treatments (T-0: non stressed, T-1: stress in early postharvest; T-2: stress in late postharvest, and T-3: stress during the whole postharvest).

	T-0	T-1	T-2	T-3	ANOVA
N_{total}	2.229 ab	2.361 a	2.176 b	1.889 c	***
C_{total}	46.086 a	44.356 ab	44.061 b	42.064 c	***
F^-	0.515 a	0.533 a	0.507 a	0.469 b	***
Cl^-	0.747 a	0.562 bc	0.547 c	0.576 b	***
NO_3^-	0.066 c	0.068 bc	0.069 ab	0.072 a	***
PO_4^{3-}	0.374	0.589	0.695	0.716	ns
SO_4^{2-}	0.806 b	1.394 a	1.530 a	1.449 a	***

Values are the mean of 30 leaves per replicate ($n = 4$) taken at the end of postharvest. ANOVA: analysis of variance. Means in each row followed by a different letter were significantly different by $LSD_{0.05}$ test. *** $p \leq 0.001$. ns = not significant.

3.4. *L-arginine, Starch, and Phosphorous Content in Roots*

The winter root reserves, including L-arginine, starch, and phosphorous, are shown in Figure 5. L-arginine showed the significantly higher value in the T-1 treatment (306.5 nmol g^{-1}), similar to T-0 (276.9 nmol g^{-1}), while it was significantly reduced by 36% and 57% with respect to T-0 treatment, in the T-2 and T-3, respectively. No significant differences were detected in the starch content among treatments with an average value of 14.35 mg g^{-1} for early maturing nectarine trees. Meanwhile, the phosphorus content was significantly higher in the T-0 and T-1 (mean values of 20 mg 100 g^{-1}), whereas in the T-2 and T-3, the phosphorous concentration was significantly reduced by 30% and 60% with respect to the T-0 treatment, respectively. Interestingly, the concentration of L-arginine in roots was well correlated with the N_{total} in leaves ($r^2 = 0.68$; $p \leq 0.001$) (Figure 6).

3.5. *Phenology, Yield, and Irrigation Water Use Efficiency (IWUE)*

The pattern of the reproductive phenological stages did not show significant differences among treatments (data not shown). However, the percentage of flowers and fruit drops were significantly higher in T-1- and T-3-stressed treatments compared to the T-0 treatment (Figure 7), highlighting the sensitivity of the early postharvest to severe water stress. Similar values of fruit drop (evaluated before the commercial hand-thinning) were observed in T-0 and T-2 treatments.

In all stressed treatments (T-1, T-2, and T-3), the total yield and number of fruits per tree were significantly lower than in the T-0 treatment, depending on the intensity of the imposed water stress (Figure 8A,B). Fruit mass was slightly higher in T-0 and T-1 (Figure 8C). It is noteworthy that the T-2 treatment recorded the highest percentage of cracked fruits (>50%), although this was not significant compared to other treatments (Figure 8D).

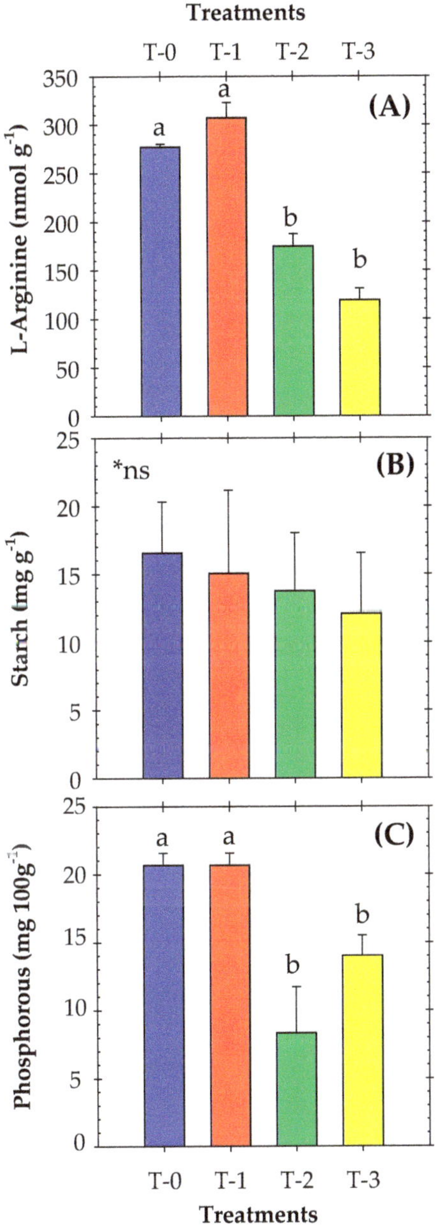

Figure 5. Nutritional root reserves: (**A**) L-arginine; (**B**) starch; and (**C**) phosphorous in early maturing nectarine trees in the different treatments (T-0: non-stressed, T-1: stress during early postharvest; T-2: stress during late postharvest; and T-3: stress during the whole postharvest) in winter dormancy. The bars are means + SE (n = 3). Different letters indicate significant differences between treatments by $LSD_{0.05}$ test. * ns: not significant.

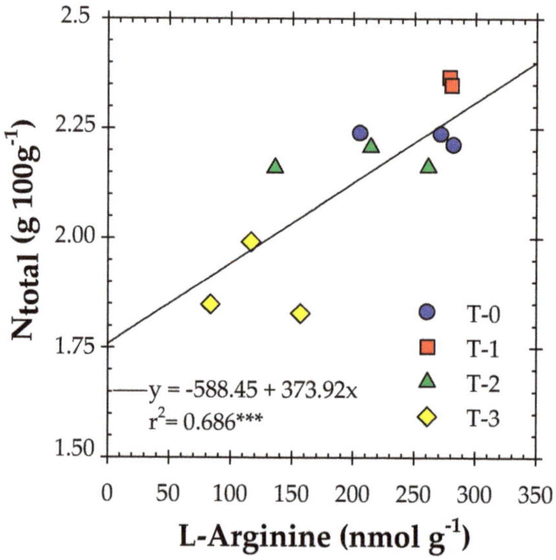

Figure 6. Relationship between leaf N_{total} and root L-arginine in early maturing nectarine trees in the different treatments (T-0: non-stressed; T-1: stress during early postharvest; T-2: stress during late postharvest; and T-3: stress during the whole postharvest). Each point corresponds to mean value of each replicate. *** significance at $p \leq 0.001$.

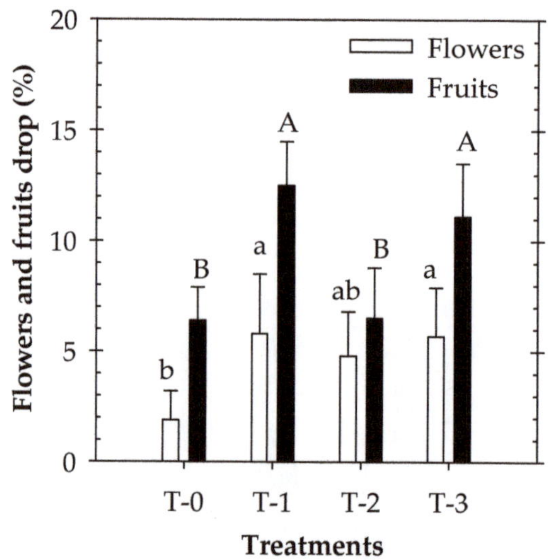

Figure 7. Flower and fruit drops (%) in the different treatments (T-0: non-stressed; T-1: stress during early postharvest; T-2: stress during late postharvest; and T-3: stress during the whole postharvest). The bars are means + SE (n = 4). Different lower and capital letters indicate significant differences between treatments for flowers and fruits by $LSD_{0.05}$ test, respectively.

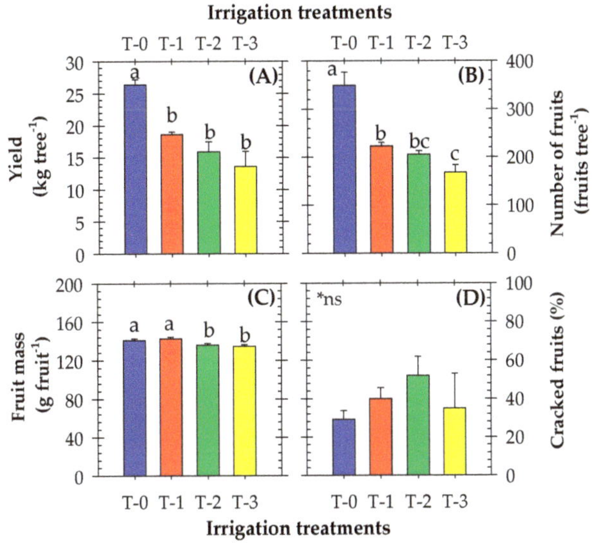

Figure 8. Yield components: (**A**) marketable total yield (kg tree^{-1}), (**B**) number of fruits per tree, (**C**) mass of individual fruit (g), and (**D**) percentage of cracked fruits, in early maturing nectarine trees in the different treatments (T-0: non-stressed, T-1: stress during early postharvest, T-2: stress during late postharvest, and T-3: stress during the whole postharvest). The bars are means + SE (n = 4). Different letters indicate significant differences between treatments by $LSD_{0.05}$ test. * ns: not significant.

All stressed treatments presented higher irrigation water use efficiency (IWUE) values than the control treatment, as shown in relative basis in Figure 9. Among them, only treatment T-3 reported significantly higher IWUE values in the experiment. The correlation between the average IWUE and the irrigation water applied showed a linear trend (y = 11.35 + 0.002x, r^2 = 0.72, $p \leq 0.001$) that predicts an increment in IWUE with respect to T-0 treatment of 129, 131, and 679% in T-1, T-2, and T-3, respectively, within the range of applied irrigation water (50–616 mm year^{-1}).

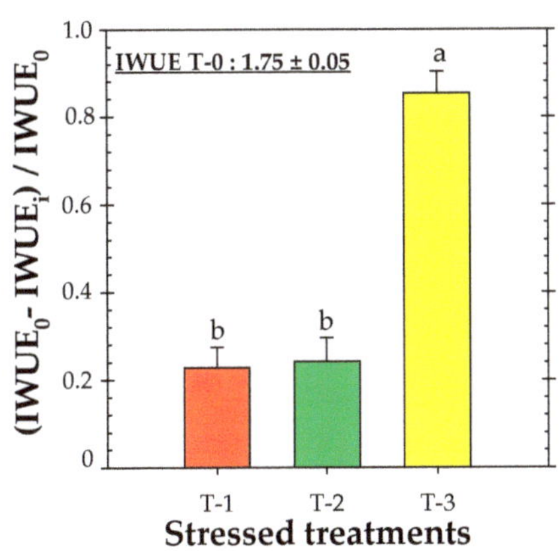

Figure 9. Relative irrigation water use efficiency (IWUE) in early maturing nectarine trees in the stressed treatments (T-1: during early postharvest, T-2: during late postharvest, and T-3: during the whole postharvest). Subscripts 'i' and '0' refer to non-stressed (T-0) and stressed treatments, respectively. The bars are means + SE (n = 4). Different letters indicate significant differences between treatments by $LSD_{0.05}$ test.

3.6. Nectarine Fruit Quality and Metabolites

No significant differences were observed between treatments in any of the physicochemical quality parameters studied (Table 2). TSS values were below the limit of marketability, mainly due to the astringent character of the nectarine cultivar. Likewise, sugar metabolite values were similar among the irrigation treatments, although the sucrose concentration tended to increase in stressed treatments respect to control (≈4% in T-1 and T-2, and ≈7% in T-3).

Table 2. Quality traits and sugars in early maturing nectarine fruits at harvest in the different treatments (T-0: non-stressed, T-1: stress during early postharvest, T-2: stress during late postharvest, and T-3: stress during the whole postharvest).

Quality Traits	T-0	T-1	T-2	T-3	ANOVA
Fruit diameter (mm)	65.56	65.76	63.89	63.96	ns
Firmness (N)	88.94	90.04	85.94	87.08	ns
L^*	36.97	37.42	37.61	37.87	ns
C^*	36.14	37.33	36.16	38.11	ns
h°	21.90	23.22	24.2	24.37	ns
TSS (°Brix)	8.60	8.71	9.08	8.26	ns
TA (mg L^{-1})	1.10	1.04	1.08	1.05	ns
MI	7.83	8.41	8.38	7.91	ns
Sugars					
Fructose (mg kg^{-1})	22.39	23.57	23.24	22.47	ns
Glucose (mg kg^{-1})	20.67	22.82	21.17	20.60	ns
myo-Inositol (mg kg^{-1})	1.08	0.92	1.00	0.85	ns
Sucrose (mg kg^{-1})	92.31	96.47	96.27	99.54	ns
UDP-glucose (mg kg^{-1})	0.04	0.04	0.04	0.04	ns
Xylose (mg kg^{-1})	0.47	0.48	0.41	0.46	ns

Values are means of 15 fruits per replicate (n = 60 fruits per treatment) for quality traits and five fruits per replicate (n = 20 fruits per treatment) for sugar. ANOVA: analysis of variance. ns: not significant. L^* = lightness; C^* = chromaticity; h° = hue angle; TSS = total soluble solids; TA = titratable acidity; MI = maturity index.

Regarding free amino acid content, there were significant differences in aspartate, proline, and valine (Table 3). Aspartate was higher in T-2 and T-3 treatments than in T-0 and T-1 treatments, whereas valine was significantly increased in nectarine fruits belonging to the T-1 treatment in comparison to the other irrigation treatments. Proline, the amino acid of prime importance for the induction of stress tolerance, had the highest value in the T-3 treatment. This finding is related to the level of water stress supported by each treatment, as indicated by the good relationship found between proline and water stress integral (Figure 10), revealing as proline concentration exponentially increased from 230 MPa day onwards. Valine increased (p = 0.039) in the stressed treatments, showing T-1 treatment the highest value (Table 3).

Table 3. Free amino acids content (mg kg^{-1}) in early maturing nectarine fruits at harvest in the different treatments (T-0: non-stressed, T-1: stress during early postharvest; T-2: stress during late postharvest, and T-3: stress during the whole postharvest).

Free Amino Acids	T-0	T-1	T-2	T-3	ANOVA
GABA	0.416	0.417	0.4064	0.449	ns
Alanine	0.327	0.349	0.336	0.351	ns
L-Arginine	nd	nd	nd	nd	

Table 3. *Cont.*

Free Amino Acids	T-0	T-1	T-2	T-3	ANOVA
Asparagine	17.702	21.908	18.763	21.257	ns
Aspartate	51.075 a	50.375 a	43.463 b	41.188 b	***
Glutamate	nd	nd	nd	nd	
Glutamine	nd	nd	nd	nd	
Isoleucine	0.057	0.088	0.070	0.062	ns
Leucine	0.058	0.066	0.052	0.050	ns
Phenylalanine	0.030	0.035	0.036	0.033	ns
Proline	0.076 c	0.145 bc	0.264 b	0.652 a	***
Threonine	0.169	0.219	0.202	0.192	ns
Tyrosine	nd	nd	nd	nd	
Valine	0.088 b	0.125 a	0.107 ab	0.102 ab	*

Values are means of five fruits per replicate (n = 20 fruits per treatment). ANOVA: analysis of variance. Means within columns followed by a different letter were significantly different by $LSD_{0.05}$. *, *** significant effect at $p \leq 0.05$ and 0.001, respectively; ns = not significant. nd: not detectable.

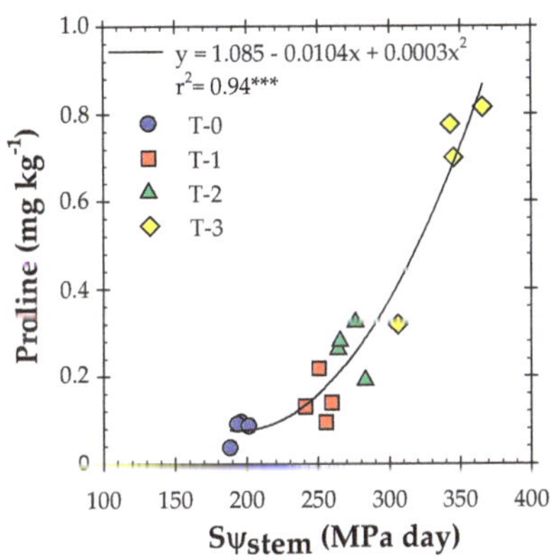

Figure 10. Relationship between proline fruit content (ppm) and the accumulated water stress integral ($S\Psi_{stem}$, MPa day) in early maturing nectarine trees in the different treatments (T-0: non-stressed, T-1: stress during early postharvest, T-2: stress during late postharvest, and T-3: stress during the whole postharvest). Each point corresponds to mean value of each replicate. *** Significance at $p \leq 0.001$.

The visual appearance of early maturing *Flariba* nectarine trees from the T-0 (fully irrigated) and T-3 (severely stressed) treatments are shown in Figure 11 from pictures taken at the end of the late postharvest period (S2) (24 September 2020). A greater defoliation can be observed in T-3 trees (Figure 11B).

(A) (B)

Figure 11. Visual appearance of *Flariba* early maturing nectarine trees: (**A**) T-0 (fully irrigated) treatment and (**B**) T-3 (severely stressed) treatment.

4. Discussion

When using RDI, it is essential not only to identify the non-critical periods but also to determine the intensity and the timing of the water stress imposed within the non-critical period to avoid adverse effects on the following harvest and on vegetative growth [48]. The role of the early postharvest period in the application of RDI irrigation strategies in early maturing peach and nectarine trees has been previously documented, ascribing the detrimental influence of water deficits during this phenological period on the floral induction and differentiation processes [6–8,12,23–25]. This statement was confirmed by data of flower and fruit drop (Figure 7). Furthermore, our experiment showed an insight not described in the literature so far: severe water deficits applied during the long postharvest period of *Flariba* nectarine trees revealed that the late postharvest stage was even more sensitive than the early postharvest stage, as winter root reserves greatly compromise yield. We here analyze step by step the effects of water deficit on soil–plant water status and fruit metabolites.

As expected, withholding irrigation water during the postharvest reduced soil water content (data not shown), predawn leaf water potential, midday stem water potential, and leaf gas exchange, along with increasing IWUE, compared to well-irrigated control trees (Figures 2, 4 and 9). In particular, the Ψ_{stem} values in the T-0 treatment ranged from −0.59 to −0.88 MPa, coinciding with values previously reported for early maturing peach and nectarine trees corresponding to non-limiting soil water conditions [7–9,25–27]. At the end of S1 (early postharvest), the stressed treatments (T-1 and T-3) showed a mean difference of 1.0 MPa with respect to the T-0 treatment. This difference became larger at the end of the S2 (late postharvest) in the T-2 and T-3 treatments, which showed a significant difference of 1.4 and 1.2 MPa compared with the T-0 treatment, respectively. The minimum Ψ_{stem} values observed during the experiment were −2.1, −2.2, and −2.0 MPa in T-1, T-2, and T-3, treatments, respectively (Figure 2B). In this sense, threshold values of Ψ_{stem} of −1.5 and −2.0 MPa during the postharvest are recommended to ensure no impairment of bloom fertility [49] and to limit the occurrence of double fruits [15], respectively.

Values of the leaf gas exchange parameters (P_n, g_s, and WUE_T) were significantly reduced by severe water deficit accordingly (Figure 4). Rahmati et al. [50] observed a reduction (>50%) in leaf gas exchange when Ψ_{stem} decreased from −1.4 to −2.0 MPa in peach trees. However, an additional Ψ_{stem} decrease ($\leq$−2.0 MPa) only led to a slight decrease in both g_s and P_n. Shackel et al. [51] noted a Ψ_{stem} threshold value of $\approx$−1.5 MPa when the decrease in P_n was compensated by a reduction in the vegetative apex growth. In our study, when irrigation was restored in stressed trees, the reduced P_n and g_s values observed at the end of S2 remained at low levels, despite the recovery observed in Ψ_{stem} values. This fact emphasizes the resilient character of this *Prunus* species [52,53]. The relative delay in stomatal opening following rewatering (Figure 4) compared with the rapid recovery shown by Ψ_{stem} (Figure 2) can also be considered as a safety mechanism, allowing plants to recover full turgor more effectively [54,55]. Moreover, greater differences between treatments were detected in P_n and g_s than in Ψ_{stem} after a rainfall episode in early maturing nectarine trees [56].

As expected, nectarine productivity was penalized in all stressed trees compared to control trees (Table 2). Despite the fact that the postharvest season is considered the most suitable period for applying RDI strategies, as previously indicated [7,8,12,23–25], drought applied during the non-critical period of postharvest affected the total nectarine yield compels us to reconsider this statement, particularly in early maturing cultivars, which bear a very long postharvest period that also coincides with the climatic period that demands the most water. Data from winter carbohydrate reserves can help us to understand this effect, as our findings indicated that the late postharvest stage was more sensitive than the early postharvest stage to severe water deficit, because the concentration of winter root reserves (mainly L-arginine and phosphorous) were significantly reduced (Figure 5), coupled with a non-significant increase in the non-reducing sugars concentration (Table 2). Carbohydrate reserves in fruit trees determine the availability of these resources to support bloom and the fruit set in the subsequent harvest season [30]. In addition, carbohydrate reserves act as a link between successive growing seasons, influencing the processes related to cropping [34]. Then, a reduction in root reserves accumulation in response to water deficit will reduce the availability of carbohydrates to support crop development (particularly flowering and fruit set) in the following growing season [31,32,34,57].

The amino acid arginine, with a high N/C ratio, serves as an important nitrogen reserve in fruit trees [58]. We were able to confirm this statement through the good relationship found between N_{total} in leaves measured at the end of the postharvest, and the L-arginine concentration in roots, measured in winter (Figure 6). In addition, L-arginine is involved in various physiological processes as plant response to stress [59]. In our study, the concentration of L-arginine in T2 and T-3 treatments was reduced compared to the T-1 treatment (also apart from the control, T-0, treatment) (Figure 5A). Moreover, the amount of N_{total} was lowest in the T-3 treatments followed by the T-2 treatment (Table 1). As indicated by Gao et al. [59], the physiological role of L-arginine could be associated with the coordinated biosynthesis of both polyamines and nitric oxide via arginine metabolism.

Phosphorus concentration was also significantly reduced in T-2 and T-3 treatments, whereas the starch concentration was similar among treatments (Figure 5B,C). Furthermore, although no significant differences were found in the TSS, MI, and sugar metabolites among fruits from the different irrigation treatments, it is noteworthy that the concentration of the sucrose increased the more the deficit imposed (Table 2). Kobashi et al. [60] on peach and Thakur and Zora [61] on nectarine trees reported that a severe water stress resulted in higher amounts of this non-reducing sugar and total sugars in the fruit. For this reason, increased carbohydrate translocation from vegetative organs to the fruit sink under water stress might be responsible for the resulting increase in sucrose in the *'Flariba'* nectarine fruits. The accumulation of higher sucrose and total sugars in fruit under RDI treatments is probably related to a higher concentration of sugars, as discussed by Stefanelli et al. [62]. A similar behavior can be observed in the more severe stress treatments (T-2 and T-3), so the water deficits during late postharvest decrease the winter root reserves of arginine and phosphorus while increasing starch and the amount of sugars in fruits (as the case of sucrose), which could be considered an adaptive mechanism of the trees to confront severe water stress.

In response to drought, amino acid catabolic enzymes rapidly increase, playing a vital role in the metabolism of amino acids under deficit irrigation conditions [63]. This behavior was observed in aspartate, proline, and valine content in stressed nectarine fruits from the stressed treatments (Table 3). Moreover, proline accumulation in response to stress (Figure 10) can also play a protective role in the photosynthetic apparatus [64]. Our results showed a significant accumulation of proline concentration in nectarine fruits from the T-3 trees (Table 3). On the other hand, the severe stress imposed in the T-2 and T-3 affected physiological behavior by promoting a decrease in leaf gas exchange values, which remained at low values despite recovering irrigation (Figure 4A,B). This situation can also be related to the amount of aspartate in *'Flariba'* nectarine fruits, which was also reduced in the T-2 and T-3 treatments (Table 3). Aspartate is involved in the regulation of many processes, such as the synthesis of protein chlorophyll biosynthesis as well as the formation of photosynthetic pigments [65]. Therefore, the fact that P_n and g_s values did not recover with the restoration of irrigation may have been due to the decrease in the aspartate amino acid in leaves, by altering the formation of chloroplasts resulting from severe water stress applied in late postharvest (Figure 4A,B).

5. Conclusions

The late postharvest period (from August to September) of early maturing *Flariba* nectarine trees grown under Mediterranean conditions was the period most sensitive to drought, since the accumulation of winter root reserves (especially L-arginine and phosphorus) was reduced, which in turn limited nectarine productivity.

Drought (severe water deficits) applied at different stages of the non-critical period of postharvest demonstrated the need to consider the intensity and duration of water stress. In our study, the rapid response of soil and plant water status indicators has allowed us to identify the sensitivity of postharvest phenological stages to severe water stress along with the implications of winter root reserves and fruit carbohydrates on the harvesting processes.

Author Contributions: Conceptualization and methodology, M.R.C. and M.C.R.-S.; software and validation, J.V. and W.C.; formal analysis and data curation, W.C. and M.R.C.; investigation and resources, all authors; writing—original draft preparation, M.R.C.; writing—review and editing, all authors; supervision, project administration, and funding acquisition, M.C.R.-S. All authors have read and agreed to the published version of the manuscript.

Funding: The Spanish State Research Agency funds (PID2019-106226RB-C21/AEI/10.13039/501100011033), and the Seneca Foundation of the Region of Murcia, Spain (19903/GERM/15) supported this research.

Institutional Review Board Statement: Not applicable.

Informed Consent Statement: Not applicable.

Data Availability Statement: The data presented in this study are available upon request from the corresponding author.

Acknowledgments: M. R. Conesa thanks the Spanish JdlC program (FJCI-2017-32045 and IJC2020-045450-I) funded by MCIN/AEI/10.13039/501100011033 and European Union NextGenerationEU/PRTR.

Conflicts of Interest: The authors declare no conflict of interest.

References

1. Varela-Ortega, C.; Blanco-Gutiérrez, I.; Esteve, P.; Bharwani, S.; Fronzek, S.; Downing, T.E. How can irrigated agriculture adapt to climate change? Insights from the Guadiana Basin in Spain. *Reg. Environ. Chang.* **2016**, *16*, 59–70. [CrossRef]
2. Fernández-García, I.; Lecina, S.; Ruiz-Sánchez, M.C.; Vera, J.; Conejero, W.; Conesa, M.R.; Domínguez, A.; Pardo, J.J.; Llélis, B.C.; Montesinos, P. Trends and challenges in irrigation scheduling in the semi-arid area of Spain. *Water* **2020**, *12*, 785. [CrossRef]
3. AEMET. State Meteorological Agency. Available online: https://www.aemet.es/en (accessed on 15 March 2022).
4. Fereres, E.; Soriano, M.A. Deficit irrigation for reducing agricultural water use. *J. Exp. Bot.* **2007**, *58*, 147–159. [CrossRef]
5. FAOSTAST. Food and Agriculture Organization Statistical Data. Available online: https://www.fao.org/statistics/en/ (accessed on 22 August 2020).
6. Alcobendas, R.; Mirás-Avalos, J.M.; Alarcón, J.J.; Pedrero, F.; Nicolás, E. Combined effects of irrigation, crop load and fruit position on size, color and firmness of fruits in an extra-early cultivar of peach. *Sci. Hortic.* **2012**, *142*, 128–135. [CrossRef]
7. De la Rosa, J.M.; Domingo, R.; Gómez-Montiel, J.; Pérez-Pastor, A. Implementing deficit irrigation scheduling through plant water stress indicators in early nectarine trees. *Agric. Water Manag.* **2015**, *152*, 207–216. [CrossRef]
8. De la Rosa, J.M.; Conesa, M.R.; Domingo, R.; Aguayo, E.; Falagán, E.; Pérez-Pastor, A. Combined effects of deficit irrigation and crop level on early nectarine trees. *Agric. Water Manag.* **2016**, *170*, 120–132. [CrossRef]
9. Conesa, M.R.; Conejero, W.; Vera, J.; Agulló, V.; García-Viguera, C.; Ruiz-Sánchez, M.C. Irrigation management practices in nectarine fruit quality at harvest and after cold storage. *Agric. Water Manag.* **2021**, *243*, 106519. [CrossRef]
10. Pérez-Pastor, A.; Ruiz-Sánchez, M.C.; Conesa, M.R. Drought stress effect on woody tree yield. In *Water Stress and Crop Plants: A Sustainable Approach*; Ahmad, P., Ed.; John Wiley & Sons Ltd.: Chichester, UK, 2016; Volume 2, Chapter 22; pp. 356–374. ISBN 9781119054368. [CrossRef]
11. Ruiz-Sánchez, M.C.; Domingo, R.; Castel, J.R. Review. Deficit irrigation in fruit trees and vines in Spain. *Span. J. Agric. Res.* **2010**, *8* (Suppl. 2), 5–20. [CrossRef]
12. Ruiz-Sánchez, M.C.; Abrisqueta, I.; Conejero, W.; Vera, J. Deficit irrigation management in early-maturing peach crop. In *Water Scarcity and Sustainable Agriculture in Semiarid Environment. Tools, Strategies, and Challenges for Woody Crops*; Elsevier: Amsterdam, The Netherlands, 2018; Chapter 6; pp. 111–126. ISBN 978-0-12-813164-0. [CrossRef]
13. Katerji, N.; Mastrorilli, M.; Rana, G. Water use efficiency of crops cultivated in the Mediterranean regions: Review and analysis. *Europ. J. Agron.* **2008**, *28*, 493–507. [CrossRef]
14. Chalmers, D.J.; Mitchell, P.D.; Van Heek, L. Control of peach tree growth and productivity by regulated water supply, tree density and summer pruning. *J. Am. Soc. Hortic. Sci.* **1981**, *106*, 307–312.
15. Naor, A.; Stern, R.; Flaishman, M.; Gal, Y.; Peres, M. Effects pf post-harvest water stress on autumnal bloom and subsequent-season productivity in mid-season 'Spadona' pear. *J. Hortic. Sci. Biotech.* **2006**, *81*, 3651–3700. [CrossRef]
16. Behboudian, M.H.; Mills, T.M. Deficit irrigation in deciduous orchards. *Hortic. Rev.* **1997**, *21*, 105–131. [CrossRef]
17. Chai, Q.; Gan, Y.; Zhao, C.; Xu, H.L.; Waskom, R.M.; Niu, Y.; Siddique, K.H.M. Regulated deficit irrigation for crop production under drought stress: A review. *Agron. Sustain. Dev.* **2016**, *36*, 1. [CrossRef]

18. Torrecillas, A.; Domingo, R.; Galego, R.; Ruiz-Sánchez, M.C. Apricot tree response to withholding irrigation at different phenological periods. *Sci. Hortic.* **2000**, *85*, 201–215. [CrossRef]
19. Uriu, K. Effect of post-harvest soil moisture depletion on subsequent yield of apricot. *Proc. Am. Soc. Hortic. Sci.* **1964**, *84*, 93–97.
20. Ruiz-Sánchez, M.C.; Egea, J.; Galego, R.; Torrecillas, A. Floral biology of 'Búlida' apricot trees subjected to postharvest drought stress. *Ann. Appl. Biol.* **1999**, *135*, 523–528. [CrossRef]
21. Girona, J.; Mata, M.; Goldhamer, D.A.; Johnson, R.S.; DeJong, T.M. Patterns of soil and tree water status and leaf functioning during regulated deficit irrigation scheduling in peach. *J. Am. Soc. Hortic. Sci.* **1993**, *118*, 580–586. [CrossRef]
22. Goldhamer, D.A.; Salinas, M.; Crisosto, C.; Day, K.R.; Soler, M.; Moriana, A. Effects of regulated deficit irrigation and partial root zone drying on late harvest peach tree performance. *Acta Hortic.* **2002**, *592*, 343–350. [CrossRef]
23. Conejero, W.; Mellisho, C.M.; Ortuño, M.F.; Moriana, A.; Moreno, F.; Torrecillas, A. Using trunk diameter sensors for regulated deficit irrigation scheduling in early maturing peach trees. *Environ. Exp. Bot.* **2011**, *71*, 409–415. [CrossRef]
24. Abrisqueta, I.; Tapia, L.M.; Conejero, W.; Sánchez-Toribio, M.I.; Abrisqueta, J.M.; Vera, J.; Ruiz-Sánchez, M.C. Response of early-maturing peach [*Prunus persica* (L.)] trees to deficit irrigation. *Span. J. Agric. Res* **2010**, *8* (Suppl. 2), 30–39. [CrossRef]
25. Vera, J.; Abrisqueta, I.; Abrisqueta, J.M.; Ruiz-Sánchez, M.C. Effect of deficit irrigation on early-maturing peach tree performance. *Irrig. Sci.* **2013**, *31*, 747–757. [CrossRef]
26. Vera, J.; Conejero, W.; Conesa, M.R.; Ruiz-Sánchez, M.C. Irrigation factor approach based on soil water content: A nectarine orchard case study. *Water* **2019**, *11*, 589. [CrossRef]
27. Conesa, M.R.; Conejero, W.; Vera, J.; Ramírez-Cuesta, J.M.; Ruiz-Sánchez, M.C. Terrestrial and remote indexes to assess moderate deficit irrigation in early-maturing nectarine trees. *Agronomy* **2019**, *9*, 630. [CrossRef]
28. Kliewer, W.M.; Cook, J.A. Arginine levels in grape canes and fruits as an indicator of nitrogen status of vineyards. *Am. J. Enol. Vitic.* **1974**, *2*, 111–118.
29. Fox, R.L. External Phosphorus requirement of crops. In *Chemistry in the Soil Environment*; Special Publication 40; Dowdy, R.H., Ed.; American Society of Agronomy: Madison, WI, USA, 1981; pp. 223–239. [CrossRef]
30. Kozolowski, T.T. Carbohydrate sources and sinks in woody plants. *Bot. Rev.* **1992**, *58*, 208–222. [CrossRef]
31. Ebel, R.; Proebsting, E.L.; Evans, R.G. Apple tree and fruit responses to early termination of irrigation in a semi-arid environment. *HortScience* **2001**, *36*, 1197–1201. [CrossRef]
32. López, G.; Mata, M.; Arbonés, A.; Solans, J.R.; Girona, J.; Marsal, J. Mitigation of effects of extreme drought during stage III of peach fruit development by summer pruning and fruit thinning. *Tree Physiol.* **2006**, *26*, 469–477. [CrossRef]
33. Abdel-Sattar, M.; Kotb, H.R.M. Nutritional status and productivity of Anna apple trees in the year following autumn irrigation determent. *Agric. Water Manag.* **2021**, *252*, 106882. [CrossRef]
34. Lopez, G.; Girona, J.; Marsal, J. Response of winter root starch concentration to severe water stress and fruit load and its subsequent effects on early peach fruit development. *Tree Physiol.* **2007**, *27*, 1619–1626. [CrossRef]
35. Vera, J.; de la Peña, J.M. *FERTIGA: Programa de Fertirrigación de Frutales*; CEBAS-CSIC: Murcia, Spain, 1994; p. 69.
36. Allen, R.G.; Pereira, L.S.; Smith, M.; Raes, D.; Wright, J.L. FAO-56 dual crop coefficient method for estimating evaporation from soil and application extensions. *J. Irrig. Drain. Eng.* **2005**, *131*, 2–13. [CrossRef]
37. Abrisqueta, I.; Abrisqueta, J.; Tapia, L.M.; Munguía, J.; Conejero, W.; Vera, J.; Ruiz-Sánchez, M.C. Basal crop coefficients for early-season peach trees. *Agric. Water Manag.* **2013**, *121*, 158–163. [CrossRef]
38. Balsalobre, N.; Koptsyukh, E.; Ruiz-Sánchez, M.C.; Vera, J.; Conejero, W.; Nicolás, M.J. Riego deficitario y raíces de nectarinos. In Proceedings of the V Congreso IDIES "I+D en Institutos de Enseñanza Secundaria", Murcia, Spain, 26 June 2018. ISBN 978-84-09-03063-7.
39. Vera, J.; Conejero, W.; Mira-García, A.B.; Conesa, M.R.; Ruiz-Sánchez, M.C. Towards irrigation automation based on dielectric soil sensors. *J. Hortic. Sci. Biotech.* **2021**, *96*, 696–707. [CrossRef]
40. Hsiao, T.C. Measurement of tree water status. In *Irrigation of Agricultural Crops. Agronomy Monograph No. 30*; Steward, B.A., Nielsen, D.R., Eds.; American Society of Agronomy: Madison, WI, USA, 1990; pp. 243–279.
41. Myers, B.J. Water stress integral-A link between short-term stress and long-term growth. *Tree Physiol.* **1988**, *4*, 315–323. [CrossRef] [PubMed]
42. Sakaguchi, S. A new method for the colorimetric determination of arginine. *J. Biochem.* **1950**, *37*, 231–236. [CrossRef]
43. Clegg, K.M. The application of the an-throne reagent to the estimation of starch in cereals. *J. Sci. Food Agric.* **1956**, *7*, 40–44. [CrossRef]
44. Teixeira, P.C.; Donagemma, A.; Fontana, W.G.G.K. *Manual de Métodos de Análise de Solos*, 3rd ed.; DF: Embrapa, Brasilia, 2017; p. 574. ISBN 978-85-7035-771-7.
45. Baggiolini, M. *Stades Repères du Cerisier. Stades Repères du Prunier. Stades Repères del'Abricotier. Stades Repères du Pêcher*; Guide Pratique de Défense des Cultures; ACTA: Paris, France, 1980.
46. Mounzer, O.; Conejero, W.; Nicolás, E.; Abrisqueta, I.; García-Orellana, Y.; Tapia, L.; Vera, J.; Abrisqueta, J.M.; Ruiz-Sánchez, M.C. Growth pattern and phenological stages of early-maturing peach trees under a Mediterranean climate. *HortScience* **2008**, *43*, 1813–1818. [CrossRef]
47. Choi, H.-K.; Choi, Y.H.; Verbene, M.; Lefeber, A.W.M.; Erkelens, C.; Verpoorte, R. Metabolic fingerprinting of wild type and transgenic tobacco plants by 1H NMR and multivariate analysis technique. *Phytochemistry* **2004**, *65*, 857–864. [CrossRef] [PubMed]

48. Pérez-Pastor, A.; Ruiz-Sánchez, M.C.; Domingo, R. Effects of timing and intensity of deficit irrigation on vegetative and fruit growth of apricot trees. *Agric. Water Manag.* **2014**, *134*, 110–118. [CrossRef]
49. Girona, J.; Gelly, M.; Mata, M.; Arbonés, A.; Rufat, J.; Marsal, J. Peach tree response to single and combined deficit irrigation regimes in deep soils. *Agric. Water Manag.* **2005**, *72*, 97–108. [CrossRef]
50. Rahmati, M.; Davarynejad, G.H.; Génard, M.; Bannayan, M.; Azizi, M.; Vercambre, G. Peach water relations, gas exchange, growth and shoot mortality under water deficit in semi-arid weather conditions. *PLoS ONE* **2015**, *10*, e0120246. [CrossRef] [PubMed]
51. Shackel, K.; Lampinen, B.; Sibbett, S.; Olson, W. The relation of midday stem water potential to the growth and physiology of fruit trees under water limited conditions. *Acta Hortic.* **2000**, *537*, 425–430. [CrossRef]
52. Abrisqueta, I.; Conejero, W.; Valdés-Vela, M.; Vera, J.; Ortuño, M.F.; Ruiz-Sánchez, M.C. Stem water potential estimation of drip-irrigated early-maturing peach trees under Mediterranean conditions. *Comput. Electron. Agric.* **2015**, *114*, 7–13. [CrossRef]
53. Conesa, M.R.; Conejero, W.; Vera, J.; Ruiz-Sánchez, M.C. Effects of postharvest water deficits on the physiological behavior of early-maturing nectarine trees. *Plants* **2020**, *9*, 1104. [CrossRef] [PubMed]
54. Mansfield, T.A.; Davies, W.J. Stomata and stomatal mechanisms. In *Physiology and Biochemistry of Drought Resistance in Plants*; Paleg, L.G., Aspinall, D., Eds.; Academic Press: New York, NY, USA, 1981; pp. 315–346.
55. Ruiz-Sánchez, M.C.; Domingo, R.; Torrecillas, A.; Pérez-Pastor, A. Water stress preconditioning to improve drought resistance in young apricot plants. *Plant Sci.* **2000**, *156*, 245–251. [CrossRef]
56. De la Rosa, J.M.; Conesa, M.R.; Domingo, R.; Pérez-Pastor, A. A new approach to ascertain the sensitivity to water stress of different plant water indicators in extra-early nectarine trees. *Sci. Hortic.* **2014**, *169*, 147–153. [CrossRef]
57. Esparza, G.; DeJong, T.M.; Weinbaum, S.A. Effects of irrigation deprivation during the harvest period on nonstructural carbohydrate and nitrogen contents of dormant, mature almond trees. *Tree Physiol.* **2001**, *21*, 1081–1086. [CrossRef] [PubMed]
58. Cantón, F.R.; Suárez, M.F.; Cánovas, F.M. Molecular aspects of nitrogen mobilization and recycling in trees. *Photosyn. Res.* **2005**, *83*, 265–278. [CrossRef]
59. Gao, H.-J.; Yang, H.-Q.; Wang, J.-X. Arginine metabolism in roots and leaves of apple (*Malus domestica Borkh.*): The tissue-specific formation of both nitric oxide and polyamines. *Sci. Hortic.* **2009**, *119*, 147–152. [CrossRef]
60. Kobashi, K.; Gemma, H.; Iwahori, S. Abscisic acid content and sugar metabolism of peaches grown under water stress. *J. Am. Soc. Hort. Sci.* **2001**, *125*, 425–428. [CrossRef]
61. Thakur, A.; Zora, S. Responses of 'Spring Bright' and 'Summer Bright' nectarines to deficit irrigation: Fruit growth and concentration of sugars and organic acids. *Sci. Horti.* **2012**, *135*, 1–226. [CrossRef]
62. Stefanelli, D.; Goodwin, I.; Jones, R. Minimal nitrogen and water use in horticulture: Effects on quality and content of selected nutrients. *Food Res. Int.* **2010**, *43*, 1833–1843. [CrossRef]
63. ElShamey, E.A.; Ali, E.F.; Selim, M.E.; ElSayed, M.A.; Ahmed, M.; Alotaibi, F.M.; Kheir, A.M. Water deficit induced physiological and amino acid responses in some rice varieties using NMR-metabolic analysis. *Agron. J.* **2021**, *113*, 4690–4704. [CrossRef]
64. Verbruggen, N.; Hermans, C. Proline accumulation in plants: A review. *Amino Acids* **2008**, *35*, 753–759. [CrossRef]
65. Kumar, V.; Sharma, N.; Kumar, K.M.; Vlaskin, M.S.; Nanda, M.; Tripathi, M.; Sanjay-Kumar, S. Microalgae with a truncated light-harvesting antenna to maximize photosynthetic efficiency and biomass productivity: Recent advances and current challenges. *Process Biochem.* **2021**, *104*, 83–91. [CrossRef]

agronomy

Article

Rice Momilactones and Phenolics: Expression of Relevant Biosynthetic Genes in Response to UV and Chilling Stresses

La Hoang Anh [1], Nguyen Van Quan [1], Vu Quang Lam [2], Akiyoshi Takami [2], Tran Dang Khanh [3,4] and Tran Dang Xuan [1,5,*]

1 Transdisciplinary Science and Engineering Program, Graduate School of Advanced Science and Engineering, Hiroshima University, Hiroshima 739-8529, Japan; hoanganh6920@gmail.com (L.H.A.); nvquan@hiroshima-u.ac.jp (N.V.Q.)

2 Division of Hematology, Department of Internal Medicine, School of Medicine, Aichi Medical University, Nagakute 480-1195, Japan; quanglamvu1991@gmail.com (V.Q.L.); takami-knz@umin.ac.jp (A.T.)

3 Agricultural Genetics Institute, Pham Van Dong Street, Hanoi 122000, Vietnam; tdkhanh@vaas.vn or tdkhanh@vnua.edu.vn

4 Center for Agricultural Innovation, Vietnam National University of Agriculture, Hanoi 131000, Vietnam

5 Center for the Planetary Health and Innovation Science (PHIS), The IDEC Institute, Hiroshima University, Higashi-Hiroshima 739-8529, Japan

* Correspondence: tdxuan@hiroshima-u.ac.jp; Tel./Fax: +81-82-424-6927

Abstract: Momilactones A (MA) and B (MB) are known as phytoalexins which principally play a role in the rice defense system against pathogens. This is the first study revealing that MA and MB contribute to rice tolerance to environmental stresses, including ultraviolet (UV) radiation and chilling conditions. The proofs were achieved by scrutinizing the responses of rice under stresses through the expression of relevant biosynthetic genes to momilactones (MRBG) and phenolics (PRBG) and their accumulation. Accordingly, the expression tendency of PRBG was in line with that of MRBGs, which increased under UV irradiation but decreased under chilling conditions. In UV-exposed rice, the proliferation of MA and MB strongly correlated to that of salicylic and chlorogenic acids, esculetin, rutin, and fisetin. In terms of increasing chilling duration, the biosynthetic propensity of MB was consistent with that of benzoic, cinnamic, ρ-coumaric, salicylic, and syringic acids, quercetin, and tricin while the syntheses of MA and other compounds were reduced. The concomitant biosyntheses of momilactones with these acknowledged stress-resistant phenolics imply that momilactones might play a role as signaling molecules in the response mechanism of rice to UV and chilling stresses. Further comprehensive studies should be conducted to validate this paradigmatic finding.

Keywords: UV radiation; chilling stress; phenolics; momilactones; gene expression; biosynthesis; ultra-performance liquid chromatography-electrospray ionization-mass spectrometry; real-time quantitative polymerase chain reaction

Citation: Hoang Anh, L.; Van Quan, N.; Quang Lam, V.; Takami, A.; Dang Khanh, T.; Dang Xuan, T. Rice Momilactones and Phenolics: Expression of Relevant Biosynthetic Genes in Response to UV and Chilling Stresses. *Agronomy* **2022**, *12*, 1731. https://doi.org/10.3390/agronomy12081731

Academic Editors: Sara Álvarez and José Ramón Acosta-Motos

Received: 29 June 2022
Accepted: 20 July 2022
Published: 22 July 2022

1. Introduction

Rice is an important food crop that plays a crucial role in ensuring food security for the whole world. Rice has a beneficial consumption as it is rich in nutrients, phytocompounds, and associated biological activities. For example, phenolics found in rice have antioxidant, antidiabetic, anti-inflammatory, anti-neurodegenerative, and anti-cardiovascular disease potentials [1]. In addition, momilactones are valuable diterpenoids of rice, which display antioxidant, antitumor (leukemia and colon cancer), antidiabetic, anti-skin aging, and anti-obesity properties [2–8]. However, with the increasing threat of climate change, many adverse conditions have occurred that negatively affect rice production worldwide [9]. Therefore, protective strategies are urgently needed for rice cultivation and maintaining food security globally. For this purpose, a deep understanding of the rice response mechanism to environmental stress is integrally required to protect rice plants from the impacts of adverse conditions.

Among stress conditions, UV radiation is one of the most challenging threats to rice cultivation. The depletion of stratospheric ozone leads to increased UV radiation reaching the Earth's surface [10]. A slight augmentation of UV irradiation can cause severe consequences to rice growth, resulting in a severe reduction in rice yields [10]. Besides UV effects, rice cultivars are also extremely susceptible to low temperature since they are distributed mainly in tropical and subtropical areas. Low temperatures occurring at critical reproductive stages can severely inhibit grain quality or impede rice yield [9]. In general, UV and chilling stresses (in separation and/or in combination) can cause various adverse effects on rice plants during physiological, biochemical, and molecular processes and ultimately impair rice yield and grain quality [11]. Therefore, the research on rice tolerance against environmental stresses is a top priority in many countries worldwide. Under UV and chilling stresses, rice cultivars can suffer from reactive oxygen species (ROS), leading to oxidative stress [12]. Oxidative stress causes damage to DNA, proteins, membranes, etc., resulting in an inhibition of rice growth [10]. Therefore, the induction of ROS and antioxidant responses are an integral part of the research on rice tolerance to stresses. Regarding antioxidants, the elevated accumulation of phenolic compounds is considered to strengthen plants' antioxidant capacity, which may help improve plant performance under stresses [13]. In recent decades, a tremendous effort has been made to understand the rice response mechanisms against the adverse environment. Several studies have been conducted to clarify the correlation between phenolic compounds and antioxidant activities of different rice varieties under various biotic and abiotic stresses such as salinity, drought, extreme temperatures, and diseases [14–17]. In addition, rice diterpenoids comprising momilactones A (MA) and B (MB) were reported to contribute to rice resistance against salt, drought, weed, UV, and heavy metal stresses [18–22]. Besides, the biosynthesis of phenolics and momilactones is closely associated with the regulation of genes encoding the major biosynthetic enzymes [23,24]. However, the expressions of the related genes to the biosynthesis of phenolics (PRBG), and especially momilactones (MRBG), in rice subjected to UV and chilling stresses have been least elucidated.

In the current research, the antioxidant responses of rice under UV and chilling effects were determined. Additionally, the accumulation of phenolics and momilactones were determined using high-performance liquid chromatography (HPLC) and ultra-performance liquid chromatography-electrospray ionization-mass spectrometry (UPLC-ESI-MS). The relative expression method applying real-time quantitative polymerase chain reaction (RT-qPCR) was utilized to evaluate the expressions of PRBG and MRBGs. In essence, comprehensive information regarding the biochemical changes and their genetic causes in rice mediated by UV and chilling effects is discussed.

2. Materials and Methods

2.1. Rice Variety, Reagents, and Standard Compounds

Seeds of a Japonica rice variety, namely Koshihikari (*Oryza sativa* L.), were provided by Japan Agricultural Cooperatives (JA), Hiroshima, Japan. The solvents, including methanol and hexane, were purchased from Junsei Chemical Co., Ltd., Tokyo, Japan, and Kanto Chemical Co., Inc., Tokyo, Japan. The chemicals sodium hypochlorite (NaClO), Folin-Ciocalteu's reagent, sodium carbonate (Na_2CO_3), aluminum chloride ($AlCl_3$), 2,2-diphenyl-1-picrylhydrazyl (DPPH), 2,2′-azinobis-(3-ethylbenzothiazoline-6-sulfonic acid) (ABTS), potassium persulfate ($K_2S_2O_8$), sodium acetate (CH_3COONa), catechol, benzoic acid, caffeic acid, chlorogenic acid, cinnamic acid, *p*-coumaric acid, *p*-hydroxybenzoic acid, gallic acid, salicylic acid, syringic acid, vanillin, esculetin, rutin, fisetin, morin, quercitin, tricin, and galagin were obtained from Kanto Chemical Co., Inc., Tokyo, Japan. Pure momilactones A (MA) and B (MB) were previously isolated from rice husk in our laboratory of the Plant Physiology and Biochemistry, Graduate School of Advanced Science and Engineering, Hiroshima University, Japan. The identification and confirmation of MA and MB were presented in the previous study by Quan et al. [2].

2.2. UV Stress Treatment

Koshihikari rice seeds were sterilized with 0.1% NaOCl for 30 min before rinsing with distilled water several times. The seeds were germinated in distilled water at 30 °C in an incubator for 48 h. Strong germinated seeds were then placed into trays (37 cm × 25 cm × 11 cm of length × width × height) containing Yoshida solution and agar 0.5%. The seedlings were grown in a growth chamber at 28 °C with a 12 h photoperiod. After 21 days, UV irradiation (253 nm, 10 μmol/m^2/s; GL15, Toshiba, Tokyo, Japan) (2 h and 4 h per day for 5 days) was provided from above with a distance of 50 cm between rice plants and UV source. Control rice plants were grown in normal conditions. All treatments were performed with 3 replications. Subsequently, rice plants were collected for further experiments.

2.3. Chilling Stress Treatment

Koshihikari rice seedlings were prepared and placed into trays with the same protocol as mentioned above. After 21 days grown in a growth chamber at 28 °C with a 12 h photoperiod, rice seedlings were subjected to chilling stress at 6 °C (4 h and 8 h per day) for 7 days. Nontreated rice plants were used as the control. All treatments were conducted in triplicate. After that, rice plants were collected for the next experiments.

2.4. Sample Preparation

The collected rice seedlings were dried in an oven at 40 °C and subsequently ground into powder. Rice samples (100 g) were then extracted with 200 mL of methanol 90% for 1 week. The obtained extracts were mixed with hexane in a separator. The methanolic phase was then collected. After that, the filtered methanolic extract was concentrated by a vacuum evaporator. The dried extract was kept in a vial for the determination of antioxidant activity and the quantification of phenolic and momilactone contents.

2.5. Antioxidant Activity

The antioxidant ability of rice seedlings under UV and chilling stresses was determined via ABTS cation decolorization and DPPH radical scavenging assays. The detailed procedures with 3 replications were described in the research of Quan et al. [25]. The inhibition of radicals was observed as the discoloration of the final solution with samples and evaluated as the decreased percentage of absorbance compared with the negative control (MeOH) at 517 nm and 734 nm for DPPH and ABTS assays, respectively. An established dose-dependent curve (linear equation) applying different concentrations of samples was used to determine the required concentration inhibiting 50% of radicals (IC_{50} values). A higher IC_{50} value means weaker inhibition.

2.6. Total Phenolic (TPC) and Flavonoid (TFC) Contents

The TPC and TFC were determined by applying the same protocols with 3 replications presented in our previous publication [26]. The calculation of TPC was based on the recorded absorbance at 765 nm and expressed as milligrams of gallic acid equivalent per gram of sample dry weight (mg GAE/g DW). TFC was measured at 430 nm and presented as milligrams of rutin equivalent per gram of sample dry weight (mg RE/g DW).

2.7. Quantification of Phenolics and Flavonoids by High-Performance Liquid Chromatography (HPLC)

The quantities of phenolic compounds were determined by the HPLC system consisting of a PU-4180 RHPLC pump, LC-Net II/ADC controller, and UV-4075 UV/Vis detector (Jasco, Tokyo, Japan). A methanolic sample (5 mg/mL) with a volume of 5.0 μL was injected into the XBridge® Shield RP18 (5 μm, 2.1 × 100 mm) column (Waters Corporation, Milford, MA, USA). Gradient elution was run with 400 μL/min of the flow rate by applying the following gradient program with 2 solvents: 95% A and 5% B over 0–2 min, followed by 30% A and 70% B over 2–12 min, subsequently changed to 0% A and 100% B over

12–22 min. From 22–34 min, the mobile phase was returned to the initial condition. In which A was 0.1% formic acid in water and B was pure acetonitrile. Phenolic compounds were determined at 280 and 350 nm. Phenolic standards (5, 10, 25, 50, and 100 μg/mL) were used to establish calibration curves for the quantification of phenolic compounds in rice based on the detected peak areas in each sample. All tests were conducted in triplicate. The outcome was presented as micrograms per gram of sample dry weight (μg/g DW).

2.8. Quantification of Momilactones by Ultra-Performance Liquid Chromatography-Electrospray Ionization-Mass Spectrometry (UPLC-ESI-MS)

MA and MB in rice extract were quantified based on the method described by Quan et al. [2]. In particular, the UPLC-ESI-MS system included LTQ Orbitrap XL mass spectrometers (Thermo Fisher Scientific, Waltham, MA, USA) and an electrospray ionization (ESI) source. For analysis, a methanolic sample (10 mg/mL) with a volume of 3.0 μL was injected into the ZORBAX Eclipse Plus C18 (1.8 μm, 2.1 × 50 mm) column (Agilent Technologies, Santa Clara, CA, USA). The column temperature was maintained at 25 °C. The gradient model with 2 solvents was set up as follows: solvent A was 0.1% trifluoroacetic acid in water, solvent B was 0.1% trifluoroacetic acid in acetonitrile. The gradient program was: 50% B during 0–5 min, then adjusted to 100% B during 5–10 min, subsequently maintained for 0.1 min, and another 5 min for equilibration. The flow rate was 300 μL/min. The operation was run in 15.1 min. A positive FTMS mode with a range of 100 to 800 m/z was applied for mass scanning. The presence of MA and MB in rice samples was confirmed by comparing their extracted ion chromatograms (EIC) and mass spectra of samples with those of standard momilactones. The calibration curves of MA and MB were established by applying different concentrations of momilactone standards (0.5, 1, 5, and 10 μg/mL) to determine the momilactone content in the rice seedlings. The peak areas of MA and MB detected in samples were used to calculate the amount of each compound by standard curves. All quantifications were performed in triplicate. The result was expressed as micrograms per gram of sample dry weight (μg/g DW).

2.9. Expression of Relevant Biosynthetic Genes to Phenolic (PRBG) and Momilactone (MRBG) by Real-Time Quantitative Polymerase Chain Reaction (RT-qPCR)

The total RNA was obtained using the RN-Sure Plant Mini Kit (Intégrale Co., Ltd., Tokushima, Japan). Single- and double-stranded DNA was removed from total RNA using DNase I, RNase-free (Thermo Scientific, Waltham, MA, USA). An amount of 1 μg of total RNA was used for synthesizing cDNA by applying a High-Capacity cDNA Reverse Transcription Kit (Applied Biosystems, Foster City, CA, USA) with an RNase Inhibitor (Applied Biosystems, Foster City, CA, USA). The obtained cDNA solution was then diluted two times for the next step. The PCR reaction was generated using KOD FX Neo (TOYOBO Co., Ltd., Osaka, Japan). The PCR program was set for initial denaturation at 96 °C for 2 min, followed by 45 cycles of denaturing at 98 °C for 15 s, annealing at 56 °C for 20 s, and extending at 72 °C for 20 s. The final extension was performed at 72 °C for 1 min.

All tests were performed in triplicate. The results of cycle threshold (Ct) values were obtained from a StepOne RT-qPCR (Applied Biosystems, Waltham, MA, USA). Based on Ct values, the relative quantifications (RQ) were calculated to compare between samples. The formula was as below:

$$RQ = 2^{-(\Delta Ct\ sample\ -\ \Delta Ct\ control)} \quad (1)$$

ΔCt sample: Ct value of tested genes in rice under stress conditions after subtracting Ct value of housekeeping gene
ΔCt control: Ct value of tested genes in rice under normal conditions after subtracting Ct value of housekeeping gene

The primers for PRBG and MRBG in the biosynthetic pathway of phenolics and momilactones [23,24] were designed by applying the NCBI Primer-BLAST program [27]. The list of primers is described in Supplementary Materials, Table S1.

2.10. Statistical Analysis

Minitab 16.0 software (Minitab Inc., State College, PA, USA) was used for data analyses. The results are expressed as mean ± standard deviation (SD) (n = 3). One-way ANOVA using Tukey's test at $p < 0.05$ was applied to indicate significant differences between samples. Pearson's correlation coefficients among parameters were determined using the same software.

3. Results and Discussion

3.1. Total Phenolic (TPC) and Flavonoid (TFC) Contents and Antioxidant Responses of Rice Seedlings against UV and Chilling Stresses

Phenolic compounds play a crucial role in the plant defense system against biotic and abiotic stresses [28]. These substances with UV-absorbing features can be effectively determined by UV-visible spectroscopy methods [29]. Preceding studies screened a wide range of wavelengths and generated the most appropriate absorbance to quantify total phenolics (TPC) (e.g., 765 nm for the Folin-Ciocalteu method) and flavonoids (TFC) (e.g., 430 nm for the aluminum chloride colorimetric protocol) in plant extracts [25,26,29,30]. Accordingly, TPC and TFC in Koshihikari (a famous Japonica model rice variety) seedlings under UV and chilling exposures were determined following the same procedures, the results are shown in Table 1. In other reports, UV stress stimulated phenolic and flavonoid contents in rice [31]. However, in this investigation, the TPC and TFC of rice seedlings were reduced under UV irradiation. This might be due to the differences in treatment conditions among studies and genetic diversity of selected rice cultivars. In fact, different rice varieties in various growing stages exhibit dissimilar mechanisms to cope with UV exposure including tolerance and susceptibility. In particular, tolerant cultivars accumulated relatively higher TPC than susceptible ones [32]. In our study, the TPC values of Koshihikari rice seedlings in UVC, UV2, and UV4 treatments were 3.23, 2.65, and 2.92 mg GAE/g DW, respectively (Table 1). While the TFC of UV-treated rice seedlings was approximately two times lower than the control. The TFC values of UVC, UV2, and UV4 were 3.23, 1.84, and 1.73 mg RE/g DW, respectively (Table 1). For the chilling treatments, the TPC of Chi4 (2.81 mg GAE/g DW) was significantly lower than the control (4.08 mg GAE/g DW) (Table 1). The TPC of Chi8 (5.57 mg GAE/g DW) remarkably increased compared to ChiC and Chi4 (Table 1). Generally, chilling conditions enhanced rice TPC in the present study, in line with the research of Rayee et al. [16]. On the other hand, Rayee et al. [16] indicated a significant upregulation of rice TFC under chilling effects, whereas TFC reduced in our report. Notably, the TFC of Chi4 (0.13 mg RE/g DW) was 2.4 times lower than the control (0.31 mg RE/g DW) (Table 1). The TFC of Chi8 (0.28 mg RE/g DW) increased compared to Chi4 (Table 1); however, it was lower than the control.

Table 1. Total phenolic (TPC) and flavonoid (TFC) contents and antioxidant response of rice seedlings against UV and chilling stresses.

Treatment	Sample	TPC (mg GAE/g DW)	TFC (mg RE/g DW)	DPPH Assay IC_{50} (mg/mL)	ABTS Assay IC_{50} (mg/mL)
UV	UVC	3.23 ± 0.22 ^{a}	3.23 ± 0.15 ^{a}	0.81 ± 0.01 ^{b}	1.05 ± 0.07 ^{a}
	UV2	2.65 ± 0.17 ^{b}	1.84 ± 0.07 ^{b}	0.91 ± 0.02 ^{a}	1.14 ± 0.06 ^{a}
	UV4	2.92 ± 0.20 ^{ab}	1.73 ± 0.04 ^{b}	0.95 ± 0.02 ^{a}	1.08 ± 0.15 ^{a}
Chilling	ChiC	4.08 ± 0.16 ^{b}	0.31 ± 0.01 ^{a}	0.49 ± 0.02 ^{b}	0.85 ± 0.02 ^{b}
	Chi4	2.81 ± 0.17 ^{c}	0.13 ± 0.01 ^{c}	0.61 ± 0.08 ^{a}	1.00 ± 0.04 ^{a}
	Chi8	5.57 ± 0.23 ^{a}	0.28 ± 0.01 ^{b}	0.48 ± 0.02 ^{b}	0.85 ± 0.08 ^{b}

Outcome is expressed as mean ± standard deviation (SD). Samples in the same column within a treatment followed by different superscript letters indicate significant differences at $p < 0.05$. TPC, total phenolic content; TFC, total flavonoid content; GAE, gallic acid equivalent; RE, rutin equivalent; DW, dry weight; DPPH, 2,2-diphenyl-1-picrylhydrazyl; ABTS, 2,2′-azinobis-(3-ethylbenzothiazoline-6-sulfonic acid); UVC, control rice seedlings without UV treatments; UV2, UV-treated rice seedlings for 2 h per day; UV4, UV-treated rice seedlings for 4 h per day; ChiC, control rice seedlings without chilling treatments; Chi4, chill-treated rice seedlings for 4 h per day; Chi8, chill-treated rice seedlings for 8 h per day.

Regarding antioxidants, this activity can help rice plants mitigate injuries from environmental stress [12]. In the current study, the antioxidant responses of rice seedlings under UV and chilling were determined via antiradical (DPPH and ABTS) assays (Table 1). In the DPPH assay, the antiradical abilities of rice seedlings decreased when the UV duration increased. The IC_{50} values of UVC, UV2, and UV4 were 0.81, 0.91, and 0.95 mg/mL, respectively (Table 1). On the other hand, rice seedlings revealed an insignificant difference in antiradical activities between the control and treatments in the ABTS assay. The IC_{50} values of UVC, UV2, and UV4 were 1.05, 1.14, and 1.08 mg/mL, respectively (Table 1). In chilling treatments, the antiradical activities of Chi4 decreased (IC_{50} = 0.61 and 1.00 mg/mL for the DPPH and ABTS assays, respectively), compared to the control (IC_{50} = 0.49 and 0.85 mg/mL for the DPPH and ABTS assays, respectively) (Table 1). However, the antiradical ability of the rice seedlings was enhanced when the level of chilling conditions was elevated (the IC_{50} values of Chi8 were 0.48 and 0.85 mg/mL for DPPH and ABTS, respectively) (Table 1). Overall, the antioxidant capacity might contribute to the response of the rice seedlings under chilling exposure. Meanwhile, UV irradiation might decrease the antioxidant capacities of rice seedlings.

3.2. Changes in Phenolic and Momilactone Contents of Rice Seedlings under UV Treatment

The changes in the chemical profiles of the rice seedlings in response to UV irradiation are shown in Table 2. Accordingly, a total of 17 compounds were detected in the rice samples. In which, there were 15 phenolic compounds belonging to the groups of simple phenols (catechol), phenolic acids (benzoic, caffeic, chlorogenic, *ρ*-coumaric, *ρ*-hydroxybenzoic, and salicylic acids), phenolic aldehydes (vanillin), coumarins (esculetin), and flavonoids (rutin, fisetin, morin, quercetin, tricin, and galangin). Other diterpenoids consisting of MA and MB were also found in the rice samples. The contents of identified compounds in the rice seedlings under UV irradiation and the control are shown in Table 2.

Among the detected phenolic compounds, the quantities of benzoic acid, caffeic acid, *ρ*-coumaric acid, morin, quercetin, and galangin reduced with the increasing period of UV treatment (Table 2). Significantly, the greatest inhibitions were found in the production of benzoic acid, caffeic acid, morin, and galangin. Notably, the benzoic acid and morin contents decreased by 1.8- and 1.4-fold, respectively, in UV4 compared to UVC (Table 2). Caffeic acid and galangin accounted for 65.49 and 69.14 μg/g DW, respectively, in the control. However, these compounds were not found in rice seedlings subjected to UV stress (Table 2). On the other hand, the impacts of UV irradiation might lead to the proliferation of compounds comprising catechol, chlorogenic acid, *ρ*-hydroxybenzoic acid, salicylic acid, esculetin, rutin, and fisetin. Catechol and *ρ*-hydroxybenzoic acid were found in UV2 with 7.77 and 41.22 μg/g DW, respectively; however, these compounds were not detected in UV4 and UVC (Table 2). Remarkably, salicylic acid was not found in UVC. However, this phenolic acid was notably one of the most abundant compounds in UV2 and UV4, with the contents of 852.78 and 853.39 μg/g DW, respectively (Table 2). These phenolic compounds were previously reported to contribute to rice antioxidant activity, especially salicylic acid, which is widely known to be an essential phytohormone involved in the proliferation of ROS scavenging enzymes in rice subjected to UV irradiation [10]. However, in the present research, a slight decrease in antioxidant activity was observed, so the high phenolic content might not be helping rice cultivars deal with oxidative stress (Table 1). Therefore, these compounds might contribute to the rice defense system through other metabolic pathways. For instance, accumulated phenols in plant cells play a role as protectors under the epidermal layer, thereby protecting cellular components and reducing damages to DNA and important functional enzymes under the influence of UV exposure [33]. Moreover, rice cultivars can mitigate UV damage by increasing UV-absorbing compounds, such as flavonoids and anthocyanins [10]. Accordingly, in this study, esculetin, rutin, and fisetin might be produced to counteract the negative effect of UV light, suggesting a photoprotective role of rice phenolics. The amounts of esculetin and rutin remarkably increased by 2.5- and 1.7-fold, respectively, in UV4 compared to UVC (Table 2). In addition to phenolics, momilactones have been considered important compounds in the rice defense system to deal with UV irradiation. In particular, the quantity of MB was significantly enhanced in UV-affected rice [21]. In agreement with the previous report, the elevated contents of MA and MB were documented (from 3.6- to 6.4-fold over the control) when the UV level increased (Table 2). The findings suggest that momilactones might play a crucial role in the rice defense system against UV stress as novel phytohormones, which requires deeper confirmation.

Table 2. Changes in chemical contents of rice seedlings in response to UV stress.

Compounds	Content (µg/g DW)		
	UVC	UV2	UV4
a. Simple phenols			
Catechol	-	7.77 ± 0.23	-
b. Phenolic acids			
Benzoic acid	2307.63 ± 54.28 ^{a}	1449.23 ± 30.11 ^{b}	1259.39 ± 41.05 ^{c}
Caffeic acid	65.49 ± 6.47	-	-
Chlorogenic acid	40.27 ± 2.11 ^{c}	61.32 ± 1.42 ^{a}	49.25 ± 0.14 ^{b}
ρ-Coumaric acid	292.69 ± 7.15 ^{a}	236.61 ± 0.57 ^{b}	234.56 ± 0.50 ^{b}
ρ-Hydroxybenzoic acid	-	41.22 ± 5.13	-
Salicylic acid	-	852.78 ± 6.22 ^{a}	853.39 ± 2.36 ^{a}
c. Phenolic aldehydes			
Vanillin	43.08 ± 4.36 ^{a}	41.56 ± 0.75 ^{a}	39.90 ± 2.18 ^{a}
d. Coumarins			
Esculetin	18.76 ± 1.19 ^{c}	62.48 ± 2.93 ^{a}	46.47 ± 1.66 ^{b}
e. Flavonoids			
Rutin	228.83 ± 10.57 ^{c}	499.02 ± 21.16 ^{a}	382.53 ± 66.04 ^{b}
Fisetin	87.08 ± 5.77 ^{b}	301.99 ± 11.39 ^{a}	92.92 ± 1.63 ^{b}
Morin	78.44 ± 4.93 ^{a}	-	54.49 ± 1.41 ^{b}
Quercetin	155.44 ± 15.97 ^{a}	144.47 ± 7.07 ^{a}	141.10 ± 3.98 ^{a}
Tricin	234.70 ± 9.54 ^{a}	201.05 ± 8.16 ^{b}	211.15 ± 6.21 ^{b}
Galangin	69.14 ± 0.85	-	-
f. Diterpenoids			
Momilactone A	10.73 ± 1.13 ^{c}	39.34 ± 0.55 ^{b}	69.14 ± 7.44 ^{a}
Momilactone B	5.48 ± 0.38 ^{c}	19.60 ± 0.73 ^{b}	26.46 ± 2.45 ^{a}

The content of each compound is presented as mean ± standard deviation (SD). Different superscript letters in a row indicate significant differences at $p < 0.05$. -, not determined; DW, dry weight; UVC, control rice seedlings without UV treatments; UV2, UV-treated rice seedlings for 2 h per day; UV4, UV-treated rice seedlings for 4 h per day.

3.3. Changes in Phenolic and Momilactone Contents of Rice Seedlings under Chilling Treatments

The influence of chilling stress on the contents of 18 detected compounds grouped into phenolic acids (benzoic, caffeic, chlorogenic, cinnamic, *ρ*-coumaric, *ρ*-hydroxybenzoic, gallic, salicylic, syringic acids), phenolic aldehydes (vanillin), coumarins (esculetin), flavonoids (rutin, fisetin, morin, quercetin, and tricin), and diterpenoids (MA and MB) of the rice seedlings is displayed in Table 3, followed by a detailed demonstration.

Table 3. Changes in chemical contents of rice seedlings in response to chilling stress.

Compounds	Content (μg/g DW)		
	ChiC	Chi4	Chi8
a. Phenolic acids			
Benzoic acid	959.20 ± 22.89 [a]	766.09 ± 3.09 [c]	829.10 ± 11.30 [b]
Caffeic acid	40.16 ± 2.19 [a]	35.13 ± 4.46 [a]	33.01 ± 1.32 [a]
Chlorogenic acid	110.78 ± 0.03 [a]	56.37 ± 6.04 [b]	57.00 ± 4.22 [b]
Cinnamic acid	44.53 ± 0.52 [a]	4.91 ± 0.27 [c]	8.99 ± 0.67 [b]
p-Coumaric acid	346.72 ± 3.36 [a]	262.69 ± 0.97 [c]	290.83 ± 2.29 [b]
p-Hydroxybenzoic acid	177.23 ± 25.85 [a]	45.12 ± 4.24 [b]	-
Gallic acid	25.64 ± 0.20	-	-
Salicylic acid	1262.93 ± 10.71 [b]	1012.81 ± 58.17 [c]	1402.41 ± 46.89 [a]
Syringic acid	189.47 ± 3.94 [b]	153.79 ± 3.25 [c]	232.67 ± 12.06 [a]
b. Phenolic aldehydes			
Vanillin	-	10.35 ± 2.71	-
c. Coumarins			
Esculetin	0.04 ± 0.00 [a]	0.03 ± 0.00 [b]	0.03 ± 0.00 [b]
d. Flavonoids			
Rutin	0.57 ± 0.01 [a]	0.43 ± 0.01 [b]	0.44 ± 0.01 [b]
Fisetin	0.18 ± 0.01 [ab]	0.19 ± 0.00 [a]	0.15 ± 0.02 [b]
Morin	0.07 ± 0.01	-	-
Quercetin	0.08 ± 0.01 [a]	0.03 ± 0.00 [c]	0.04 ± 0.00 [b]
Tricin	0.16 ± 0.01 [a]	0.06 ± 0.00 [c]	0.10 ± 0.00 [b]
e. Diterpenoids			
Momilactone A	27.28 ± 1.10 [a]	7.90 ± 0.40 [b]	6.14 ± 0.49 [b]
Momilactone B	8.75 ± 0.50 [a]	4.04 ± 0.08 [c]	5.68 ± 0.24 [b]

The content of each compound is presented as mean ± standard deviation (SD). Different superscript letters in a row indicate significant differences at $p < 0.05$. -, not determined; DW, dry weight; ChiC, control rice seedlings without chilling treatments; Chi4, chill-treated rice seedlings for 4 h per day; Chi8, chill-treated rice seedlings for 8 h per day.

Based on the data presented in Table 3, the quantities of almost all detected phenolic acids in the rice plants decreased under chilling conditions in which the amounts of chlorogenic acid and cinnamic acid were dramatically reduced by 1.9 and 5.0 times, respectively, in Chi8 compared to ChiC. Similarly, the *p*-hydroxybenzoic acid content in Chi4 was 3.9-fold lower than ChiC. However, this compound was not detected in Chi8 (Table 3). Gallic acid was only determined in the control, whereas this compound was not detected in Chi4 or Chi8 (Table 3). In the study of Rayee et al. [16], the contents of all phenolic acids found in chilling-affected rice plants were also reduced, while the antioxidant activity was enhanced. Rayee et al. [16] hypothesized that phenolic compounds other than phenolic acids, such as phenols, polyphenols, and flavonoids, may be involved in the rice resistance to chilling conditions. However, all flavonoid quantities were reduced in our study under the impact of chilling (Table 3). Among them, the quantities of quercetin and tricin critically diminished by 2.0 and 1.6 times, respectively, in Chi8 compared to ChiC (Table 3). In addition, morin was found in the control, but not in either Chi4 or Chi8 (Table 3). The results show that flavonoids might also play a negligible role in the defense mechanisms of rice against chilling stress. On the other hand, salicylic acid had enhanced quantities in Chi8 (1402.41 μg/g DW) compared to ChiC (1262.93 μg/g DW) (Table 3). In previous reports, low temperature promoted the accumulation of endogenous salicylic acid in numerous plant species, for example, *Arabidopsis*, cereals, wheat, and grape, suggesting that this phenolic acid is involved in regulating plant responses to chilling conditions [34]. However, apart from a slight increase in the levels of endogenous salicylic acid in rice subjected to chilling stress, no direct correlation between salicylic acid treatment and the chilling tolerance of rice has been announced. Rice plants also responded negatively to chilling conditions with the exogenous application of salicylic acid by increasing electrolyte leakage, lipid peroxidation, and decreasing the antioxidant enzyme activity [35,36]. Moreover, in this study, the content of syringic acid in Chi8 (232.67 μg/g DW) was elevated compared to the control (189.47 μg/g DW) (Table 3). However, no reports have explained how syringic acid contributes to rice resistance to chilling effects. In addition to phenolics, momilactones, as mentioned above, play an important role in the ability of rice plants to cope with different stresses [18,19,21]. This study indicated, for the first time, that MA might not contribute to the chilling tolerance of rice because of its reduced accumulation (3.5 and 4.4 times compared to the control) (Table 3). The quantity of MB in Chi4 was 2.2 times lower than ChiC. However, the amount was enhanced with the increasing duration of the chilling treatment, which was consistent with the proliferation of benzoic acid, cinnamic acid, ρ-coumaric acid, salicylic acid, syringic acid, quercetin, and tricin (Table 3). This result implies that MB may contribute to the physiological responses of rice to chilling stress. Therefore, forthcoming studies should be carried out to clarify this mechanism.

3.4. Transcriptional Response Involved in Biosynthesis of Phenolics and Momilactones in Rice Seedlings under UV and Chilling Stresses

According to the scheme presented by Anh et al. [23], the *PAL* gene encodes an important enzyme, namely phenylalanine ammonia-lyase, in the biosynthetic pathway of phenolic compounds. The major enzymes involved in the formation of momilactones consist of syn-copalyl diphosphate synthase-like; syn-pimara-7,15-diene synthase-like; 9-beta-pimara-7,15-diene oxidase-like; and momilactone A synthase-like, which are encoded by the *OsCPS4, OsKSL4, CYP99A3, OsMAS, OsMAS2* genes, respectively [24]. The expressions of targeted genes were determined by relative quantification (RQ) methods using the housekeeping gene. Particularly, selected housekeeping genes should exhibit stable expression in different tissue types, development stages, and experimental treatments. In this study, two housekeeping genes including *actin* and *eIF-4A* were examined, which are among the most stable genes commonly used for rice research with different treatment conditions [37–40]. After screening, we obtained that *actin* (Ct values ranged from 23.28 to 29.27) revealed a stronger expression than *eIF-4A* (Ct ranged from 27.40 to 31.93) in tested samples (Supplementary Materials, Table S2 and Figure S1). Ideally, an optimal housekeeping gene should have an average expression with Ct value of between 15 and 30 [41]. Hereby, we selected *actin* as the internal reference gene to evaluate the expressions (RQ values) of phenolic (PRBG)-and momilactone (MRBG)-relevant biosynthetic genes in rice seedlings treated with UV and chilling conditions (Supplementary Materials, Table S3). The changes in expression levels of the tested genes with the increasing effects of UV and chilling stresses are displayed in Figure 1.

Based on the results in Figure 1 and Supplementary Materials, Table S3, the expressions of all tested genes elevated in rice under UV effects. Remarkably, *CYP99A3* and *OsMAS* had the highest level of expression, followed by *OsMAS2, OsKSL4, OsCPS4,* and *PAL*. The RQ values of *PAL, OsCPS4, OsKSL4, CYP99A3, OsMAS,* and *OsMAS2* in UV2 were 2.047, 3.710, 5.851, 48.217, 54.932, and 33.265, respectively. While in UV4, the RQ values of these genes were 2.582, 3.852, 17.887, 58.781, 48.148, and 42.267, respectively. Of which, the expressions of *OsKSL4, CYP99A3,* and *OsMAS* in rice subjected to the increased level of UV irradiations were previously evaluated in an Indica rice variety Basmati 1 [39]; however, the expressions of *OsCPS4* and *OsMAS2* and the quantities of MA and MB were not presented. Additionally, the response mechanisms to UV stress may vary among different rice varieties. Therefore, this is the first study indicating the changes of MA and MB syntheses in the Japonica model rice variety Koshihikari based on their accumulation and relevant gene expressions under UV exposure.

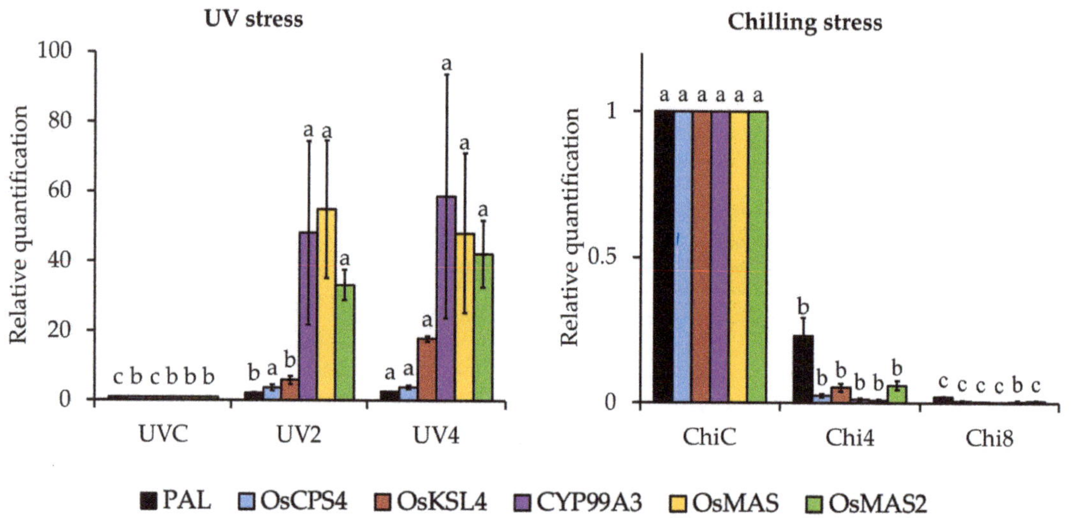

Figure 1. Expressions of relevant genes to the biosynthesis of phenolics and momilactones in rice seedlings against UV and chilling stresses. UVC, control rice seedlings without UV treatments; UV2, UV-treated rice seedlings for 2 h per day; UV4, UV-treated rice seedlings for 4 h per day; ChiC, control rice seedlings without chilling treatments; Chi4, chill-treated rice seedlings for 4 h per day; Chi8, chill-treated rice seedlings for 8 h per day; *PAL*, gene encoding phenylalanine ammonia-lyase; *OsCPS4*, gene encoding syn-copalyl diphosphate synthase-like; *OsKSL4*, gene encoding syn-pimara-7,15-diene synthase-like; *CYP99A3*, gene encoding 9-beta-pimara-7,15-diene oxidase-like; *OsMAS*, gene encoding momilactone A synthase-like; *OsMAS2*, gene encoding momilactone A synthase-like. Different letters (a,b,c) enclosed with columns (same colors) express significant differences at $p < 0.05$.

On the other hand, the rice responded to the increasing duration of the chilling treatment by a severe decrease in expressions of PRBG and MRBG. Particularly, the RQ values of *PAL* were 0.230 and 0.021 in Chi4 and Chi8, respectively. Furthermore, the RQ values of *OsCPS4*, *OsKSL4*, *CYP99A3*, *OsMAS*, and *OsMAS2* were 0.026, 0.054, 0.012, 0.010, and 0.061, respectively, in Chi4. Meanwhile, these values were 0.008, 0.004, 0.002, 0.006, 0.007, respectively, in Chi8 (Figure 1 and Supplementary Materials, Table S3). Notably, there has been no previous report indicating the effects of low temperature on the expressions of MRBG in correlation with their components in rice.

3.5. Pearson's Correlation Coefficients between Antioxidant Activity, Chemical Profiles, and Relevant Gene Expressions of Rice Seedlings under UV Treatments

The effects of UV on phenolic content and the expression of relevant genes in the phenylpropanoid pathway were previously indicated [31,42]. Of which, *PAL* is an important enzyme in the formation of cinnamic acid from phenylalanine which is the first step in a general phenylpropanoid pathway [23]. Thousands of phenolic compounds such as phenolic acids, flavonoids, anthocyanins, and lignins are derived from phenylpropanoids [43]. Therefore, enhanced activity of *PAL* is expected to increase the TPC of targeted organisms [43]. However, preceding studies showed different contributions of *PAL* to TPC which may depend on various factors such as plant species, treatment conditions, and sample preparations. For example, the upregulated *PAL* activity was consistent with the elevated TPC in cut carrots exposed to UV radiation [44]. In contrast, TPC did not show any correlation with PAL activity during the cold storage of lettuce [45], similar to the results shown in our report. This suggests that PAL plays an essential role but may not be the only factor determining the TPC in rice. Other enzymes in the biosynthetic pathway of phenolic compounds, such as cinnamic acid 4-hydroxylase (*C4H*), 4-coumaroyl:CoA-ligase

(*4CL*), chalcone synthase (*CHS*), chalcone reductase (*CHR*), etc., may also be important contributors to TPC in targeted organisms [23]. On the other hand, *PAL* may strongly correlate with specific phenolic compounds instead of TPC in plant response to different treatment conditions [46]. Accordingly, this research is the first indication that the expression of *PAL* was tightly linked to the accumulation of chlorogenic acid, salicylic acid, esculetin, and rutin in UV-exposed rice (Supplementary Materials, Table S4). Meanwhile, the proliferation of MA and MB might be controlled by the stimulated expressions of MRBG, including *OsCPS4*, *OsKSL4*, *CYP99A3*, *OsMAS*, and *OsMAS2* (Supplementary Materials, Table S4). Notably, the regulation of PRBG expression was consistent with that of MRBG expressions (Supplementary Materials, Table S4). Generally, this report confirms the contribution of chlorogenic acid, salicylic acid, esculetin, rutin, MA, and MB to the response mechanisms of rice against UV stress. However, the proliferation of these compounds had a low association with the antioxidant activities of rice treated with UV irradiation (correlation values of below 0.24), implying that they might contribute to rice tolerance against UV irradiation through other biological pathways (Supplementary Materials, Table S4). Based on a strong correlation between the biosyntheses of momilactones and the mentioned stress-resistant phenolics, MA and MB might play a role as signaling compounds in the physiological response of rice against UV irradiation.

3.6. Pearson's Correlation Coefficients between Antioxidant Activity, Chemical Profiles, and Relevant Gene Expressions of Rice Seedlings under Chilling Treatments

Regarding chilling stress, the antioxidant activity was closely associated with the changes in rice TPC and TFC (Table 1 and Supplementary Materials, Table S5) which may contribute to rice tolerance against low temperature during 8 h. In previous reports, the serious occurrence of oxidative stress in rice during the first 8 h of chilling stress was implicated, especially, a burst of ROS accumulation was recorded after the initial 4 h [47,48]. Whereas there was a sudden decrease in antioxidant enzymes of rice seedlings during the first 2 h of chilling treatment [47]. Under the generation of ROS during chilling exposure, rice plants may first use their existing antioxidant molecules (e.g., phenolics and flavonoids), followed by producing more of these metabolites to eliminate the damages [49]. These mentioned facts may explain the reduction in rice TPC and TFC and associated antioxidant activities under chilling conditions during the first 4 h in the current study. However, rice plants possibly started to produce more phenolics and flavonoids which may enhance the antioxidant capacities to deal with chilling stress after 8 h. On the other hand, the decreased expression of *PAL* was tightly linked to the diminished contents of almost all detected phenolics, except for syringic acid and salicylic acid (Supplementary Materials, Table S5). For momilactone biosynthesis, the reduced quantities of momilactones might be caused by the impeded expressions of MRBG (Supplementary Materials, Table S5). Furthermore, there was a linear relationship between the expressions of PRBG and MRBG (Supplementary Materials, Table S5). We suppose that chilling conditions might impede the expression of these genes in rice, resulting in the decreased quantities of almost all phenolics and momilactones. With the increasing level of the chilling duration, we found that the elevated contents of salicylic acid, syringic acid, benzoic acid, *ρ*-coumaric acid, quercitin, tricin, and MB might be involved in the strengthened antioxidant capacity (DPPH and ABTS assays) of rice seedlings against chilling stress (Supplementary Materials, Table S5). Additionally, the strong correlation between the proliferation of salicylic acid and syringic acid suggests that the synergic effect of these compounds might contribute to rice resistance to low temperatures.

3.7. Concluding Remarks

This is the first study to interpret the critical role of concomitant syntheses of phenolics and momilactones in rice under UV and chilling stresses based on the evidence obtained from both secondary metabolite and gene expression levels. The schematic representation of rice responses to UV and chilling influences via the accumulation of major compounds and the expressions of relevant biosynthetic genes is summarized in Figure 2. Accordingly, the upregulated expression of *PAL* was closely associated with the enhanced contents of chlorogenic acid, salicylic acid, esculetin, and rutin under UV exposure. Meanwhile, the decreased expression of *PAL* was tightly linked to the reduced accumulation of almost all phenolic compounds, except for syringic acid and salicylic acid in rice treated with chilling conditions. Although known to be potent radical scavengers, phenolic compounds may not always represent strong antioxidant capacity of plants under adverse conditions due to the lack of effective substitution of hydroxyl (-OH) groups [50]. On the other hand, plants may utilize their endogenous phenolic compounds to combat environmental stresses through various biological pathways [10,33]. For instance, phenols accumulated under the epidermal layer can protect cellular components, DNA, and important enzymes from UV damages [33]. Flavonoids are also proliferated to absorb UV radiations, thereby helping plants mitigate injuries [33]. Some compounds, such as salicylic acid, act as phytohormones contributing to the promotion of antioxidant enzymes in the response mechanism of rice plants against unfavorable conditions [10]. In addition to phenolics, this study documented a potential role of momilactones in the rice defense system since their biosyntheses were regulated under increasing levels of UV and chilling exposures. Remarkably, MA and MB biosyntheses are enhanced under UV effects, whereas they are decreased in rice treated with 6 °C exposure. There was a linear relationship between the expressions of PRBG and MRBG in rice exposed to UV and chilling conditions. Furthermore, the proliferation of MA and MB was in line with that of chlorogenic acid, esculetin, rutin, fisetin, and salicylic acid under UV treatment. In chill-treated rice, the quantity of MB was consistent with that of benzoic acid, cinnamic acid, *ρ*-coumaric acid, salicylic acid, syringic acid, quercetin, and tricin, while the contents of MA and the remaining detected compounds were reduced. Probably, momilactones might contribute to the antioxidant ability of rice plants by neutralizing the oxidative effects of free radicals. However, these compounds revealed a moderate anti-free radical ability (IC_{50} = 2.84 and 1.28 mg/mL for MA and MB, respectively against ABTS radical cations) [3]. In addition, the minor concentrations of momilactones in rice suggest that momilactones may contribute a secondary role in the total antioxidant activity of rice. Therefore, the actual role of momilactones in the antioxidant response of rice to unfavorable conditions requires further confirmation. On the other hand, the biosyntheses of momilactones were closely correlated to that of stress-resistant phenolics and flavonoids in the present research (Supplementary Materials, Tables S4 and S5). Most likely, they appear to be signaling substances inducing protection in rice against UV and chilling stresses. Thus, the potential functions of momilactones and their relationship with resistance-associated molecules in rice under stresses, including phytohormones (e.g., salicylic acid, abscisic acid, ethylene, and jasmonic acid) and functional enzymes (e.g., antioxidant enzymes and DNA damage repair enzymes), may be a promising approach that should be comprehensively scrutinized by future studies.

Our findings may contribute to future research on protecting rice cultivation. Of which, the exogenous application of interested compounds should be thoroughly evaluated to confirm their direct contributions to rice tolerance against biotic and abiotic stresses [38,51]. Preceding reports indicated the effects of salicylic acid on rice responses to UV and chilling exposures. Accordingly, rice responded positively to UV radiations with salicylic acid treatment through the enhancement of antioxidant enzymes, photosynthesis, pollen viability, leaf phenolic content, and yield [11]. In contrast, although the endogenous amount of salicylic acid was increased in rice under low temperature, the exogenous application of this substance caused negative impacts on rice tolerance against chilling conditions by increasing oxidative stress and decreasing antioxidant capacity [35,36]. On the other hand, momilactones may play an integral role in rice physiological responses to various stresses [18,19,52]. However, there has been no research on the effects of momilactone treatment due to the confined availability on the market as well as the difficulty in isolation and purification of these compounds [2]. Therefore, next investigations should focus on the contribution of exogenous application of momilactones to rice resistance against UV and chilling stresses. Moreover, because adverse conditions simultaneously activate a combination of responsive molecules, the synergistic action of compounds, such as phenolics and momilactones, is a promising field of future research.

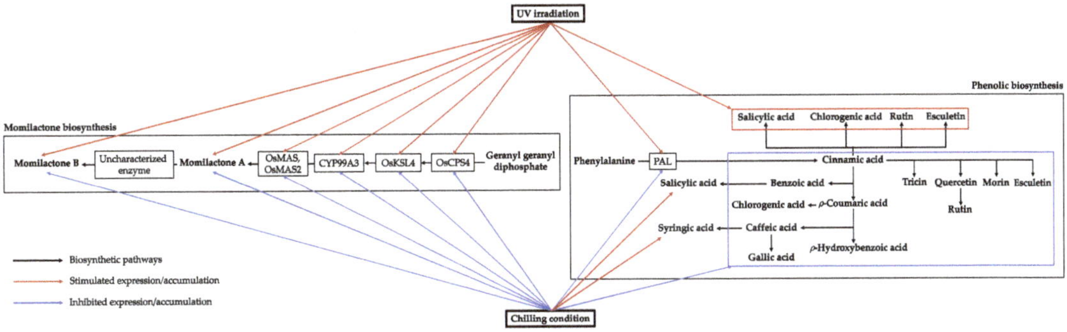

Figure 2. The effects of UV and chilling stresses on the accumulation of rice phenolics and momilactones and the expression of relevant biosynthetic genes. The biosynthetic pathways are aggregated following the schemes described by Anh et al. [23] and Shimura et al. [24]. *PAL*, gene encoding phenylalanine ammonia-lyase; *OsCPS4*, gene encoding syn-copalyl diphosphate synthase-like; *OsKSL4*, gene encoding syn-pimara-7,15-diene synthase-like; *CYP99A3*, gene encoding 9-beta-pimara-7,15-diene oxidase-like; *OsMAS*, gene encoding momilactone A synthase-like; *OsMAS2*, gene encoding momilactone A synthase like.

Furthermore, genetic engineering with the support of advanced techniques and a comprehensive genome database of rice is another effective approach to improve rice tolerance against stresses focusing on the biosynthesis of secondary metabolites such as phenolics and momilactones. Many major enzymes in the biosynthetic pathway of phenolic compounds have been described [23]. Although momilactones were identified over the last several decades to exhibit valuable biological activities, the catalytic steps in the biosynthetic pathway of these compounds remain to be least elucidated. In particular, the genes encoding the hydrolyzation of MA to MB in rice are uncharacterized [53]. Clarifying these unclear aspects may lead to new research strategies for improving rice tolerance against unfavorable conditions. Moreover, expanding research on the proliferation of secondary compounds and the relevant gene expressions will not only protect rice production from the impacts of environmental stresses, but also promise other valuable achievements. For instance, phenolics and momilactones present a potential for treating various human diseases [2–5]. Therefore, the simultaneous accumulation of phenolics and momilactones in rice is promising for medicinal and pharmaceutical purposes. This study is expected to support the sustainable development goals (SDGs) of ensuring healthy lives and ending poverty and hunger globally, especially in developing countries.

Supplementary Materials: The following supporting information can be downloaded at: https://www.mdpi.com/article/10.3390/agronomy12081731/s1, Figure S1: A boxplot for the Ct values of candidate housekeeping genes (*eIF-4A* and *actin*) in rice under UV and chilling conditions; Table S1: Primer sequences of the tested genes; Table S2: Expression (Ct value) of candidate housekeeping genes (*eIF-4A* and *actin*) in rice under UV and chilling stresses; Table S3: Relative quantification (RQ) of relevant genes to the biosynthesis of phenolics and momilactones in rice response to UV and chilling stresses; Tables S4 and S5: Pearson's correlation coefficients between antioxidant activity, chemical profiles, and relevant gene expressions of rice seed-lings under UV and chilling stresses.

Author Contributions: Conceptualization, T.D.X., N.V.Q., T.D.K. and A.T.; methodology, N.V.Q., V.Q.L., and L.H.A.; investigation, L.H.A. and V.Q.L.; data curation, L.H.A. and V.Q.L.; writing—original draft preparation, L.H.A.; writing—review and editing, T.D.X., N.V.Q., A.T., V.Q.L., T.D.K. and L.H.A.; supervision, T.D.X., N.V.Q., T.D.K. and A.T.; project administration, T.D.X., N.V.Q. and A.T. All authors have read and agreed to the published version of the manuscript.

Funding: This research received no external funding.

Institutional Review Board Statement: Not applicable.

Informed Consent Statement: Not applicable.

Data Availability Statement: Not applicable.

Conflicts of Interest: The authors declare no conflict of interest.

References

1. Wongsa, P. Phenolic compounds and potential health benefits of pigmented rice. In *Recent Advances in Rice Research*; Ansari, M.U.R., Ed.; IntechOpen: London, UK, 2021.
2. Quan, N.V.; Tran, H.-D.; Xuan, T.D.; Ahmad, A.; Dat, T.D.; Khanh, T.D.; Teschke, R. Momilactones A and B are α-amylase and α-glucosidase inhibitors. *Molecules* **2019**, *24*, 482. [CrossRef]
3. Quan, N.V.; Thien, D.D.; Khanh, T.D.; Tran, H.-D.; Xuan, T.D. Momilactones A, B, and tricin in rice grain and by-products are potential skin aging inhibitors. *Foods* **2019**, *8*, 602. [CrossRef] [PubMed]
4. Quan, N.V.; Xuan, T.D.; Tran, H.-D.; Ahmad, A.; Khanh, T.D.; Dat, T.D. Contribution of momilactones A and B to diabetes inhibitory potential of rice bran: Evidence from in vitro assays. *Saudi Pharm. J.* **2019**, *27*, 643–649. [CrossRef] [PubMed]
5. Kim, S.-J.; Park, H.-R.; Park, E.; Lee, S.-C. Cytotoxic and antitumor activity of momilactone B from rice hulls. *J. Agric. Food Chem.* **2007**, *55*, 1702–1706. [CrossRef] [PubMed]
6. Okada, A.; Shimizu, T.; Okada, K.; Kuzuyama, T.; Koga, J.; Shibuya, N.; Nojiri, H.; Yamane, H. Elicitor induced activation of the methylerythritol phosphate pathway toward phytoalexins biosynthesis in rice. *Plant Mol. Biol.* **2007**, *65*, 177–187. [CrossRef]
7. Park, C.; Jeong, N.Y.; Kim, G.-Y.; Han, M.H.; Chung, I.-M.; Kim, W.-J.; Yoo, Y.H.; Choi, Y.H. Momilactone B induces apoptosis and G1 arrest of the cell cycle in human monocytic leukemia U937 cells through downregulation of pRB phosphorylation and induction of the cyclin-dependent kinase inhibitor $p21^{Waf1/Cip1}$. *Oncol. Rep.* **2014**, *31*, 1653–1660. [CrossRef]
8. Lee, S.C.; Chung, I.-M.; Jin, Y.J.; Song, Y.S.; Seo, S.Y.; Park, B.S.; Cho, K.H.; Yoo, K.S.; Kim, T.-H.; Yee, S.-B.; et al. Momilactone B, an allelochemical of rice hulls, induces apoptosis on human lymphoma cells (Jurkat) in a micromolar concentration. *Nutr. Cancer* **2008**, *60*, 542–551. [CrossRef]
9. Singh, R.; Jwa, N.-S. Understanding the responses of rice to environmental stress using proteomics. *J. Proteome Res.* **2013**, *12*, 4652–4669. [CrossRef]
10. Hidema, J.; Kumagai, T. Sensitivity of rice to ultraviolet-B radiation. *Ann. Bot.* **2006**, *97*, 933–942. [CrossRef]
11. Khan, M.I.R.; Fatma, M.; Per, T.S.; Anjum, N.A.; Khan, N.A. Salicylic acid-induced abiotic stress tolerance and underlying mechanisms in plants. *Front. Plant Sci.* **2015**, *6*, 462. [CrossRef]
12. Apel, K.; Hirt, H. Reactive oxygen species: Metabolism, oxidative stress, and signal transduction. *Annu. Rev. Plant Biol.* **2004**, *55*, 373–399. [CrossRef] [PubMed]
13. Walter, M.; Marchesan, E. Phenolic compounds and antioxidant activity of rice. *Braz. Arch. Biol. Technol.* **2011**, *54*, 371–377. [CrossRef]
14. Minh, L.T.; Khang, D.T.; Ha, P.T.T.; Tuyen, P.T.; Minh, T.N.; Quan, N.V.; Xuan, T.D. Effects of salinity stress on growth and phenolics of rice (*Oryza sativa* L.). *Int. Lett. Nat. Sci.* **2016**, *57*, 1–10. [CrossRef]
15. Quan, N.T.; Anh, L.H.; Khang, D.T.; Tuyen, P.T.; Toan, N.P.; Minh, T.N.; Minh, L.T.; Bach, D.T.; Ha, P.T.T.; Elzaawely, A.A.; et al. Involvement of secondary metabolites in response to drought stress of rice (*Oryza sativa* L.). *Agriculture* **2016**, *6*, 23. [CrossRef]
16. Rayee, R.; Xuan, T.D.; Tran, H.-D.; Fakoori, N.A.; Khanh, T.D.; Dat, T.D. Responses of flavonoids, phenolics, and antioxidant activity in rice seedlings between Japonica and Indica subtypes to chilling stress. *Int. Lett. Nat. Sci.* **2020**, *77*, 41–50. [CrossRef]
17. Toan, N.P.; Ha, P.T.T.; Xuan, T.D. Effects of rice blast fungus (*Pyricularia grisea*) on phenolics, flavonoids, antioxidant capacity in rice (*Oryza sativa* L.). *Int. Lett. Nat. Sci.* **2017**, *61*, 1–7. [CrossRef]
18. Xuan, T.D.; Minh, T.N.; Anh, L.H.; Khanh, T.D. Allelopathic momilactones A and B are implied in rice drought and salinity tolerance, not weed resistance. *Agron. Sus. Dev.* **2016**, *36*, 52. [CrossRef]
19. Kato-Noguchi, H. Stress-induced allelopathic activity and momilactone B in rice. *Plant Growth Regul.* **2009**, *59*, 153–158. [CrossRef]
20. Kato-Noguchi, H.; Peters, R.J. The role of momilactones in rice allelopathy. *J. Chem. Ecol.* **2013**, *39*, 175–185. [CrossRef]
21. Kato-Noguchi, H.; Kujime, H.; Ino, T. UV-induced momilactone B accumulation in rice rhizosphere. *J. Plant Physiol.* **2007**, *164*, 1548–1551. [CrossRef]
22. Quan, N.V.; Xuan, T.D.; Tran, H.-D.; Thuy, N.T.D. Inhibitory activities of momilactones A, B, E, and 7-ketostigmasterol isolated from rice husk on paddy and invasive weeds. *Plants* **2019**, *8*, 159. [CrossRef] [PubMed]
23. Anh, L.H.; Quan, N.V.; Nghia, L.T.; Xuan, T.D. Phenolic allelochemicals: Achievements, limitations, and prospective approaches in weed management. *Weed Biol. Manag.* **2021**, *21*, 37–67.
24. Shimura, K.; Okada, A.; Okada, K.; Jikumaru, Y.; Ko, K.-W.; Toyomasu, T.; Sassa, T.; Hasegawa, M.; Kodama, O.; Shibuya, N.; et al. Identification of a biosynthetic gene cluster in rice for momilactones. *J. Biol. Chem.* **2007**, *282*, 34013–34018. [CrossRef]
25. Quan, N.V.; Xuan, T.D.; Tran, H.-D.; Thuy, N.T.D.; Trang, L.T.; Huong, C.T.; Andriana, Y.; Tuyen, P.T. Antioxidant, α-amylase and α-glucosidase inhibitory activities and potential constituents of *Canarium tramdenum* Bark. *Molecules* **2019**, *24*, 605. [CrossRef]

26. Anh, L.H.; Quan, N.V.; Lam, V.Q.; Iuchi, Y.; Takami, A.; Teschke, R.; Xuan, T.D. Antioxidant, anti-tyrosinase, anti-α-amylase, and cytotoxic potentials of the invasive weed *Andropogon virginicus*. *Plants* **2020**, *10*, 69. [CrossRef] [PubMed]
27. National Center for Biotechnology Information (NCBI). Available online: https://www.ncbi.nlm.nih.gov (accessed on 18 June 2021).
28. Chowdhary, V.; Alooparampil, S.; Pandya, R.V.; Tank, J.G. Physiological function of phenolic compounds in plant defense system. In *Phenolic Compounds—Chemistry, Synthesis, Diversity, Non-Conventional Industrial, Pharmaceutical and Therapeutic Applications*; Badria, F.A., Ed.; IntechOpen: London, UK, 2021.
29. Aleixandre-Tudo, J.L.; du Toit, W. The role of UV-visible spectroscopy for phenolic compounds quantification in winemaking. In *Frontiers and New Trends in the Science of Fermented Food and Beverages*; Solís-Oviedo, R.L., Pech-Canul, A.D.L.C., Eds.; IntechOpen: London, UK, 2019.
30. Anh, L.H.; Xuan, T.D.; Thuy, N.T.D.; Quan, N.V.; Trang, L.T. Antioxidant and α-amylase inhibitory activities and phytocompounds of *Clausena indica* fruits. *Medicines* **2020**, *7*, 10.
31. Cho, M.-H.; Lee, S.-W. Phenolic phytoalexins in rice: Biological functions and biosynthesis. *Int. J. Mol. Sci.* **2015**, *16*, 29120–29133. [CrossRef]
32. Caasi-Lit, M.; Whitecross, M.I.; Nayudu, M.; Tanner, G.J. UV-B irradiation induces differential leaf damage, ultrastructural changes and accumulation of specific phenolic compounds in rice cultivars. *Funct. Plant Biol.* **1997**, *24*, 261. [CrossRef]
33. Sharma, A.; Shahzad, B.; Rehman, A.; Bhardwaj, R.; Landi, M.; Zheng, B. Response of phenylpropanoid pathway and the role of polyphenols in plants under abiotic stress. *Molecules* **2019**, *24*, 2452. [CrossRef]
34. Miura, K.; Tada, Y. Regulation of water, salinity, and cold stress responses by salicylic acid. *Front. Plant Sci.* **2014**, *5*, 4. [CrossRef]
35. Wang, D.H.; Li, X.X.; Su, Z.K.; Ren, H.X. The role of salicylic acid in response of two rice cultivars to chilling stress. *Biol. Plant* **2009**, *53*, 545–552. [CrossRef]
36. Nadarajah, K.; Hamid, N.W.A.; Rahman, N.A. SA-mediated regulation and control of abiotic stress tolerance in rice. *Int. J. Mol. Sci.* **2021**, *22*, 5591. [CrossRef] [PubMed]
37. Narsai, R.; Ivanova, A.; Ng, S.; Whelan, J. Defining reference genes in *Oryza sativa* using organ, development, biotic and abiotic transcriptome datasets. *BMC Plant Biol.* **2010**, *10*, 56. [CrossRef] [PubMed]
38. Xuan, T.D.; Khang, T.D. Effects of exogenous application of protocatechuic acid and vanillic acid to chlorophylls, phenolics and antioxidant enzymes of rice (*Oryza sativa* L.) in submergence. *Molecules* **2018**, *23*, 620. [CrossRef]
39. Wankhede, D.P.; Kumar, K.; Singh, P.; Sinha, A.K. Involvement of mitogen activated protein kinase kinase 6 in UV induced transcripts accumulation of genes in phytoalexin biosynthesis in rice. *Rice* **2013**, *6*, 35. [CrossRef]
40. Jain, N.; Vergish, S.; Khurana, J.P. Validation of house-keeping genes for normalization of gene expression data during diurnal/circadian studies in rice by RT-qPCR. *Sci. Rep.* **2018**, *8*, 3203. [CrossRef]
41. Li, J.; Zhang, Z.; Xu, C.; Wang, D.; Lv, M.; Xie, H. Identification and validation of reference genes for real-time RT-PCR in *Aphelenchoides besseyi*. *Mol. Biol. Rep.* **2020**, *47*, 4485–4494. [CrossRef]
42. He, Y.; Li, X.; Zhan, F.; Xie, C.; Zu, Y.; Li, Y.; Yue, M. Resistance-related physiological response of rice leaves to the compound stress of enhanced UV-B radiation and *Magnaporthe oryzae*. *J. Plant Interact.* **2018**, *13*, 321–328. [CrossRef]
43. Hsieh, L.-S.; Yeh, C.-S.; Pan, H.-C.; Cheng, C.-Y.; Yang, C.-C.; Lee, P.-D. Cloning and expression of a phenylalanine ammonia-lyase gene (BoPAL2) from *Bambusa oldhamii* in *Escherichia coli* and *Pichia pastoris*. *Protein Expr. Purif.* **2010**, *71*, 224–230. [CrossRef]
44. Surjadinata, B.; Jacobo-Velázquez, D.; Cisneros-Zevallos, L. UVA, UVB and UVC light enhances the biosynthesis of phenolic antioxidants in fresh-cut carrot through a synergistic effect with wounding. *Molecules* **2017**, *22*, 668. [CrossRef]
45. Materska, M.; Olszówka, K.; Chilczuk, B.; Stochmal, A.; Pecio, L.; Pacholczyk-Sienicka, B.; Piacente, S.; Pizza, C.; Masullo, M. Polyphenolic profiles in lettuce (*Lactuca sativa* L.) after $CaCl_2$ treatment and cold storage. *Eur. Food Res. Technol.* **2019**, *245*, 733–744. [CrossRef]
46. Morales, L.O.; Tegelberg, R.; Brosche, M.; Keinanen, M.; Lindfors, A.; Aphalo, P.J. Effects of solar UV-A and UV-B radiation on gene expression and phenolic accumulation in *Betula pendula* leaves. *Plant Physiol.* **2010**, *30*, 923–934. [CrossRef] [PubMed]
47. Bonnecarrère, V.; Borsani, O.; Díaz, P.; Capdevielle, F.; Blanco, P.; Monza, J. Response to photoxidative stress induced by cold in japonica rice is genotype dependent. *Plant Sci.* **2011**, *180*, 726–732. [CrossRef] [PubMed]
48. Yun, K.Y.; Park, M.R.; Mohanty, B.; Herath, V.; Xu, F.; Mauleon, R.; Wijaya, E.; Bajic, V.B.; Bruskiewich, R.; de los Reyes, B.G. Transcriptional regulatory network triggered by oxidative signals configures the early response mechanisms of japonica rice to chilling stress. *BMC Plant. Biol.* **2010**, *10*, 16. [CrossRef]
49. Yildirim, A.B. Ultraviolet-B-induced changes on phenolic compounds, antioxidant capacity and HPLC profile of in vitro-grown plant materials in *Echium orientale* L. *Ind. Crops Prod.* **2020**, *153*, 112584. [CrossRef]
50. Indradi, R.B.; Fidrianny, I.; Wirasutisna, K.R. DPPH scavenging activities and phytochemical content of four Asteraceae plants. *Int. J. Pharmacogn. Phytochem. Res.* **2017**, *9*, 755–759. [CrossRef]
51. Quan, N.T.; Xuan, T.D. Foliar application of vanillic and *p*-hydroxybenzoic acids enhanced drought tolerance and formation of phytoalexin momilactones in rice. *Arch. Agron. Soil Sci.* **2018**, *64*, 1831–1846. [CrossRef]
52. Kato-Noguchi, H.; Hasegawa, M.; Ino, T.; Ota, K.; Kujime, H. Contribution of momilactone A and B to rice allelopathy. *J. Plant Physiol.* **2010**, *167*, 787–791. [CrossRef]
53. Nguyen, T.-D.; Dang, T.-T.T. Demystifying the momilactone pathway. *Nat. Chem. Biol.* **2021**, *17*, 126–128. [CrossRef]

agronomy

MDPI

Article

Salicylic Acid Improves the Salt Tolerance Capacity of *Saponaria officinalis* by Modulating Its Photosynthetic Rate, Osmoprotectants, Antioxidant Levels, and Ion Homeostasis

Lingxin Xu [1], Hong Chen [1], Tingting Zhang [1], Yanan Deng [1], Junxin Yan [1,2,*] and Lei Wang [1,2,*]

1 College of Landscape Architecture, Northeast Forestry University, Harbin 150040, China; xulingxin@nefu.edu.cn (L.X.); fjylchenhong@nefu.edu.cn (H.C.); zhangtingting@nefu.edu.cn (T.Z.); 1290079578@nefu.edu.cn (Y.D.)

2 Consulting and Design Institute, Northeast Forestry University, Harbin 150040, China

* Correspondence: yanjunxin@nefu.edu.cn (J.Y.); wanglei@nefu.edu.cn (L.W.)

Abstract: Salicylic acid (SA) plays an important role in regulating salt stress tolerance in plants. However, there are no studies on the effect of exogenous SA on *Saponaria officinalis* under salt stress. To study the effectiveness of SA on mitigating salt stress, *S. officinalis* were used in a pot experiment of salt stress simulated with an NaCl solution (100, 200, and 300 mmol L^{-1}), while an SA solution (0, 0.2, 0.4, 0.6, 0.8, 1.0 mmol L^{-1}) was sprayed on leaves. Under salt stress, spraying SA caused an increase in the salt damage index, electrolyte leakage, and a reduction in malondialdehyde and Na^+ content, but an increase in the rate of photosynthesis, chlorophyll, soluble sugar, soluble protein, free proline, K^+, Mg^{2+}, Ca^{2+} content, the K^+/Na^+ ratio, superoxide dismutase, peroxidase, catalase, ascorbate peroxidase activity, and the comprehensive score. The results show that SA improves the salt tolerance capacity of *S. officinalis* by modulating its photosynthetic rate, osmoprotectants, antioxidant levels, and ion homeostasis. However, the effectiveness of SA was not linearly related to its concentration. In summary, our findings reveal the protective roles of SA against salinity in *S. officinalis* and suggest that the use of 0.6 mmol L^{-1} of SA in salt stress conditions could be an effective approach to reduce the damage caused by saline soil in *S. officinalis*.

Keywords: antioxidase; factor analysis; ion homeostasis; osmoregulatory substance; salicylic acid; salt stress

Citation: Xu, L.; Chen, H.; Zhang, T.; Deng, Y.; Yan, J.; Wang, L. Salicylic Acid Improves the Salt Tolerance Capacity of *Saponaria officinalis* by Modulating Its Photosynthetic Rate, Osmoprotectants, Antioxidant Levels, and Ion Homeostasis. *Agronomy* **2022**, *12*, 1443. https://doi.org/10.3390/agronomy12061443

Academic Editors: Sara Álvarez and José Ramón Acosta-Motos

Received: 24 April 2022
Accepted: 13 June 2022
Published: 16 June 2022

1. Introduction

Salinization is one of the top ten threats to the world's soil resources. Salinity has affected more than 3600 Mha of soil, and the total area is expanding by around 1.5 Mha of land each year [1]. As an example, more than 70% of Northeast China is currently salinized land [2]. Improving the level of salinity in soil using physical or chemical methods is costly, of limited effectiveness, and can cause secondary soil salinity [3]. Plant repair is one of the best ways to improve the salt levels in soil. *Saponaria officinalis* is widespread in its distribution in the saline areas of northeast area of China, and has the potential to become a plant material for improved soil salinity. Additionally, the leaves of *S. officinalis* can be used to extract active surface agents, creating economic value [4]. However, salt stress may produce harmful effects on *S. officinalis* by osmotic stress, peroxidative damage, and ion toxicity [5]. This will not only affect the improvement of the saline soil, but also produce economic losses. Under such situations, it is necessary to explore ways to improve the salt tolerance capacity of *S. officinalis*.

The use of exogenous hormones to alleviate the harmful effects of salt stress on plants is providing new ideas for the use of various plants in saline soil [6]. Salicylic acid (SA) is a widely found phytohormone that stimulates various responses in plant cells under adverse conditions [7]. Researchers believe that SA is a potential growth regulator that can improve plant salt tolerance [7,8]. The positive effects of SA may stem from its involvement in activating the osmotic regulation system and antioxidant system [8,9], which reduce the adverse effects of salt stress, water deficiency, structural cell damage (disrupted chlorophyll, proteins, and lipids), disordered metabolic activity (blocked photosynthesis), and maintain ion homeostasis [10]. However, there are currently no studies on the effect of exogenous SA on *S. officinalis* under salt stress, and no reports of SA application on *S. officinalis* under salt stress. Whether or not SA can improve salt tolerance capacity and how SA improves salt tolerance capacity of *S. officinalis* are still unknown.

We speculated that SA could improve the salt tolerance capacity of *S. officinalis* by modulating the photosynthetic rate, osmoprotectants, antioxidant levels, and ion homeostasis. Thus, the following physiological indexes of *S. officinalis* were determined after SA treatment and the exposure to salt stress. The comprehensive score, salt damage index, electrolyte leakage, and malondialdehyde content were used to evaluate the degree of salt toxicity; the photosynthesis rate and chlorophyll content were used to evaluate the photosynthetic capabilities; soluble sugar, soluble protein, and free proline content were used to evaluate the accumulation of osmoprotectants; superoxide dismutase, peroxidase, catalase, and ascorbate peroxidase activity were used to evaluate the antioxidant levels; Na^+, K^+, Mg^{2+}, Ca^{2+} content and the K^+/Na^+ ratio were used to evaluate ion homeostasis. We hope that exploring the effectiveness of SA in the mitigation of the adverse effects of salinity and investigating the possible mechanisms of SA enhancement of salt tolerance in the *S. officinalis* can provide a feasible method and an effective concentration to promote the growth of *S. officinalis* under salt stress.

2. Materials and Methods

2.1. Plant Materials and Growth Conditions

The experiment was conducted in a randomized complete block design. Each group had 24 pots of two-year-old *S. officinalis*. An amount of 1000 g dray mixture of cultivated soil (N, 2.8%; P_2O_5, 1.9%; K_2O, 0.8%; organic matter, 60%; humic acid, 35%) and vermiculite at 3:1 (*v:v*) was packed in a plastic pot (170 × 123 × 153 mm). The experiment was started 30 days after *S. officinalis* were transplanted. The experiment start date was defined as day 0. Six treatments were set up (SA; 0, 0.2, 0.4, 0.6, 0.8, and 1.0 mmol L^{-1}). SA (Tianjin Yongda Chemical Reagent Co., Ltd., Tianjin, China) was dissolved with a small amount of absolute ethanol (Tianjin Yongda Chemical Reagent Co., Ltd., Tianjin, China) and diluted with distilled water to specific concentrations (final solutions contained 0.05% pure ethanol). After the SA solutions were sprayed evenly (one drop of Tween-20 was added to each liter of the solution to ensure that the SA solution was evenly attached to the leaf surface; 15 mL to each pot; Tween-20, Tianjin Yongda Chemical Reagent Co., Ltd., Tianjin, China), 250 mL of NaCl solution (100, 200, 300 mmol L^{-1}; Tianjin Yongda Chemical Reagent Co., Ltd., Tianjin, China) was injected into each pot to simulate salt stress (mild, moderate, and severe). As a control, distilled water (contained 0.05% pure ethanol; 15 mL to each pot) was sprayed on the leaf surface, and 250 mL of distilled water was injected into the pots. Leaves of *S. officinalis* were harvested on day 30 under the above-mentioned conditions to determine various physiological and biochemical attributes.

2.2. Determination of Assays and Contents

A grade scale for salt damage index (SDI) was outlined as follows [11]. Level 1: No damage; Level 2: the leaf edge and leaf tip of old leaves turn yellow and dry, while the new leaves show no symptoms; Level 3: 1/3 to 2/3 leaves show yellowing and drying symptoms; Level 4: more than 2/3 of the leaves appear yellowing and drying symptoms; Level 5: Severe damage, almost all leaves have withered or plant death occurs (Figures 1 and 2).

$$\text{SDI (\%)} = \sum (X \times N)/(5N) \times 100\%$$

Figure 1. The effects of salinity and salicylic acid on *Saponaria officinalis*.

Figure 2. The representative images of the leaves considered for calculating the salt damage index.

X (index of each damage grade), N (number of each damage grade).

Fresh samples were placed in 20 mL of deionized water in a closed vial and incubated at 25 °C for 24 h. The conductivity of the liquid was measured with a DDS-11A conductivity meter (Shanghai INESA Scientific Instrument Co., Ltd., Shanghai, China). The conductivity of the liquid was measured again after being heated at 100 °C for 20 min.

$$EL\ (\%) = (C_1/C_2) \times 100\%$$

C_1 (electrical conductivity before heating), C_2 (electrical conductivity after heating).

Fresh samples were used to determine malondialdehyde (MDA) content according to the method of thiobarbituric acid [12]. The optical density was measured with a Cary 60 UV-Vis Spectrophotometer (Agilent Technologies Inc., Santa Clara, CA, USA).

Chlorophyll (Chl) content was measured using the acetone extraction method and calculated using the optical density [12]. An amount of 0.3 g of leaves was homogenized with 10 mL of 80% pure acetone (Tianjin Yongda Chemical Reagent Co., Ltd., Tianjin, China), and the homogenate was centrifuged (4 °C, 14,000 r min^{-1}, 15 min) to obtain the supernatant solution. The absorbance at 663 and 645 nm was recorded. Photosynthesis rate (Pn) was measured with a Li-6400 Portable Photosynthesis System (LI-COR Biotechnology, Lincoln, NE, USA). The conditions in the leaf chamber included a light intensity of 1000 μmol m^{-2} s^{-1}, a leaf temperature of 25 °C, the CO_2 mole fraction of 450 μmol mol^{-1}, and a flow rate of 500 mL min^{-1}, with a relative humidity of 50–70% [13].

The quick-frozen samples were used to determinate the osmoregulatory material content. Soluble sugar (SS) content was quantified by the anthrone sulphuric acid method [14]. Soluble protein content (SP) content was quantified by the Coomassie brilliant blue method. The leaves were homogenized in 10 mL distilled water, and the supernatant was obtained after centrifugation. The supernatant was mixed with 5 mL Coomassie brilliant blue G-250

(Tianjin Yongda Chemical Reagent Co., Ltd., Tianjin, China), and the change in absorbance at 595 nm was recorded. Bovine serum protein was used to obtain the standard curve by the same method [15]. Free proline (FP) content was determined by the sulfosalicylic acid method [12].

An amount of 0.3 g leaves were homogenized with 10 mL of 0.05 mol L^{-1} precooled phosphate buffer (pH 7.8; Tianjin Yongda Chemical Reagent Co., Ltd., Tianjin, China). The homogenate was centrifuged (4 °C, 5000 r min^{-1}, 10 min) to obtain the supernatant enzyme solution. The enzyme solution (0.05 mL) was mixed with 0.05 mol L^{-1} phosphate buffer (1.5 mL), 130 mmol L^{-1} Met (0.3 mL; Sigma-Aldrich, Shanghai, China), 750 μmol L^{-1} NBT (0.3 mL; Sigma-Aldrich, Shanghai, China), 100 μmol L^{-1} EDTA-Na_2 (0.3 mL; Tianjin Yongda Chemical Reagent Co., Ltd., Tianjin, China), 2.0 μmol L^{-1} riboflavin (0.3 mL; Tianjin Yongda Chemical Reagent Co., Ltd., Tianjin, China), and distilled water (0.25 mL) to obtain a superoxide dismutase (SOD) reaction mixture. This reaction mixture was exposed to light for a color reaction, and the absorbance at 560 nm was recorded [16]. The enzyme solution (0.1 mL) was mixed with 0.05 mol L^{-1} phosphate buffer (1.5 mL), 0.05 mol L^{-1} guaiacol solution (1 mL; Tianjin Yongda Chemical Reagent Co., Ltd., Tianjin, China), and 2% hydrogen peroxide (H_2O_2; 1 mL; Tianjin Yongda Chemical Reagent Co., Ltd., Tianjin, China) to obtain a peroxidase (POD) reaction mixture. The change in absorbance at 470 nm was recorded [15]. The enzyme solution (0.1 mL) was placed in the test tube. An amount of 0.05 mol L^{-1} phosphate buffer (1 mL) was added to distilled water (1.7 mL) to obtain a catalase (CAT) reaction mixture. The mixture rested at 25 °C for 3 min. Then, 200 mM H_2O_2 (0.2 mL) was added to the test tube. The change in absorbance at 240 nm was recorded [17]. The enzyme solution (0.1 mL) was mixed with 0.05 mol L^{-1} phosphate buffer (2.7 mL), 15 mmol L^{-1} ASA (1 mL), and H_2O_2 (0.1 mL) to produce the ascorbate peroxidase (APX) reaction mixture. Then, the change in absorbance at 290 nm was recorded [18].

Na^+, K^+, Mg^{2+}, and Ca^{2+} content were determined from oven-dried leaf samples, after wet-digesting the samples in an HNO_3–$HClO_4$ acid mixture (4:1 v/v), by ICP-OES Optima 8300 (Perkins Elmer, Norwalk, CT, USA) [19].

2.3. Statistical Analysis

The statistical chart was made with Origin 2019 (OriginLab, Northampton, MA, USA). Data analysis was performed with SPSS 22.0 (Statistical Product and Service Solutions, IBM, Amonk, NY, USA). A two-way analysis and one-way analysis of variance (ANOVA) was performed, and the differences between the mean values were analyzed for significance following the Tukey method. Correlation analysis was performed by the Pearson method. Data were standardized according to the correlation analysis results. When the metric is positively correlated with the SDI, standardized data = $(X - X_{min})/(X_{max} - X_{min})$; When the metric is negatively correlated with the SDI, standardized data = $(X_{max} - X)/(X_{max} - X_{min})$. Factor analysis was performed on the standardized data.

3. Results

3.1. Salt Damage

Two-way ANOVA tests indicated that the salt data index (SDI) was affected by salt stress and SA treatment but was not affected by the interaction of these two factors, while electrolyte leakage (EL) and malondialdehyde (MDA) content were affected by salt stress and SA treatment and the interaction of these two factors (Table 1). The SDI increased with the increase of salt stress. For example, the SDI increased by 75% when compared with the control under severe salt stress (Table 2). A similar trend was also found for the EL and MDA content (Figure 3). The SDI of the SA treatment were lower than SA0 at each salt stress, but the difference was insignificant ($p > 0.05$, Table 2). However, the EL and MDA content of SA0.6 were significantly lower than for SA0 under each salt stress ($p < 0.05$, Figure 3). When compared with SA0, the EL of SA0.6 decreased by 16.33, 35.20, and 26.58% respectively, and the MDA content decreased by 21.72, 25.45, and 24.40% respectively. Notably, the EL of SA0.2 and SA1.0 was not significantly different from that of SA0

($p > 0.05$, Figure 3a). The MDA content of SA1.0 was even higher than that of SA0 under severe salt stress (Figure 3b).

3.2. Chlorophyll Content and Photosynthesis Rate

Two-way ANOVA tests indicated that the chlorophyll (Chl) content and photosynthesis rate (Pn) were affected by salt stress and SA treatments, but the interactions of these two factors were significant only for Chl (Table 1). The Chl content and Pn decreased with an increase in salt stress, and SA treatment was shown to alleviate this reduction (Figure 4). For example, the Chl of SA0.6 was increased by 14.72, 18.94, and 0.25% when compared with that of SA0 at mild, moderate, and severe salt stresses, respectively (Figure 4a). The Pn of SA0.6 was significantly higher than SA0 at mild and moderate salt stress ($p > 0.05$), but the difference was not significant between SA0.6 and SA0 at severe salt stress ($p < 0.05$, Figure 4b).

Table 1. Two-way analysis of variance (ANOVA) of the effects of salinity (S), salicylic acid (SA), and their interactions on the salt damage index, electrolyte leakage, malondialdehyde content, chlorophyll content, photosynthesis rate, soluble sugar content, soluble protein content, free proline content, superoxide dismutase activity, peroxidase activity, catalase activity, ascorbate peroxidase activity, Na^+ content, K^+ content, Mg^{2+} content, Ca^{2+} content, the and K^+/Na^+ ratio of *Saponaria officinalis.*

Source of Variation	Variable		
	Salt	SA	Salt × SA
Salt damage index (%)	120.522 ***	4.366 **	0.578 NS
Electrolyte leakage (%)	1593.293 ***	227.847 ***	32.059 ***
Malondialdehyde content (mmol g^{-1} FW)	890.767 ***	134.467 ***	17.678 ***
Chlorophyll content (mg g^{-1} FW)	1001.17 ***	34.134 ***	1.739 NS
Photosynthesis rate (μmol CO_2 m^{-2} s^{-1})	2543.288 ***	30.735 ***	2.572 *
Soluble sugar content (μmol g^{-1} FW)	62.318 ***	3.202 *	0.993 NS
Soluble protein content (mg g^{-1} FW)	302.128 ***	23.867 ***	6.927 ***
Free proline content (μg g^{-1} FW)	2218.661 ***	44.961 ***	8.897 ***
Superoxide dismutase activity (U min^{-1} g^{-1} protein)	2459.838 ***	134.081 ***	21.971 ***
Peroxidase activity (U min^{-1} g^{-1} protein)	1789.024 ***	40.996 ***	11.225 ***
Catalase activity (U s^{-1} g^{-1} protein)	284.676 ***	32.282 ***	3.409 **
Ascorbate peroxidase activity (U s^{-1} g^{-1} protein)	91.259 ***	14.435 ***	4.164 ***
Na^+ content (mg g^{-1} DW)	3929.347 ***	102.827 ***	5.435 ***
K^+ content (mg g^{-1} DW)	3178.515 ***	119.37 ***	5.187 ***
Mg^{2+} content (mg g^{-1} DW)	996.112 ***	42.766 ***	3.997 **
Ca^{2+} content (mg g^{-1} DW)	930.038 ***	27.836 ***	4.026 **
K^+/Na^+ ratio	3298.456 ***	78.773 ***	8.833 ***

The data represent F-values at the 0.05 level. *, **, *** and NS indicate significance at $p < 0.05$, $p < 0.01$, $p < 0.001$, and $p > 0.05$, respectively. FW, fresh weight; DW, dry weight.

Table 2. The effects of salinity and salicylic acid on the salt damage index of *Saponaria officinalis.*

Treatment	Salt Damage Index (%)		
	Mild Salt Stress	Moderate Salt Stress	Severe Salt Stress
Control	20.00 ± 0.00 b	20.00 ± 0.00 b	20.00 ± 0.00 b
SA0	51.67 ± 10.41 a	66.67 ± 10.41 a	80.00 ± 5.00 a
SA0.2	41.67 ± 11.55 a	65.00 ± 8.66 a	78.33 ± 7.64 a
SA0.4	33.33 ± 7.64 ab	63.33 ± 5.77 a	71.67 ± 7.64 a
SA0.6	31.67 ± 7.64 ab	58.33 ± 2.89 a	68.33 ± 5.77 a
SA0.8	35.00 ± 5.00 ab	63.33 ± 2.89 a	76.67 ± 2.89 a
SA1.0	43.33 ± 2.89 a	65.00 ± 5.00 a	78.33 ± 7.64 a

The data are presented as means ± SEs (n = 3). One-way analysis of variance (ANOVA) was used to test the differences between different treatments under mild, moderate, and severe salt stress. The differences were considered significant when $p < 0.05$, as indicated by the different letters.

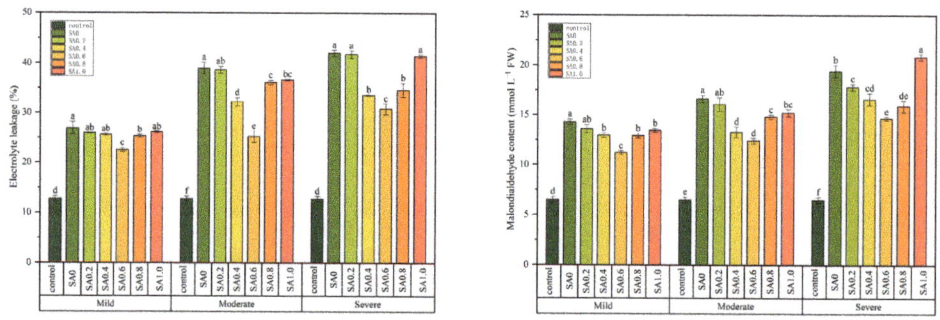

(a) (b)

Figure 3. The effects of salinity and salicylic acid on the electrolyte leakage (**a**) and malondialdehyde content (**b**) of *Saponaria officinalis*. One-way analysis of variance (ANOVA) was used to test the differences between different treatments under mild, moderate, and severe salt stress. The differences were considered significant when $p < 0.05$, as indicated by the different letters ($n = 3$). FW, fresh weight.

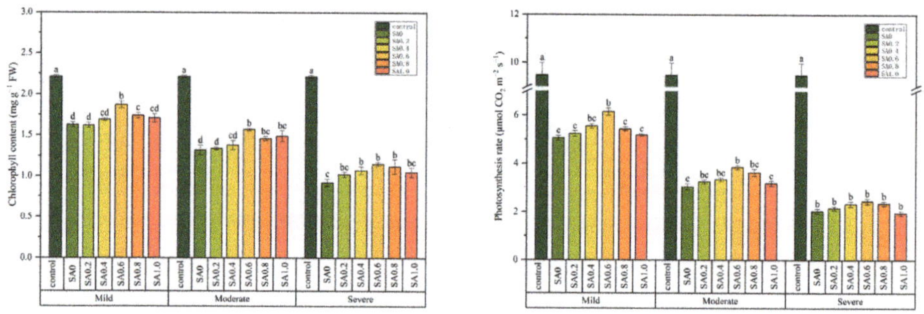

(a) (b)

Figure 4. The effects of salinity and salicylic acid on the chlorophyll content (**a**) and photosynthesis rate (**b**) of *Saponaria officinalis*. One-way analysis of variance (ANOVA) was used to test the differences between different treatments under mild, moderate, and severe salt stresses. The differences were considered significant when $p < 0.05$, as indicated by the different letters ($n = 3$). FW, fresh weight.

3.3. Osmoregulatory Material Content

Two-way ANOVA tests indicated that the soluble sugar (SS), soluble protein (SP), and free proline (FP) content were affected by salt stress and SA treatment, but the interactions of these two factors were significant only for SP and FP content (Table 1). The SS, SP and FP content increased with the increase of salt stress, and SA treatment was shown to alleviate this increase (Table 3). There were few differences between the SS content of SA treatment and that of SA0 (Table 3). However, when compared with SA0, the SP and FP content were higher. For example, the SP of SA0.6 was increased by 10.89, 5.47, and 10.94%, and the FP content was increased by 11.28, 4.66, and 8.34% when compared with SA0 (Table 3).

Table 3. The effects of salinity and salicylic acid on the soluble sugar, soluble protein, and free proline content of *Saponaria officinalis*.

Treatment		Soluble Sugar (μmol g^{-1} FW)	Soluble Protein (mg g^{-1} FW)	Free Proline (μg g^{-1} FW)
Mild salt stress	Control	124.64 ± 0.92 ^{b}	4.80 ± 0.15 ^{c}	116.88 ± 2.32 ^{d}
	SA0	157.56 ± 0.41 ^{a}	5.51 ± 0.23 ^{b}	174.14 ± 1.33 ^{c}
	SA0.2	161.50 ± 6.62 ^{a}	5.60 ± 0.08 ^{b}	184.66 ± 0.96 ^{b}
	SA0.4	163.64 ± 4.73 ^{a}	6.05 ± 0.05 ^{a}	190.91 ± 0.56 ^{a}
	SA0.6	164.63 ± 7.86 ^{a}	6.11 ± 0.25 ^{a}	193.78 ± 1.22 ^{a}
	SA0.8	159.37 ± 2.88 ^{a}	5.52 ± 0.03 ^{b}	182.60 ± 2.11 ^{b}
	SA1.0	159.22 ± 3.79 ^{a}	5.75 ± 0.08 ^{ab}	174.83 ± 1.48 ^{c}
Moderate salt stress	Control	124.64 ± 0.92 ^{b}	4.80 ± 0.15 ^{b}	116.88 ± 2.32 ^{c}
	SA0	157.56 ± 0.41 ^{a}	6.22 ± 0.20 ^{a}	210.26 ± 2.65 ^{b}
	SA0.2	161.50 ± 6.62 ^{a}	6.38 ± 0.13 ^{a}	211.37 ± 1.30 ^{b}
	SA0.4	163.64 ± 4.73 ^{a}	6.45 ± 0.02 ^{a}	216.34 ± 1.01 ^{ab}
	SA0.6	164.63 ± 7.86 ^{a}	6.56 ± 0.10 ^{a}	220.05 ± 5.16 ^{a}
	SA0.8	159.37 ± 2.88 ^{a}	6.47 ± 0.09 ^{a}	219.94 ± 0.72 ^{a}
	SA1.0	159.22 ± 3.79 ^{a}	6.36 ± 0.11 ^{a}	218.58 ± 0.52 ^{a}
Severe salt stress	Control	124.64 ± 0.92 ^{c}	4.80 ± 0.15 ^{d}	116.88 ± 2.32 ^{d}
	SA0	164.54 ± 1.36 ^{b}	6.49 ± 0.07 ^{c}	231.57 ± 2.25 ^{c}
	SA0.2	169.70 ± 1.36 ^{ab}	6.88 ± 0.15 ^{b}	234.51 ± 1.17 ^{bc}
	SA0.4	171.56 ± 4.83 ^{ab}	7.19 ± 0.05 ^{ab}	244.11 ± 2.94 ^{ab}
	SA0.6	172.49 ± 3.00 ^{ab}	7.20 ± 0.15 ^{ab}	250.88 ± 3.40 ^{a}
	SA0.8	178.55 ± 6.73 ^{a}	7.27 ± 0.20 ^{a}	251.43 ± 7.25 ^{a}
	SA1.0	168.74 ± 4.58 ^{ab}	6.46 ± 0.02 ^{c}	239.40 ± 2.39 ^{bc}

The data are presented as means ± SEs (n = 3). One-way analysis of variance (ANOVA) was used to test the differences between different treatments under mild, moderate, and severe salt stresses. The differences were considered significant when $p < 0.05$, as indicated by the different letters. FW, fresh weight.

3.4. Antioxidase Activity

Two-way ANOVA tests indicated that superoxide dismutase (SOD), peroxidase (POD), catalase (CAT), and ascorbate peroxidase (APX) activity were affected by salt stress and SA treatment and the interaction of these two factors (Table 1). The SOD, POD, CAT, and APX activity decreased with the increase in salt stress. For example, the SOD, POD, CAT, and APX activity decreased by 65.50, 68.86, 27.28, and 21.14% when compared with the control under severe salt stress, respectively (Figure 5). After application of the SA spray, the SOD, POD, CAT, and APX activity was higher than that of SA0. The SOD, POD, CAT, and APX activity of SA0.6 was significantly higher than that of SA0 at each salt stress ($p < 0.05$, Figure 5). However, under mild salt stress, the difference of SOD and POD activity between SA0.2, SA1.0, and SA0 did not reach a significant level (($p > 0.05$, Figure 5a,b).

3.5. Ion Homeostasis

Two-way ANOVA tests indicated that Na^+, K^+, Mg^{2+}, and Ca^{2+} content was affected by salt stress, SA treatment, and the interaction of these two factors (Table 1). The Na^+ content increased with the increase in salt stress, while K^+, Mg^{2+}, Ca^{2+} content, and the K^+/Na^+ ratio decreased with the increase in salt stress (Table 4). The K^+, Mg^{2+}, Ca^{2+} content, and the K^+/Na^+ ratio was higher than that of SA0, and the Na^+ content was lower than that of SA0 after SA application. The Na^+, K^+, Mg^{2+}, Ca^{2+} content, and K^+/Na^+ ratio of SA0.6 were significantly different from that of SA0 at each salt stress ($p < 0.05$, Table 4). However, under severe salt stress, the difference of Na^+, Mg^{2+}, Ca^{2+} content, and the K^+/Na^+ ratio between SA0.2, SA1.0, and SA0 did not reach a significant level ($p > 0.05$, Table 4).

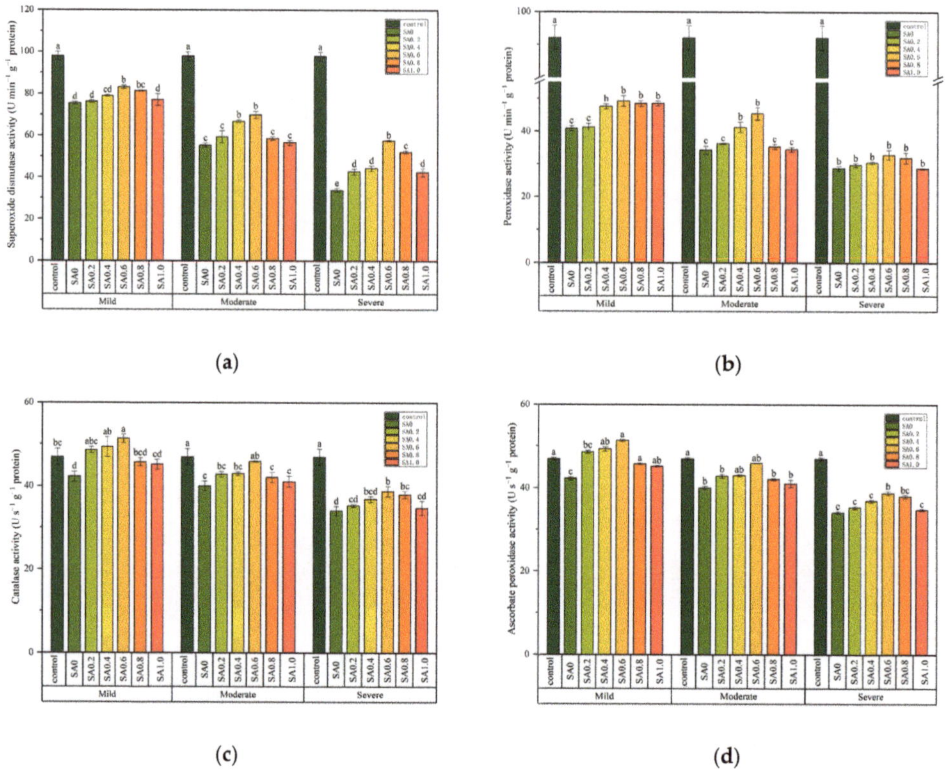

Figure 5. The effects of salinity and salicylic acid on superoxide dismutase activity (**a**), Peroxidase activity (**b**), catalase activity (**c**), ascorbate peroxidase activity (**d**) of *Saponaria officinalis*. One-way analysis of variance (ANOVA) was used to test the differences between different treatments under mild, moderate, and severe salt stresses. The differences were considered significant when $p < 0.05$, as indicated by the different letters ($n = 3$).

Table 4. The effects of salinity and salicylic acid on Na^+ content, K^+ content, Mg^{2+} content, Ca^{2+} content, and the K^+/Na^+ ratio of *Saponaria officinalis*.

Treatment		Na^+ Content (mg g^{-1} DW)	K^+ Content (mg g^{-1} DW)	Mg^+ Content (mg g^{-1} DW)	Ca^+ Content (mg g^{-1} DW)	K^+/Na^+ Ratio
Moderate salt stress	Control	5.461 ± 0.243 ^{e}	23.101 ± 0.166 ^{a}	9.919 ± 0.192 ^{a}	12.651 ± 0.186 ^{a}	4.237 ± 0.220 ^{a}
	SA0	12.286 ± 0.081 ^{a}	17.896 ± 0.058 ^{f}	5.904 ± 0.173 ^{e}	9.473 ± 0.123 ^{d}	1.457 ± 0.005 ^{d}
	SA0.2	11.545 ± 0.205 ^{b}	18.202 ± 0.159 ^{ef}	6.658 ± 0.095 ^{cd}	9.593 ± 0.288 ^{d}	1.577 ± 0.03 ^{cd}
	SA0.4	11.098 ± 0.273 ^{bc}	19.074 ± 0.141 ^{bc}	7.138 ± 0.34 ^{bc}	10.170 ± 0.291 ^{bc}	1.719 ± 0.038 ^{bc}
	SA0.6	9.992 ± 0.116 ^{d}	19.298 ± 0.028 ^{b}	7.217 ± 0.048 ^{b}	10.500 ± 0.027 ^{b}	1.932 ± 0.024 ^{b}
	SA0.8	10.978 ± 0.082 ^{c}	18.802 ± 0.119 ^{cd}	6.916 ± 0.072 ^{bcd}	9.793 ± 0.182 ^{cd}	1.713 ± 0.012 ^{bc}
	SA1.0	11.435 ± 0.140 ^{bc}	18.519 ± 0.323 ^{de}	6.428 ± 0.128 ^{d}	9.549 ± 0.170 ^{d}	1.619 ± 0.009 ^{cd}
Mild salt stress	Control	5.461 ± 0.243 ^{d}	23.101 ± 0.166 ^{a}	9.919 ± 0.192 ^{a}	12.651 ± 0.186 ^{a}	4.237 ± 0.220 ^{a}
	SA0	9.443 ± 0.169 ^{a}	20.230 ± 0.111 ^{f}	7.912 ± 0.205 ^{e}	10.458 ± 0.125 ^{c}	2.143 ± 0.027 ^{f}
	SA0.2	8.742 ± 0.109 ^{b}	21.520 ± 0.169 ^{d}	8.531 ± 0.301 ^{d}	11.411 ± 0.257 ^{b}	2.462 ± 0.040 ^{de}
	SA0.4	7.935 ± 0.050 ^{c}	22.102 ± 0.075 ^{c}	9.163 ± 0.019 ^{bc}	11.678 ± 0.175 ^{b}	2.786 ± 0.019 ^{bc}
	SA0.6	7.840 ± 0.216 ^{c}	22.614 ± 0.088 ^{b}	9.779 ± 0.292 ^{a}	11.716 ± 0.409 ^{b}	2.886 ± 0.083 ^{b}
	SA0.8	8.187 ± 0.091 ^{c}	21.239 ± 0.140 ^{de}	9.324 ± 0.274 ^{ab}	11.685 ± 0.108 ^{b}	2.594 ± 0.012 ^{cd}
	SA1.0	9.026 ± 0.263 ^{ab}	20.973 ± 0.289 ^{e}	8.582 ± 0.051 ^{cd}	11.492 ± 0.145 ^{b}	2.324 ± 0.055 ^{ef}

Table 4. *Cont.*

Treatment		Na^+ Content (mg g^{-1} DW)	K^+ Content (mg g^{-1} DW)	Mg^+ Content (mg g^{-1} DW)	Ca^+ Content (mg g^{-1} DW)	K^+/Na^+ Ratio
Severe salt stress	Control	5.461 ± 0.243 ^{c}	23.101 ± 0.166 ^{a}	9.919 ± 0.192 ^{a}	12.651 ± 0.186 ^{a}	4.237 ± 0.220 ^{a}
	SA0	15.055 ± 0.331 ^{a}	14.918 ± 0.105 ^{e}	5.044 ± 0.337 ^{c}	8.033 ± 0.122 ^{c}	0.991 ± 0.025 ^{c}
	SA0.2	14.672 ± 0.281 ^{a}	15.683 ± 0.206 ^{d}	5.531 ± 0.421 ^{bc}	8.322 ± 0.092 ^{bc}	1.069 ± 0.012 ^{bc}
	SA0.4	13.897 ± 0.123 ^{b}	16.343 ± 0.182 ^{c}	5.362 ± 0.084 ^{bc}	8.543 ± 0.141 ^{b}	1.176 ± 0.008 ^{bc}
	SA0.6	13.545 ± 0.177 ^{b}	17.081 ± 0.238 ^{b}	5.974 ± 0.121 ^{b}	8.662 ± 0.126 ^{b}	1.261 ± 0.015 ^{b}
	SA0.8	14.692 ± 0.086 ^{a}	16.131 ± 0.151 ^{cd}	5.452 ± 0.058 ^{bc}	8.532 ± 0.234 ^{b}	1.098 ± 0.015 ^{bc}
	SA1.0	14.953 ± 0.224 ^{a}	15.743 ± 0.385 ^{cd}	5.311 ± 0.230 ^{bc}	8.071 ± 0.182 ^{c}	1.053 ± 0.040 ^{bc}

The data are presented as means ± SEs (n = 3). One-way analysis of variance (ANOVA) was used to test the differences between different treatments under mild, moderate, and severe salt stresses. The differences were considered significant when $p < 0.05$, as indicated by the different letters. DW, dry weight.

3.6. Factor Analysis

We speculate that those plants with more osmolators, a higher antioxidant enzyme activity, and a stable ionic concentration suffer less from salinity under equivalent salt stress. However, we could not accurately evaluate the degree of salt damage in plants by a single indicator. For example, the SDI differences among the individual treatment groups did not reach significant levels under salt stress (Table 2). In addition, under severe salt stress, in contrast to the SA0, SA1.0 had less SDI, but a higher MDA content, less SP content, but a higher SOD activity (Tables 2 and 3, Figures 3 and 5). Therefore, we considered a comprehensive evaluation of the salt damage of each treatment group by factor analysis. The results of the factor analysis indicate that there was no significant difference between SA1.0 and SA0 under severe salt stress (p> 0.05, Table 5). Additionally, SA0.2, SA0.4, SA0.6, SA0.8, and SA1.0 were significantly different from the SA0 (p <0.05) under the three kinds (mild, moderate, and severe) of salt stress (Table 5). Moreover, the combined score of SA0.6 was the smallest among all three conditions (Table 5).

Table 5. Comprehensive score of *Saponaria officinalis* in different treatment groups.

Treatment	Comprehensive Score		
	Mild Salt Stress	Moderate Salt Stress	Severe Salt Stress
Control	−0.809 ± 0.085 ^{bc}	−0.809 ± 0.085 ^{d}	−0.809 ± 0.085 ^{e}
SA0	−0.085 ± 0.155 ^{a}	0.803 ± 0.113 ^{a}	1.681 ± 0.064 ^{a}
SA0.2	−0.731 ± 0.063 ^{b}	0.264 ± 0.215 ^{b}	1.16 ± 0.1 ^{bc}
SA0.4	−1.219 ± 0.301 ^{c}	−0.238 ± 0.095 ^{c}	0.759 ± 0.162 ^{c}
SA0.6	−1.733 ± 0.221 ^{d}	−0.708 ± 0.16 ^{d}	0.032 ± 0.231 ^{d}
SA0.8	−0.941 ± 0.072 ^{bc}	0.242 ± 0.092 ^{b}	0.336 ± 0.093 ^{d}
SA1.0	−0.665 ± 0.065 ^{b}	0.426 ± 0.102 ^{b}	1.427 ± 0.195 ^{ab}

The data are presented as means ± SEs. One-way analysis of variance (ANOVA) was used to test the differences between different treatments. The differences were considered significant when $p < 0.05$, as indicated by the different letters.

4. Discussion

Based on the extrinsic morphology of plants, a grade scale for SDI can be used to simply describe the salt damage in plants, while Pn also reflects plant physiological activities to some extent. The results achieved in this work showed that salinity makes the SDI of *S. officinalis* increase while SA (0.4, 0.6, 0.8 mmol L^{-1}) treatment makes the SDI stay at relatively low levels (Table 2). The opposite trend was observed for Pn (Figure 4).The increasing of SDI and the decrease of Pn can be attributed to cellular oxidative damage and Chl decomposition (Figures 3 and 4). EL and the cellular peroxidation product MDA can reflect the extent of oxidative damage in plant cells [20]. Changes in Chl content are associated with visual symptoms and photosynthetic efficiency changes during the disease incidence [21]. SA may improve the salt tolerance capacity of *S. officinalis* by alleviating cellular oxidative damage and avoiding Chl breakdown, since the *S. officinalis* treated with SA had a lower EL, lower MDA content, and higher Chl content (Figures 3 and 4). Similarly, SA (100 μmol L^{-1}) reduced the MDA content in *Oryza sativa* under salt conditions (200 mmol L^{-1}) [5]. SA leaf treatment (0.5 and 1.0 mM) reduced the electrolyte leakage in *Ocimum basilicum* under different levels of NaCl stress (60, 120 mmol L^{-1}) [22]. SA treatment significantly alleviated the decrease of photosynthetic pigment content in *Trigonella foenum-graecum* caused by salt stress [23]. Meanwhile, the application of 2 mmol L^{-1} SA to the leaves of *Raphanus sativus* significantly enhanced the levels of photosynthetic pigments under salt stress [12]. SA also promoted the Pn of *Brassica juncea* [24] and cucumber (*Cucumis sativus* cv. 'Zhongnong 26') [25] under salt stress. It seems no doubt that SA can alleviate oxidative damage, avoid Chl breakdown, and promote photosynthesis. Therefore, SA may improve the salt tolerance capacity of *S. officinalis* by modulating the photosynthetic rate.

SA may mitigate the oxidative damage of *S. officinalis* by maintaining high antioxidant enzyme activity (Figure 5). *S. officinalis* sprayed with SA (0.4, 0.6 mmol L^{-1}) also has higher SOD, POD, CAT, and FP activity when compared with *S. officinalis* that is sprayed with distilled water (Figure 5). In *Brassica parachinensis* under salt stress, treatment with 1 mmol L^{-1} SA enhanced the SOD, CAT, and APX activity [26]. Additionally, SA treatment increased the SOD and POD activity and reduced lipid peroxidation levels, subsequently reducing the toxicity of salt stress in *Cucumis sativus* [27]. Faghih et al. found that SA enhanced the APX, POD, and SOD activity of *Lycopersicon esculentum* under salt stress [28]. It is worth mentioning that protective effect of SA on Chl may be achieved by enhancing the activity of APX, since APX is present in the chloroplasts (Figures 4 and 5). Therefore, SA may improve the salt tolerance capacity of *S. officinalis* by modulating antioxidant levels.

Salinity affects plant growth and development through osmotic stress, while plants resist salt stress by accumulating osmotic regulatory substances. In this study, *S. officinalis* leaves accumulated a lot of Na^+ under salt stress (Tables 3 and 4). Although the accumulation of Na^+ is able to increase the cellular water potential, excessive Na^+ can cause ion toxicity in plants [13]. We found that the K^+ content and K^+/Na^+ ratio of *S. officinalis* were decreased (Table 4), and the reduction in $^{K+}$ caused by Na^+ is a well-known competitive process in plant tissues [29]. Furthermore, we found that salt stress decreased the Mg^{2+} and Ca^{2+} contents in *S. officinalis* (Table 4). SA may maintain the ionic balance within plant leaves by limiting the excessive accumulation of Na^+ (Table 4). SA promotes the absorption of K^+ and Ca^{2+} because SA has an inhibitory effect on the excess accumulation of Na^+, alleviating the competitive effect of Na^+ on K^+ [29], and keeping the excess Na^+ from replacing Ca^{2+} on the membrane binding site [30]. However, the promoting effects of SA on Mg^{2+} need to be further studied. Studies on *Mentha pulegium* revealed that 1 mmol L^{-1} SA reduced the accumulation of Na^+ in the leaves under salt stress, promoted the accumulation of K^+ in leaves, and improved the K^+/Na^+ ratio [31]. It also found that SA contrasted the massive entry of Na^+, in favor of K^+, Ca^{2+}, and Mg^{2+} accumulation in maize [32]. The protective effect of SA on Chl may be also related to promoting the accumulation of Mg^{2+}, since Mg^{2+} is an important constituent element of Chl (Table 4). Therefore, SA may improve the salt tolerance capacity of *S. officinalis* by modulating ion homeostasis.

In addition to inorganic salt ions, organic matter such as SS and SP and FP have important roles in plant osmotic regulation. In this study, *S. officinalis* sprayed with SA (0.4, 0.6, 0.8 mmol L^{-1}) also had higher SS, SP, and FP content when compared with the *S. officinalis* that was sprayed with distilled water, but the difference in the SS content was not significant (Table 3). The results show that for *S. officinalis*, SA mainly alleviates salt damage by regulating its accumulation of FP and SP. Similarly, 0.5 mmol L^{-1} SA increased the FP content of lentils under 100 mmol L^{-1} NaCl stress [33]. However, Studies in *O. basilicum* showed that foliar treatment with SA at 0.5 and 1.0 mmol L^{-1} concentrations under saline soil conditions (60 and 120 mM) increased the SP content but decreased the FP content [20]. The above phenomenon shows that, under salt stress, SA may alleviate osmotic damage by inducing plants to accumulate more organic matter. However, the change rules of different osmotic substances are different according to different plant species. SA may act via the accumulation of osmosis modulators, which is achieved by regulating the activity of certain enzymes. For example, SA promotes the activity of sucrose synthase and enhances the content of soluble sugars in plant organs [34]. Meanwhile, SA leaf spray improved pyrroline-5-carboxylic acid reductase activity while it inhibited the proline oxidase activity in Mentha pulegium under saline soil conditions [35]. Since FP can protect Rubisco (the key enzyme that determines carbon assimilation rate in photosynthesis), promoting the accumulation of FP is also one of the reasons why SA promotes Pn under saline stress [35]. In summary, SA may improve the salt tolerance capacity of *S. officinalis* by modulating osmoprotectants.

The effects of SA on plant physiological and biochemical properties vary depending on the experimental conditions, such as the intensity and stress duration and SA application method [7]. Some scholars have suggested that the effects of SA depend on the application dose; low concentrations of SA reduce plant damage while high concentrations exacerbate plant damage [7]. The absorption and utilization of SA varied based on the plant species, genotype, developmental stage, and stress strength [35]. It was found in this study that high concentrations of SA (1.0 mmol L^{-1}) had little effect on the antioxidant enzymes of *S. officinalis* (Figure 5), whereas 0.1 mol L^{-1} SA slightly promoted APX and SOD activity and decreased the POX activity of Strawberry [28]. Moderate concentrations of SA (0.6 mmol L^{-1}) were more effective in alleviating the harmful effects on *S. officinalis* under salt stress, while higher (1.0 mmol L^{-1}) concentrations of SA were not as effective in alleviating harmful effects on *S. officinalis* under saline stress (Table 5). The positive effect of SA was not enhanced with its increased concentration.

5. Conclusions

Exogenous SA can effectively alleviate the harmful effect of salt stress on *S. officinalis*, improving the osmotic stress, oxidative damage, and ion poisoning caused by salinity. (1) SA promotes the growth of *S. officinalis* by modulating the photosynthetic rate; (2) SA enhances the osmoregulatory ability by promoting SP and FP synthesis in *S. officinalis*; (3) SA enhances the antioxidant capacity by maintaining SOD, POD, CAT, and APX activity; (4) The positive effect of SA was not enhanced with increasing concentrations. Moderate concentrations of SA (0.6 mmol L^{-1}) are more effective in alleviating the harmful effects on *S. officinalis* under salt stress. Thus, the external administration of SA can be used as a feasible method to improve plant salt tolerance, but the concentration of SA needs to be standardized based on the degree of salt damage and the plant type.

Author Contributions: Data curation, H.C., T.Z. and Y.D.; Writing—original draft, L.X.; Writing—review & editing, J.Y. and L.W. All authors have read and agreed to the published version of the manuscript.

Funding: This research was funded by the National Key R&D program of China under grant (2021YFD1500600); Fundamental Research Funds for the Central Universities under grant (2572021BK01); National Natural Science Foundation of China under grant (31800546).

Institutional Review Board Statement: Not applicable.

Informed Consent Statement: Not applicable.

Data Availability Statement: Not applicable.

Conflicts of Interest: The authors declare no conflict of interest.

References

1. Rahman, M.M.; Mostofa, M.G.; Keya, S.S.; Siddiqui, M.N.; Ansary, M.M.U.; Das, A.K.; Rahman, M.A.; Tran, L.S.-P. Adaptive Mechanisms of Halophytes and Their Potential in Improving Salinity Tolerance in Plants. *Int. J. Mol. Sci.* **2021**, *22*, 10733. [CrossRef] [PubMed]
2. Wang, Y.; Jie, W.; Peng, X.; Hua, X.; Yan, X.; Zhou, Z.; Lin, J. Physiological Adaptive Strategies of Oil Seed Crop Ricinus communis Early Seedlings (Cotyledon vs. True Leaf) Under Salt and Alkali Stresses: From the Growth, Photosynthesis and Chlorophyll Fluorescence. *Front. Plant Sci.* **2019**, *9*, 1939. [CrossRef] [PubMed]

3. Kulak, M.; Gul, F.; Sekeroglu, N. Changes in growth parameter and essential oil composition of sage (*Salvia officinalis* L.) leaves in response to various salt stresses. *Ind. Crops Prod.* **2020**, *145*, 112078. [CrossRef]
4. Smulek, W.; Zdarta, A.; Pacholak, A.; Zgola-Grzeskowiak, A.; Marczak, L.; Jarzebski, M.; Kaczorek, E. *Saponaria officinalis* L. extract: Surface active properties and impact on environmental bacterial strains. *Colloids Surf. B-Biointerfaces* **2017**, *150*, 209–215. [CrossRef]
5. Mostofa, M.G.; Fujita, M.; Tran, L.-S.P. Nitric oxide mediates hydrogen peroxide- and salicylic acid-induced salt tolerance in rice (*Oryza sativa* L.) seedlings. *Plant Growth Regul.* **2015**, *77*, 265–277. [CrossRef]
6. Mostofa, M.G.; Fujita, M. Salicylic acid alleviates copper toxicity in rice (*Oryza sativa* L.) seedlings by up-regulating antioxidative and glyoxalase systems. *Ecotoxicology* **2013**, *22*, 959–973. [CrossRef]
7. Horvath, E.; Szalai, G.; Janda, T. Induction of abiotic stress tolerance by salicylic acid signaling. *J. Plant Growth Regul.* **2007**, *26*, 290–300. [CrossRef]
8. Hayat, Q.; Hayat, S.; Irfan, M.; Ahmad, A. Effect of exogenous salicylic acid under changing environment: A review. *Environ. Exp. Bot.* **2010**, *68*, 14–25. [CrossRef]
9. Abdoli, S.; Ghassemi-Golezani, K.; Alizadeh-Salteh, S. Responses of ajowan (Trachyspermum ammi L.) to exogenous salicylic acid and iron oxide nanoparticles under salt stress. *Environ. Sci. Pollut. Res.* **2020**, *27*, 36939–36953. [CrossRef]
10. Oueslati, S.; Karray-Bouraoui, N.; Attia, H.; Rabhi, M.; Ksouri, R.; Lachaal, M. Physiological and antioxidant responses of *Mentha pulegium* (Pennyroyal) to salt stress. *Acta Physiol. Plant.* **2010**, *32*, 289–296. [CrossRef]
11. Zhong, Y.-p.; Qi, X.-j.; Chen, J.-y.; Li, Z.; Bai, D.-f.; Wei, C.-g.; Fang, J.-b. Growth and physiological responses of four kiwifruit genotypes to salt stress and resistance evaluation. *J. Integr. Agric.* **2019**, *18*, 83–95. [CrossRef]
12. Bukhat, S.; Manzoor, H.; Athar, H.-u.-R.; Zafar, Z.U.; Azeem, F.; Rasul, S. Salicylic Acid Induced Photosynthetic Adaptability of Raphanus sativus to Salt Stress is Associated with Antioxidant Capacity. *J. Plant Growth Regul.* **2020**, *39*, 809–822. [CrossRef]
13. Wang, Y.; Xu, Y.; Peng, X.; Yan, J.; Yan, X.; Zhou, Z.; Lin, J. Cotyledon removal decreases salt tolerance during seedling establishment of Ricinus communis, an oilseed energy crop species. *Ind. Crops Prod.* **2019**, *142*, 111857. [CrossRef]
14. Bai, J.-h.; Liu, J.-h.; Zhang, N.; Yang, J.-h.; Sa, R.-l.; Wu, L. Effect of Alkali Stress on Soluble Sugar, Antioxidant Enzymes and Yield of Oat. *J. Integr. Agric.* **2013**, *12*, 1441–1449. [CrossRef]
15. Tuna, A.L.; Kaya, C.; Dikilitas, M.; Yokas, I.; Burun, B.; Altunlu, H. Comparative effects of various salicylic acid derivatives on key growth parameters and some enzyme activities in salinity stressed maize (*Zea mays* L.) plants. *Pak. J. Bot.* **2007**, *39*, 787–798. [CrossRef]
16. Giannopolitis, C.N.; Ries, S.K. Superoxide dismutases: I. Occurrence in higher plants. *Plant Physiol.* **1977**, *59*, 309–314. [CrossRef]
17. Seckin, B.; Sekmen, A.H.; Turkan, I. An Enhancing Effect of Exogenous Mannitol on the Antioxidant Enzyme Activities in Roots of Wheat Under Salt Stress. *J. Plant Growth Regul.* **2009**, *28*, 12–20. [CrossRef]
18. Nakano, Y.N.; Asada, K. Hydrogen Peroxide is Scavenged by Ascorbate-specific Peroxidase in Spinach Chloroplasts. *Plant Cell Physiol.* **1981**, *22*, 867–880. [CrossRef]
19. Souana, K.; Taibi, K.; Abderrahim, L.A.; Amirat, M.; Achir, M.; Boussaid, M.; Mulet, J.M. Salt-tolerance in *Vicia faba* L. is mitigated by the capacity of salicylic acid to improve photosynthesis and antioxidant response. *Sci. Hortic.* **2020**, *273*, 109641. [CrossRef]
20. Gunes, A.; Inal, A.; Alpaslan, M.; Eraslan, F.; Bagci, E.G.; Cicek, N. Salicylic acid induced changes on some physiological parameters symptomatic for oxidative stress and mineral nutrition in maize (*Zea mays* L.) grown under salinity. *J. Plant Physiol.* **2007**, *164*, 728–736. [CrossRef]
21. Selem, E.E.; Hamed, R.E.A.; Kamel, H.A.; Hegazy, H.S. Physiological and Biochemical Response of Gamma Irradiated *Sesamum indicum* L. Seed Grown in Heavy Metal Contaminated Soil. *Biosci. Res.* **2018**, *15*, 1063–1072.
22. Elhindi, K.M.; Al-Amri, S.M.; Abdel-Salam, E.M.; Al-Suhaibani, N.A. Effectiveness of salicylic acid in mitigating salt-induced adverse effects on different physio-biochemical attributes in sweet basil (*Ocimum basilicum* L.). *J. Plant Nutr.* **2017**, *40*, 908–919. [CrossRef]
23. Abdelhameed, R.E.; Abdel Latef, A.A.H.; Shehata, R.S. Physiological Responses of Salinized Fenugreek (*Trigonella foenum-graecum* L.) Plants to Foliar Application of Salicylic Acid. *Plants* **2021**, *10*, 657. [CrossRef]
24. Nazar, R.; Umar, S.; Khan, N.A. Exogenous salicylic acid improves photosynthesis and growth through increase in ascorbate-glutathione metabolism and S assimilation in mustard under salt stress. *Plant Signal. Behav.* **2015**, *10*, e1003751. [CrossRef]
25. Miao, Y.; Luo, X.; Gao, X.; Wang, W.; Li, B.; Hou, L. Exogenous salicylic acid alleviates salt stress by improving leaf photosynthesis and root system architecture in cucumber seedlings. *Sci. Hortic.* **2020**, *272*, 109577. [CrossRef]
26. Kamran, M.; Xie, K.; Sun, J.; Wang, D.; Shi, C.; Lu, Y.; Gu, W.; Xu, P. Modulation of growth performance and coordinated induction of ascorbate-glutathione and methylglyoxal detoxification systems by salicylic acid mitigates salt toxicity in choysum (*Brassica parachinensis* L.). *Ecotoxicol. Environ. Saf.* **2020**, *188*, 109877. [CrossRef]
27. Shi, Q.; Zhu, Z. Effects of exogenous salicylic acid on manganese toxicity, element contents and antioxidative system in cucumber. *Environ. Exp. Bot.* **2008**, *63*, 317–326. [CrossRef]
28. Faghih, S.; Ghobadi, C.; Zarei, A. Response of Strawberry Plant cv. 'Camarosa' to Salicylic Acid and Methyl Jasmonate Application Under Salt Stress Condition. *J. Plant Growth Regul.* **2017**, *36*, 651–659. [CrossRef]
29. Botella, M.A.; Martinez, V.; Pardines, J.; Cerda, A. Salinity induced potassium deficiency in maize plants. *J. Plant Physiol.* **1997**, *150*, 200–205. [CrossRef]

30. Cachorro, P.; Ortiz, A.; Cerda, A. Implications of Calcium Nutrition on the Response of *Phaseolus vulgaris* L. to Salinity. *Plant Soil* **1994**, *159*, 205–212. [CrossRef]
31. Ghassemi-Golezani, K.; Farhadi, N. The Efficacy of Salicylic Acid Levels on Photosynthetic Activity, Growth, and Essential Oil Content and Composition of Pennyroyal Plants Under Salt Stress. *J. Plant Growth Regul.* **2021**. [CrossRef]
32. Sultan, I.; Khan, I.; Chattha, M.U.; Hassan, M.U.; Barbanti, L.; Calone, R.; Ali, M.; Majid, S.; Ghani, M.A.; Batool, M.; et al. Improved salinity tolerance in early growth stage of maize through salicylic acid foliar application. *Ital. J. Agron.* **2021**, *16*, 1810. [CrossRef]
33. Misra, N.; Saxena, P. Effect of salicylic acid on proline metabolism in lentil grown under salinity stress. *Plant Sci.* **2009**, *177*, 181–189. [CrossRef]
34. Sha, H.; Liu, H.; Wang, J.; Jia, Y.; Wang, X.; Zou, D.; Zhao, H. Physiological mechanism of salicylic acid regulating salt tolerance of crops. *J. Northeast. Agric. Univ.* **2017**, *48*, 80–88. [CrossRef]
35. Farhadi, N.; Ghassemi-Golezani, K. Physiological changes of Mentha pulegium in response to exogenous salicylic acid under salinity. *Sci. Hortic.* **2020**, *267*, 109325. [CrossRef]

agronomy

MDPI

Article

Physiological, Metabolic and Transcriptional Responses of Basil (*Ocimum basilicum* Linn. var. *pilosum* (Willd.) Benth.) to Heat Stress

Lei Qin [1,2,†], Chengyuan Li [1,3,†], Dongbin Li [4], Jiayan Wang [1], Li Yang [1], Aili Qu [1,*] and Qingfei Wu [1,*]

1 School of Biological and Chemical Engineering, NingboTech University, Ningbo 315100, China; qinlei2012shandong@163.com (L.Q.); lcysigoto@163.com (C.L.); wangjiayan107@163.com (J.W.); yangli@nit.net.cn (L.Y.)
2 College of Agronomic Sciences, Shandong Agricultural University, Tai'an 271018, China
3 College of Chemical and Biological Engineering, Zhejiang University, Hangzhou 310058, China
4 Ningbo Forest Farm, Ningbo 315440, China; comeonldb@163.com
* Correspondence: quaili@nit.zju.edu.cn (A.Q.); feiqw1234@163.com (Q.W.)
† These authors contributed equally to this work.

Abstract: As a medicinal and edible plant, basil (*Ocimum basilicum* Linn. var. *pilosum* (Willd.) Benth.) has rich nutrition and significant economic value. The increase in heat stress caused by global warming adversely affects the growth and yield of plants. However, the response mechanism of basil to heat stress is poorly understood. This work investigated the changes in phenotype, metabolome, and transcriptome in basil under heat stress. The results showed that heat stress triggered severe oxidative damage and photosynthesis inhibition in basil. Metabonomic analysis showed that, compared to the control group, 29 significantly differentially accumulated metabolites (DAMs) were identified after 1 d of heat treatment, and 37 DAMs after the treatment of 3 d. The DAMs were significantly enriched by several pathways such as glycolysis or gluconeogenesis; aminoacyl-tRNA biosynthesis; and alanine, aspartate, and glutamate metabolism. In addition, transcriptomic analysis revealed that 15,066 and 15,445 genes were differentially expressed after 1 d and 3 d of heat treatment, respectively. Among them, 11,183 differentially expressed genes (DEGs) were common response genes under 1 d and 3 d heat treatment, including 5437 down-regulated DEGs and 6746 up-regulated DEGs. All DEGs were significantly enriched in various KEGG (Kyoto Encyclopedia of Genes and Genomes) pathways, most dominated by glyoxylate and dicarboxylate metabolism, followed by starch and sucrose metabolism, and by the biosynthesis and metabolism of other secondary metabolites. Overall, all the above results provided some valuable insights into the molecular mechanism of basil in response to heat stress.

Keywords: basil; heat stress; metabolome; transcriptome; molecular mechanism

Citation: Qin, L.; Li, C.; Li, D.; Wang, J.; Yang, L.; Qu, A.; Wu, Q. Physiological, Metabolic and Transcriptional Responses of Basil (*Ocimum basilicum* Linn. var. *pilosum* (Willd.) Benth.) to Heat Stress. *Agronomy* **2022**, *12*, 1434. https://doi.org/10.3390/agronomy12061434

Academic Editors: Sara Álvarez and José Ramón Acosta-Motos

Received: 26 May 2022
Accepted: 14 June 2022
Published: 15 June 2022

1. Introduction

Ocimum basilicum (basil) is an annual aromatic plant belonging to the Lamiaceae family, widely cultivated in many countries, and has significant economic value [1]. Many important compounds in basil have been isolated and identified, such as volatile oils, and various polyphenols and flavonoids with antioxidant, anti-inflammatory, and other pharmacological activities [2–4]. Therefore, basil plants are often used as seasoned vegetables in cooking and as ingredients in some medical products industry [5]. The herb *Ocimum basilicum* Linn. var. *pilosum* (Willd.) Benth., as a variety of *Ocimum basilicum*, is widely distributed throughout China [6,7], contains many kinds of essential oils, and has been used as medicine to treat some diseases in folk prescriptions [7]. To the best of our knowledge, however, very few studies have focused on the heat tolerance of this plant.

As one of the common and serious abiotic interferences, heat stress imposes negative effects on the normal growth of plants, and damages agricultural production, which is becoming a worldwide problem [8,9]. It is estimated that the global air temperature will rise by about 0.2 °C per decade and thus increase another 1.8–4 °C by the year 2100 [10–12], so the world has to face more severe challenges of heat stress. By limiting various physiological and biochemical responses, heat stress could adversely impact plant development. In general, heat stress reduces photosynthetic efficiency and injures photochemical reactions and carbon metabolism, thereby diminishing the plant productivity and shortening the life cycle [13–15]. The functions and integrity of cell membranes are also highly sensitive to heat stress, and cellular membrane permeability and cellular electrolytes would be increased under heat stress, leading to membrane instability [16–19]. In addition, heat stress could induce oxidative stress to produce excessive reactive oxygen species (ROS) in plant cells, while ROS formation would give rise to various negative effects, such as lipid peroxidation, nucleic acid damage, enzyme inactivation, and protein oxidation [20]. All these factors would degenerate physiological and biochemical processes of a plant under heat stress, limit plant growth, reduce its yield, and even cause its death.

On the other hand, plants have evolved a series of strategies in adaption to heat stress. There are two ways to mitigate deleterious effects due to heat stress. One is to prevent heat stress by reducing the absorption of solar radiation; the other is to enhance plant tolerance to heat stress [11]. Generally, plants can avoid short-term heat stress through transpirational cooling, altering membrane lipid compositions, or changing leaf orientation. For the plants grown in hot climates, small hairs (tomentose) have always been found to form a thick coat on the surface of the leaf and cuticles, contributing to reduce the absorption of solar radiation. Moreover, plants can deal with heat stress by regulating a lot of physiological processes. For instance, plants under heat stress would accumulate many more osmotic protection substances including proline, polyols, and soluble carbohydrates, which could lower the osmotic potential of cells and attenuate the stress effects [21]. In addition, plants have formed a defense system of antioxidation to scavenge excessive ROS, which contain nonenzymatic antioxidants, such as ascorbic acid (vitamin C), phenolic compounds, flavonoids, carotenoids, and other secondary metabolites, as well as enzymatic antioxidants including superoxide dismutase (SOD), peroxidase (POD), and catalase (CAT) [22]. Furthermore, the responses of a plant to heat stress are also mediated by its own hormones, including abscisic acid (ABA), jasmonic acid (JA), ethylene (ET), cytokinins (CKs), brassinosteroids (BRs), and salicylic acid (SA) [23], and these hormones participate in responses to heat stress mainly through triggering stress-adaptive signaling cascades [24,25].

At the molecular level, it has been reported that a large number of genes such as heat shock protein (HSP) genes, osmosensors, mitogen-activated protein kinase (MAPK) cascades, transcription factors, and second messengers would participate in the plant response to heat stress [26,27]. Currently, RNA-sequencing (RNA-Seq) is becoming a fast and effective way to uncover gene expression profiles, so it has been widely used to explore stress response mechanisms. A series of studies have revealed basil's response to high temperatures. For example, when subjected to a short-term heat stress, the contents of phenolic compounds and activities of antioxidant enzymes were remarkably increased in *Ocimum basilicum* L. cv. 'Genovese' [28]. However, when 'Genovese' was grown under heat stress and elevated CO_2 concentrations for a long time, the content of phenolic acid in the leaves decreased, the wax decreased, and the photosynthesis increased [9]. Due to basil displaying a significant number of varieties with different genetic backgrounds, their resistance to heat stress may also be different, and the molecular mechanism of basil's response to heat stress also needs further investigation. In this study, the changes in physiological features and metabolites in basil (*Ocimum basilicum* Linn. var. *pilosum* (Willd.) Benth.) under heat stress were investigated and analyzed, and RNA-Seq was used to identify differentially expressed unigenes. The results will be helpful for cloning and analyzing heat-resistant genes in basil, theoretically providing some insights for molecular breeding in the future.

2. Materials and Methods

2.1. Plant Materials and Treatment

The seeds of *Ocimum basilicum* Linn. var. *pilosum* (Willd.) Benth. were provided by Zhejiang Academy of Agricultural Science (Hangzhou, China). Seeds were sown in plastic pots filled with turfy soil, which was rich in incompletely decomposed plant residues, humus, and some minerals, and the pots were placed in an artificial illumination incubator. The cultivation conditions were set as follows: a photoperiod of 12 h light (30 °C)/12 h dark (24 °C), air relative humidity of 60%, and an illumination of 300 μmol m^{-2} s^{-1}. To ensure the healthy and normal growth of plants, half-strength Hoagland nutrient solution was irrigated daily. For heat treatment, the seedlings of eight weeks were exposed to a day/night temperature of 42/36 °C, and the leaves were collected as heat-stressed samples at 0, 1, 3, and 5 d after treatment. For each treatment, three biological replicates were performed, and each biological replicate contained 15 individual plants. Then, the collected samples were immediately frozen in liquid nitrogen and stored at −80 °C before use. To ensure the consistency of the experiment, the leaves of each sample were taken from the top to bottom of the third and fourth leaves, respectively.

2.2. Physiological Analysis

After heat treatment with different times (0, 1, 3, and 5 d), the contents of oxidizing substances including malonaldehyde (MDA), H_2O_2, and O_2^- were determined according to the instructions of reagent kits (Beijing Solarbio Science and Technology, Beijing, China). In brief, for MDA content measurement, 0.1 g of leaf tissue was weighed and 1 mL of extract buffer was added for ice bath homogenization. Centrifuging at 8000× *g* for 10 min at 4 °C, the supernatant was taken and placed on ice for testing. Finally, the absorbance of the supernatant at 532 nm (A532) and 600 nm (A600) was measured, and MDA content = 32.258 × (ΔA532–ΔA600) ÷ weight. The H_2O_2 content was calculated based on the absorbance of the supernatant at 415 nm, and the content = Δ Asample ÷ Δ Astandard ÷ weight. As the purple-red azo compound has a characteristic absorption peak at 530 nm, the O_2^- content in the sample can be calculated according to the absorbance at 530 nm (A530) and standard curve. Photosynthetic parameters were measured using a portable photosynthesis system (Licor-6400, LI-COR, Lincoln, NE, USA), and all measurements were performed at a photon flux density of 1000 μmol m^{-2} s^{-1}, CO_2 concentration of 400 μmol mol^{-1}, and leaf temperature of 30 °C. Chlorophyll fluorescence parameters were measured using a Dual-PAM 100 chlorophyll fluorescence analyzer (Heinz Walz, Effeltrich, Germany) by the manufacturer's instructions. Before measurement, plants need to be dark-adapted for at least half an hour. The measurement was repeated six times and averaged.

2.3. Extraction and Detection of Metabolites

The extraction of metabolites was conducted following a previous method [29]. In detail, the sampled leaves were quickly ground into powder in a mortar cooling with liquid nitrogen, and the accurately weighed 50 mg powder was put into a 2 mL centrifuge tube. Next, 0.5 mL of the mixed solution at acetonitrile: isopropanol: water (3:3:2, *v*/*v*/*v*) and several zirconium beads of 2 mm were added in turn into the tube, which was placed into the high-throughput tissue grinding instrument, shaken at 30 Hz for 20 s, and allowed to stand for 10 s. This was repeated 8 times and then sonicated in an ice-water bath for 5 min. Once again, 0.5 mL of the mixed solution of acetonitrile: isopropanol: water (3:3:2, *v*/*v*/*v*) was added and then sonicated in an ice-water bath for 5 min. After centrifugation at 14,000 rpm for 2 min, the supernatant solution of 500 μL was pipetted into a new 2 mL tube and concentrated to dryness with a vacuum concentrator for 8–10 h, while the rest of the supernatant was placed into a refrigerator at −80 °C for backup. The as-concentrated sample was redissolved with 80 μL of methoxypyridine solution of 20 mg/mL concentration, and followed by 30 s of vortex vibration and 60 min of incubation at 60 °C. Sequentially, 100 μL of BSTFA-TMCS (99:1) derivatization reagent was added, vortexed for 30 s, incubated at 70 °C for 90 min, and centrifuged at 14,000 rpm for 3 min, and the supernatant of 100 μL was put into the detection bottle. The final samples were temporarily stored in a sealed cup for testing, and GC-TOF detection would be completed within 24 h.

Samples were analyzed by gas chromatography with a DB-5MS capillary column (Agilent, Santa Clara, CA, USA). In the analysis, the constant flow of helium was at 1 mL/min. A 1 μL sample was injected through the autosampler in a split ratio of 1:10, helium was used as the carrier gas at constant flow rate of 1 mL/min, the inlet temperature was 280 °C, and the transfer line and ion source temperatures were 320 °C and 230 °C, respectively. The heating program initially started at 50 °C for 0.5 min, increased to 320 °C at a rate of 15 °C/min, and remained at 320 °C for 9 min. Mass spectrometry was carried out with a full scope at a scan rate of 10 spec/s, an electron energy of −70 V, and a solvent delay of 3 min.

2.4. Metabolomics Analysis

The raw data of mass spectrometry were converted into net CDF format by Agilent MSD Chem Station for peak identification, peak filtering, and peak alignment, and the data matrix comprised the mass to charge ratio (m/z), retention time, and intensity. Metabolites were annotated in conjunction with the AMDIS program using the National Institute of Standards and Technology (NIST) commercial database and Wiley Registry metabolome database. Among them, the alkane retention index was used for further substance characterization according to the retention index provided by The Golm Metabolome Database (GMD) (http://gmd.mpimp-golm.mpg.de/ (12 October 2021). Then, the metabolomic data were analyzed using the MetaboAnalystR package in R language [30].

2.5. cDNA Library Construction and Sequencing

The leaves of each sample were fully ground into powder under liquid nitrogen cooling. Then, about 0.2 g of powder was weighed for total RNA extraction using the RNAprep Pure Plant Plus Kit (Tiangen, Beijing, China). The RNA purity and concentration were measured using an Agilent 2100 Bioanalyzer (Agilent, Santa Clara, CA, USA). Approximately 1.5 μg of RNA per sample was used to construct the sequencing library according to the previously described method [31]. Then, the library preparations were sequenced to generate 150 bp paired-end reads using the Illumina HiSeq 4000 platform (Illumina, San Diego, CA, USA). The raw sequencing data were uploaded to the NCBI BioProject database, given a project number of PRJNA833588.

2.6. Transcriptomic Analysis

The raw data obtained by sequencing included a small number of reads with sequencing adapters or with low sequencing quality. In order to ensure the quality and reliability of data analysis, the original data should be filtered, and the adaptor contamination and low-quality reads were removed using Trimmomatic software (version 0.38). After removing unclear reads (unknown nucleotides > 5%) and low-quality reads (Q-value < 20), a clean sequence was obtained. The clean reads were assembled de novo using Trinity software, and the redundancies in the assembled transcript sequences were removed using CD-HIT-EST software with the global sequence identity threshold cut-off setting at 90%.

Uni genes were functionally annotated with alignments to the KOG, NT, R, Uniprotand Swiss-Prot databases. The DESeq R package (1.10.1) was used to analyze the differentially expressed genes (DEGs) with an adjusted p-value of <0.05. GO annotation and KEGG annotation were completed by a Python script and the GhostKOALA website (https://www.kegg.jp/ghostkoala/ (accessed on 9 November 2021), respectively. GO and KEGG enrichment analysis was performed by TBtools and Mapman [32].

2.7. Quantitative Real-Time PCR (qRT-PCR) Analysis

First-strand cDNA was synthesized from 1 μg of RNA using the PrimeScript RT reagent Kit with gDNA Eraser (Takara, Dalian, China). qRT-PCR analyses were conducted using SYBR® Premix Ex Taq™ II (Takara, Dalian, China) on an ABI Prism 7700 (Applied Biosystems, Foster City, CA, USA). The PCR procedures went through at 95 °C for 10 s, followed by 40 cycles at 95 °C for 5 s, and then at 60 °C for 30 s. The gene-specific primers were obtained in Primer Premier 5.0, and all primer sequences are listed in Table S1. The comparative CT method ($2^{-\Delta\Delta CT}$ method) was used to quantify the relative expression of genes, and the actin gene was selected as an internal control to normalize the data. Each experiment was performed in three technical replicates.

3. Results

3.1. Physiological Responses to Heat Stress

In order to explore the effect of heat stress on basil growth, after treating at high temperature (42/36 °C, light/dark, 12 h/12 h) with different times, the phenotypes of basil were analyzed. As shown in Figure 1, the plants as the control group (CK) grew well and the leaves unfurled (Figure 1a). The leaves appeared slightly curly and saggy after 1 d of heat treatment (Figure 1b), became more curly and drooping after 3 d of heat treatment (Figure 1c), and the leaves as well as the plants were severely withered to approach death after 5 d of heat treatment (Figure 1d). In addition, the heat-induced oxidation damage to basil growth was investigated. It was found that compared with the CK, the MDA content as an indicator of cell membrane damage significantly increased by 1.3-fold after 1 d of heat treatment, and increased continuously with the extension of treatment time (Figure 2a). Similarly, the ROS content including H_2O_2 and O_2^- was increased dramatically with increasing heat treatment time (Figure 2b,c), and subsequent experiments only analyzed plants that had been heat-treated for 1 d and 3 d, as the plants were close to death after 5 d of heat treatment.

Figure 1. Phenotype of basil under heat stress. (**a**) Phenotype of control group (CK), (**b**) phenotype of high-temperature treatment for 1 d, (**c**) phenotype of high-temperature treatment for 3 d, (**d**) phenotype of high temperature treatment for 5 d.

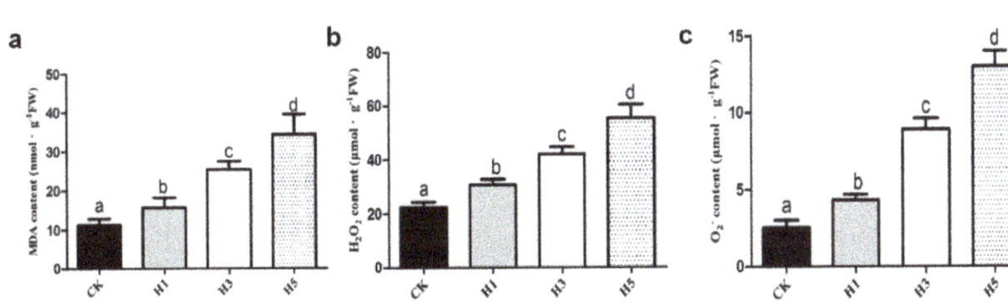

Figure 2. Determination of oxidizing substance content. (**a**) MDA content, (**b**) H_2O_2 content, (**c**) O_2 content. CK, control group; H1, group with heat stress treatment for 1 d; H3, group with heat stress treatment for 3 d; H5, group with heat stress treatment for 5 d.

Moreover, after being exposed to heat stress for 1 d and 3 d, the photosynthetic capacity and photochemical activity of basil seedlings were also measured. The net photosynthetic rate (Pn) and the maximum photochemical efficiency of PSII (Fv/Fm) showed no significant difference compared with the CK after 1 d of treatment, but dramatically declined after 3 d of treatment (Figure 3a,b). Furthermore, under heat stress, the value of nonphotochemical quenching (NPQ) indicated that the change in plant heat dissipation capacity increased gradually with treatment time, while the value of photochemical quenching coefficient (qP) indicated that the openness of the PSII active center gradually decreased with time (Figure 3c,d).

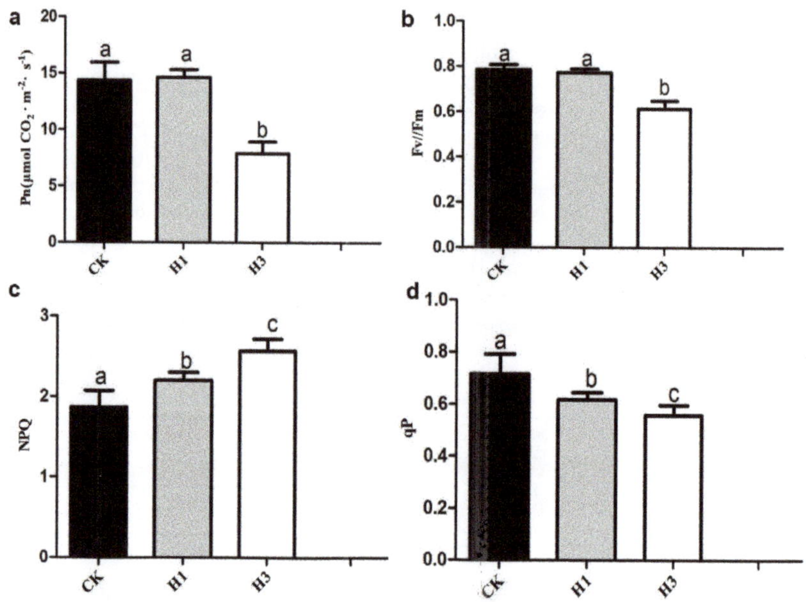

Figure 3. Measurement of photosynthetic parameters and chlorophyll fluorescence parameters under heat stress. (**a**) Pn, the net photosynthetic rate, (**b**) Fv/Fm, the maximum photochemical efficiency of PSII, (**c**) NPQ, the nonphotochemical quenching, (**d**) qP, the photochemical quenching coefficient. CK, control group; H1, group with heat stress treatment for 1 d; H3, group with heat stress treatment for 3 d.

3.2. *Metabolomics Analysis of the Leaves under Heat Stress*

As an aromatic plant, basil comprises various volatiles with important medicinal activities. To further understand the changes in volatiles under heat stress, the metabolites in basil leaves after 1 d and 3 d of heat treatment were analyzed by GC–MS technology, and a total of 222 metabolites were identified, including various nucleic acids, organic acids, peptides, and steroids (Table S2, Figure 4a). Principal component analysis (PCA) showed the internal structures of several variables related to different principal components. In the PCA score plot (Figure 4b), all biological replicates were clustered together, indicating a good correlation among core samples within groups, and also verifying the data reliability. In addition, there was an obvious distinction between the CK and the heat-stressed groups, where the latter almost clustered together and thus possessed similar metabolic profiles and phenotypes, which were significantly different from the former. Overall, the above results revealed that heat stress could significantly influence the metabolism of volatile substances in basil.

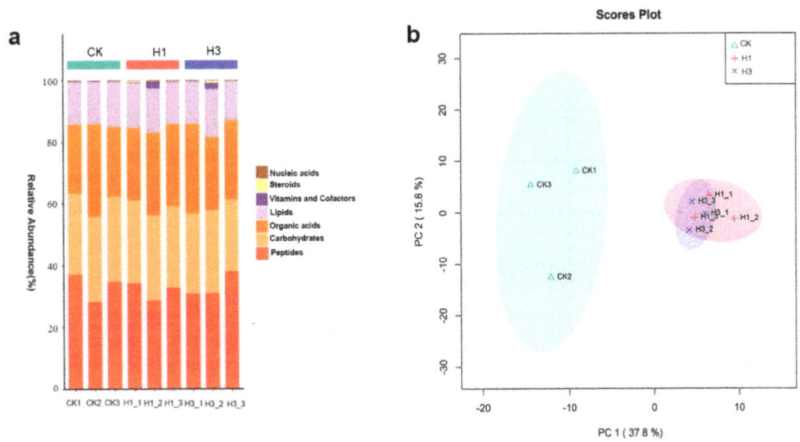

Figure 4. Statistical analysis of metabolome data. (**a**) Classification of identified metabolites, (**b**) principal component analysis (PCA) score plot. CK, control group; H1, group with heat stress treatment for 1 d; H3, group with heat stress treatment for 3 d.

3.3. Differentially Accumulated Metabolites (DAMs) Induced by Heat Stress

Orthogonal projections to latent structures discriminant analysis (OPLS-DA) as a method of multivariate statistical analysis with monitoring pattern recognition can effectively eliminate irrelevant effects on the research, and scientifically screen different metabolites. As shown in Figure 5, distinct metabolic differentiations were observed between the CK and treatment 1 (H1), between the CK and treatment 2 (H3), and between treatment 1 (H1) and treatment 2 (H3). All Q2 values in the three models were greater than 0.5, suggesting that all models were adequately reliable. Subsequently, the fold change (FC) and variable importance in project (VIP) values of the OPLS-DA model were combined to further screen the DAMs. By setting FC $\geq$ 2 or $\leq$ 0.5 together with VIP $\geq$ 1.0 as thresholds for significant differences, the significantly DAMs of 29 between CK and H1 (H1 had 16 up-regulated and 13 down-regulated), 37 between CK and H3 (H3 had 17 up-regulated and 20 down-regulated), and 9 between H1 and H3 (H1 had 5 up-regulated and 4 down-regulated) were determined, respectively, as presented in Table S3. Furthermore, Venn diagram analysis revealed that although no common DAMs were found in each comparison group, 18 DAMs were observed in CK-H1 and CK-H3 (Figure S1, Table S3), indicating that these metabolites may play crucial functions in response to heat stress.

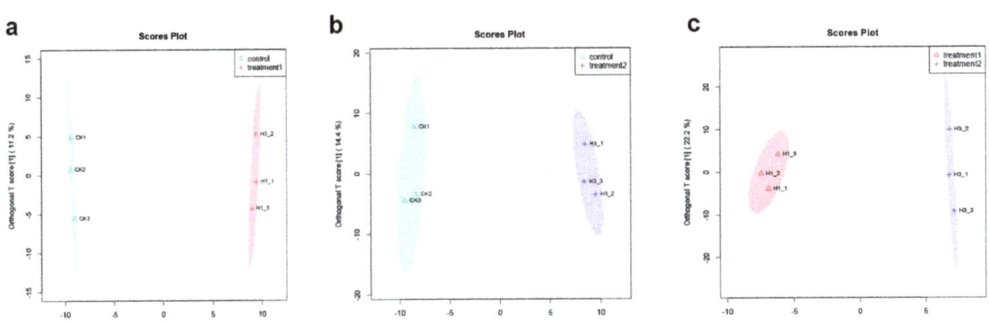

Figure 5. The score plots of orthogonal projections to latent structures discriminant analysis (OPLS-DA) pairwise comparisons of differentially accumulated metabolites (DAMs). (**a**) Comparison groups of CK vs. H1, (**b**) comparison groups of CK vs. H3, (**c**) comparison groups of H1 vs. H3. CK, control group; H1, group with heat stress treatment for 1 d; H3, group with heat stress treatment for 3 d.

3.4. Pathway Enrichment Analysis of DAMs under Heat Stress

Based on the Kyoto Encyclopedia of Genes and Genomes (KEGG) database, DAMs from each comparison group were mapped to metabolic pathways. As displayed in Figure 6, KEGG pathway enrichment analysis showed that DAMs in the CK-H1 pair were mainly enriched in alanine, aspartate, and glutamate metabolism; glycolysis or gluconeogenesis; tyrosine metabolism; amino sugar and nucleotide sugar metabolism; and another 3 pathways. DAMs in the CK-H3 pair were mainly enriched in 7 pathways including aminoacyl-tRNA biosynthesis; alanine, aspartate, and glutamate metabolism; phenylalanine, tyrosine, and tryptophan biosynthesis; and nitrogen metabolism. DAMs in the H1-H3 comparative pair were enriched in the citrate cycle (TCA cycle); pantothenate and CoA biosynthesis; and valine, leucine, and isoleucine biosynthesis. All these enriched pathways in the above were associated with the heat tolerance of basil.

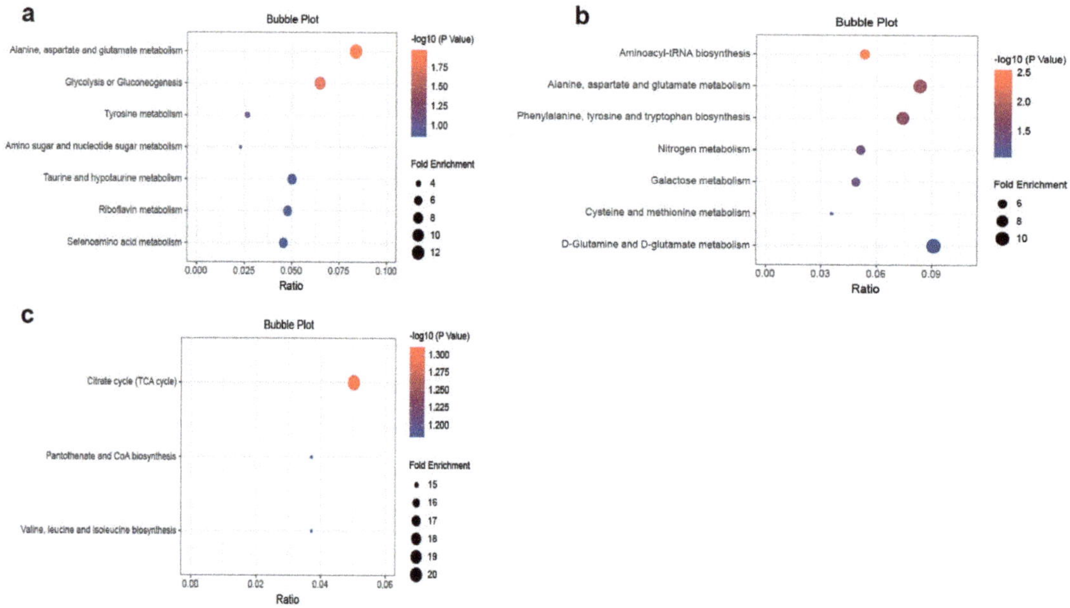

Figure 6. Koto Encyclopedia of Genes and Genomes (KEGG) pathway classification of comparison groups. (**a**) Comparison groups of CK vs. H1, (**b**) comparison groups of CK vs. H3, (**c**) comparison groups of H1 vs. H3. CK, control group; H1, group with heat stress treatment for 1 d; H3, group with heat stress treatment for 3 d.

3.5. Transcriptomic Analysis of Basil Responses to Heat Stress

To dissect the molecular response of basil to heat stress, the global gene expression profiles were detected by RNA-seq. A total of 9 cDNA libraries were constructed from CK, H1, and H3 groups. Sequencing analysis revealed that about 145.2 Gb of reads were generated. The clean reads were assembled de novo using Trinity software, and the redundancies in the assembled transcript sequences were removed using CD-HIT-EST software. In total, 205,848 unigenes were obtained with an average length of 419 bp and an N50 length of 977 bp. Whereafter, all sequences of the assembled unigenes were searched in the KOG, NT, NR, Uniprot, and Swiss-Prot databases and then annotated using the BLASTX program. In order to obtain the reliable profiles of gene expressions, the transcripts per million (TPM) values were used to quantify gene level. Differentially expressed genes (DEGs) were selected to annotate the differential expression between the CK and the heat-stressed groups based on the filter criteria of FC (a ratio of stress/control) values ≥ 2 and *p*-value < 0.05. In the result, a total of 19,400 DEGs were identified from the three comparison groups (WT-H1, WT-H3, and H1-H3), including 15,066, 15,445, and 192 DEGs in WT-H1, WT-H3, and H1-H3 pairs, respectively (Table S4). Among them, 11,183 DEGs were common response genes after heat stress of 1 d and 3 d, including 5437 down-regulated DEGs and 5746 up-regulated DEGs (Figure 7a,b). Interestingly, only 17 of 11,183 common DEGs showed a significant difference in the H1-H3 pair (Figure 7a). In addition, all DEGs were classified into 6 sub-groups with different numbers using K-means cluster analysis (Figure S2), indicating that DEGs could respond to heat stress in specific ways. All above results showed significant differences in gene expression between the CK and a heat-stressed group, while the differences among the heat-stressed groups were indistinctive.

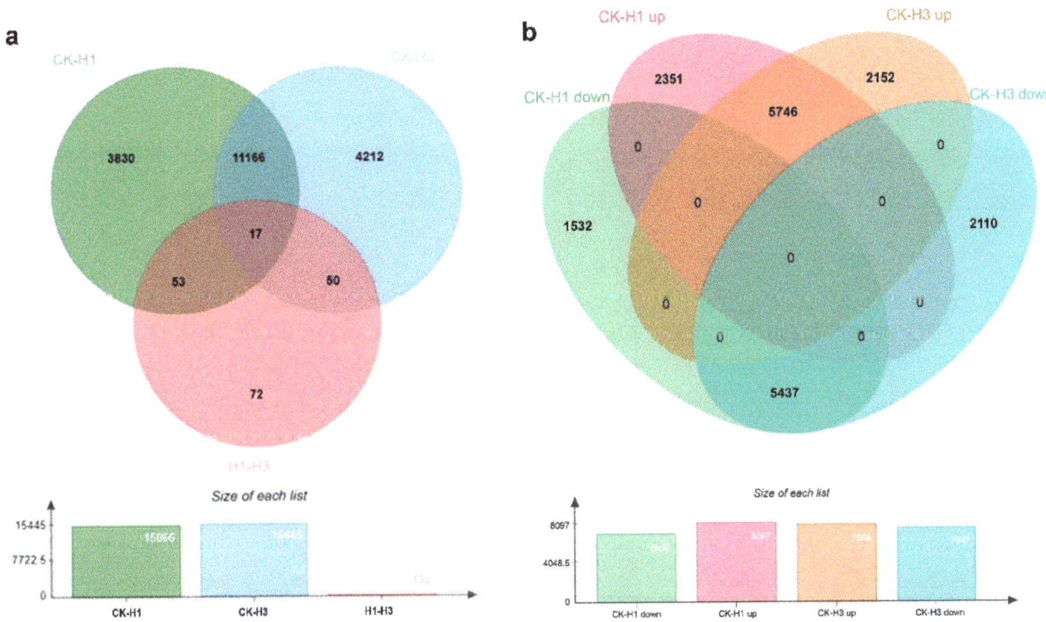

Figure 7. Venn diagram analysis of differentially expressed genes (DEGs). (**a**) Venn diagram showing the overlapping and unique DEGs amongst the comparison groups, (**b**) Venn diagram showing the up-regulated and down-regulated DEGs in CK vs. H1 and CK vs. H3. CK, control group; H1, group with heat stress treatment for 1 d; H3, group with heat stress treatment for 3 d.

3.6. GO and KEGG Analysis of DEGs under Heat Stress

The DEGs were further classified and characterized according to the functional category defined by GO and KEGG. To identify various GO terms from the significantly enriched DEGs, the function of DEGs was analyzed using TBtools software after heat stress. All 19,400 DEGs were classified into three main GO groups including molecular function, cellular component, and biological process. The 30 most enriched GO terms were shown in Figure 8, where the DEGs annotated in the molecular function group included glyceraldehyde-3-phosphate dehydrogenase (NAD+) (phosphorylating) activity, glyceraldehyde-3-phosphate dehydrogenase (NAD(P) +) (phosphorylating) activity, disordered domain specific binding, catalytic activity and hydrolase activity, and hydrolyzing O-glycosyl compounds; the DEGs in the cellular component consisted of 20 GO terms such as stromule, plastid membrane, plastid thylakoid, cell wall, chloroplast, and another 15 GO terms; and the DEGs in the biological process involved the small-molecule biosynthetic process, carbohydrate metabolic process, carbohydrate catabolic process, glucan catabolic process, and polysaccharide metabolic process.

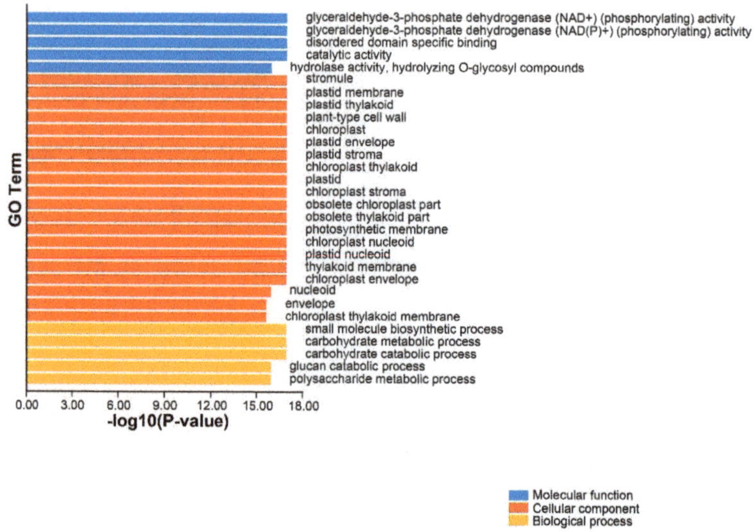

Figure 8. GO enrichment analysis of all differentially expressed genes (DEGs) in basil under heat stress.

To further uncover the effect of heat stress on the biological pathway and signal transduction pathway, all DEGs were mapped to the KEGG database. As a result, the 30 most enriched KEGG pathways were identified and are shown in Figure 9, of which glyoxylate and dicarboxylate metabolism was the most dominant pathway, followed by starch and sucrose metabolism, the biosynthesis of other secondary metabolites, and metabolism. In addition, based on the Mapman results (Figure 10), most of the DEGs were significantly down-regulated in photosynthetic pathways relating to lipids, ascorbate, glutathione, starch, and sucrose, indicating that these metabolic pathways were suppressed by heat stress, while most of the DEGs were significantly up-regulated in the pathways relating to terpenes, phenylpropanoids, phenolics, and nucleotides, suggesting that these metabolic pathways were enhanced by heat stress.

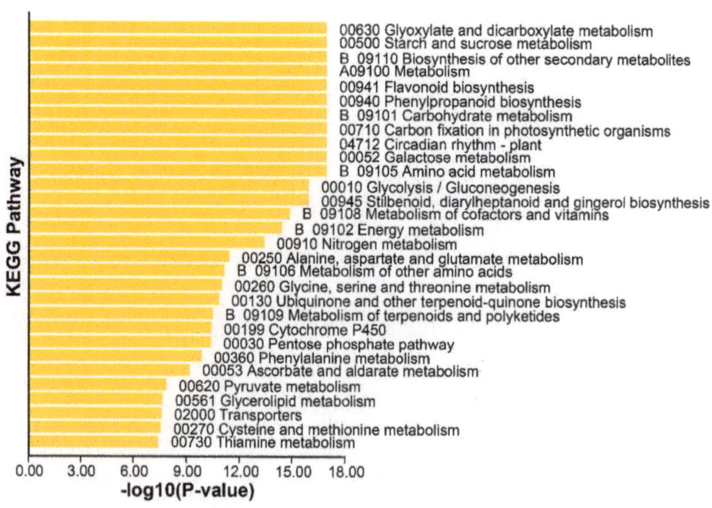

Figure 9. KEGG enrichment analysis of all differentially expressed genes (DEGs) in basil under heat stress.

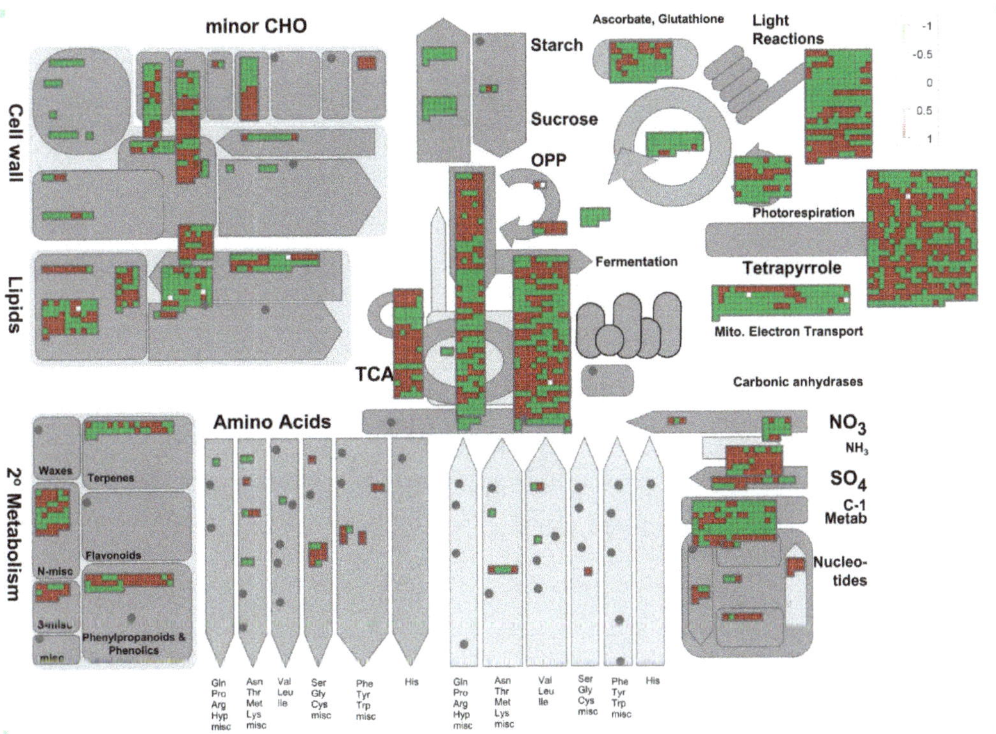

Figure 10. Comprehensive view of DEGs involved in multiple metabolic pathways analyzed by MapMan.

3.7. The DEGs Participating in Alanine, Aspartate, and Glutamate Metabolism

By combining transcriptome analysis with metabolome analysis, it was found that the metabolism of alanine, aspartate, and glutamate was significantly affected under heat stress. As presented in Table S3, after 1 d and 3 d of heat treatment, the L-alanine content dramatically reduced to 4.79% and 5.9% of its original content in the CK, respectively, and the oxalacetic acid content was similar to the original content after 1 d of heat treatment but decreased to 17% of the original content after 3 d of heat treatment (Table S3). A total of 84 DEGs were found to be involved in this pathway through KEGG analysis (Table S5). Alanine-glyoxylate aminotransferase (AGT) could catalyze the conversion of glyoxylate to glycine, and owing to the largest proportion of all DEGs in this pathway, 12 genes encoding AGT were significantly decreased after heat stress. Moreover, the expression levels of glutamate-glyoxylate aminotransferase (GGT1) and glutamine synthetase 2 (GS2) were also down-regulated after heat stress. On the contrary, many other genes such as glutamine-dependent asparagine synthase 1 (AS1), asparagine synthetase 3 (AS3), and carbamoyl phosphate synthetase A (carbamoyl phosphate synthetase A) were up-regulated after heat stress (Figure 11a). In addition, K-means cluster analysis showed that all the 84 DEGs could be classified into 4 sub-groups (Figure 11b), implying that they had specific expression patterns under heat stress.

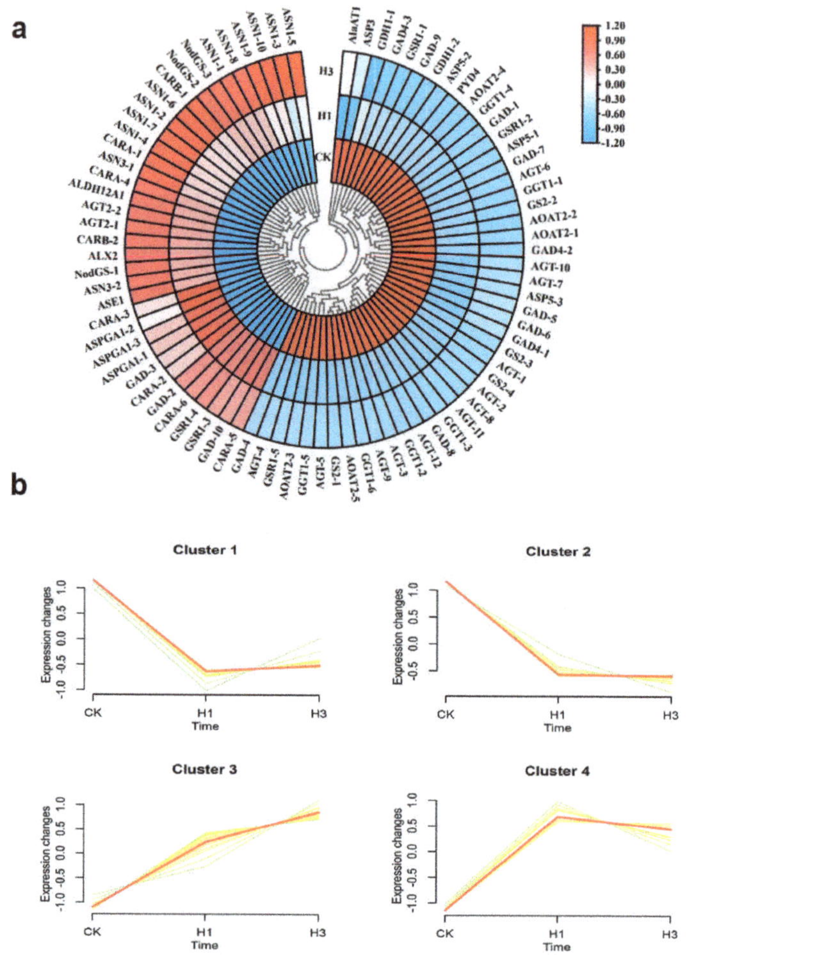

Figure 11. DEGs related to alanine, aspartate, and glutamate metabolism. (**a**) Hierarchical clustering of expression profile of DEGs, (**b**) K-means cluster analysis of DEGs.

3.8. Differentially Expressed Transcription Factors in Response to Heat Stress

Transcription factors (TFs) play important roles in plant growth and development, hormone production, as well as the response to abiotic and biotic stress [33]. In this study, 206 differentially expressed TFs were found in the DEGs, including 139 genes being up-regulated and 67 genes being down-regulated under heat treatment (Table S6). As shown in Figure 12, all these differentially expressed TFs could be divided into 17 families, including bZIP, bHLH, WRKY, and TCP. With holding the most members in all TF families, 26 DEGs were identified to belong to the heat-shock transcription factor (HSF) gene family. Interestingly, DEGs in E2F/DP, ERF, and Trihelix families only showed an upregulated expression trend under heat treatment, while DEGs in C2H2 and SBP families only showed a down-regulated expression trend. It was also worth mentioning that most of the differentially expressed WRKYs were up-regulated except for one gene (Cluster-9139.92087). All the above results suggested that basil would adapt to heat stress by altering the expression of its own TFs.

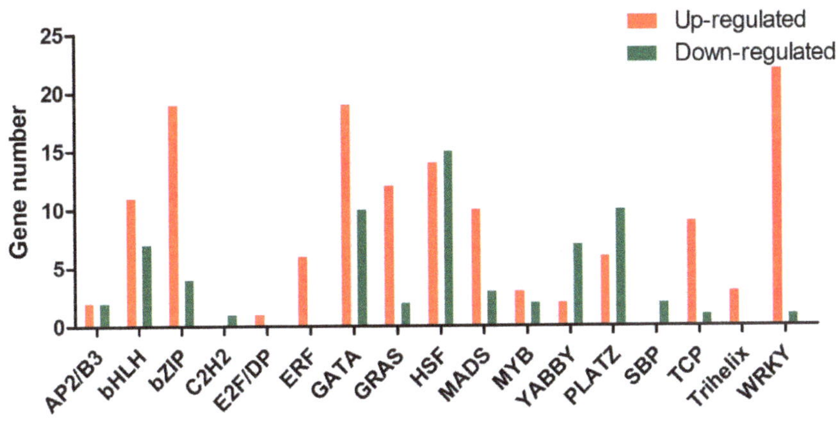

Figure 12. Differentially expressed transcription factors involved in heat stress.

3.9. Validation of RNA-seq Data by qRT-PCR

To validate the reliability of the transcriptome analysis results, 20 genes related to hormone metabolism, transcription factors, and photosynthesis were selected for qRT-PCR analysis to detect their expression levels after heat stress treatments. Overall, the two analysis methods demonstrated basically consistent trends of gene expression (Figure 13), confirming that the RNA-seq data were reliable.

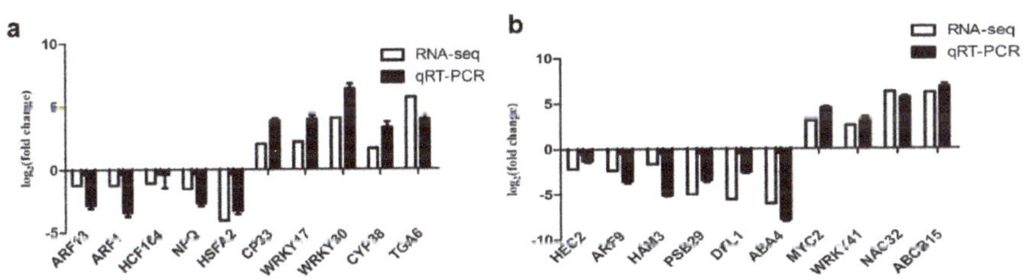

Figure 13. Validation of RNA-seq by qRT-PCR. (**a**) qRT-PCR analysis of DEGs in comparison groups of CK vs. H1, (**b**) qRT-PCR analysis of DEGs in comparison groups of CK vs. H3. CK, control group; H1, group with heat stress treatment for 1 d; H3, group with heat stress treatment for 3 d.

4. Discussion

4.1. The Ability of Basil to Resist Heat Stress

With the intensification of global warming, heat stress is becoming a severe abiotic stress to restrict the quality and yield of plants. Uncovering the mechanisms of plants in response to heat stress will facilitate the breeding of heat-resistant vegetable and crop cultivars. In this work, physiological changes combined with multi-omics analysis were used to investigate basil's response to heat stress of different times. Generally, photosynthesis is often considered as one of the most heat-sensitive physiological processes [34–36]. Heat stress commonly causes serious thermal damages to thylakoid structures, dramatically inhibits carbon metabolism and photochemical reactions, and reduces the rate of photosynthesis [11]. For instance, after being stressed at 43 °C for 2 h, tobacco would show an obvious decrease in net photosynthetic rate (Pn), stomatal conductance, and carboxylation efficiency (CE) of photosynthesis [37]. Heat stress would damage chloroplast membranes and thylakoid membranes, thus reducing the capacity of photosynthesis in soybean [38].

Consistent with previous studies, photosynthesis in basil was also suppressed under heat stress, as the values of Pn and Fv/Fm were severely decreased after 3 d of heat treatment, and most of the photosynthetic pathway-related genes were significantly down-regulated under heat stress (Figure 3a,b and Figure 10). In addition, it is worth pointing out that the temperatures in the treatment conditions used by this study (42/36 °C, light/dark, 12 h/12 h) were abnormal in South China, and it was rare to have such a high temperature continuously for several days. Interestingly, the values of Pn and Fv/Fm showed no significant difference compared with CK after 1 d of heat treatment (Figure 3a,b), and the leaves just showed slightly curling (Figure 1b). Meanwhile, under heat stress, NPQ and qP displayed the trend to increase and decrease, respectively (Figure 3c,d). These results revealed the photosynthetic thermal adaptation of basil to short-term heat stress by regulating heat dissipation and the openness of the PSII active center. Previous studies have shown that heat stress often caused oxidative stress in plants [8]. The contents of H_2O_2, O_2^-, and MDA were increased dramatically with increasing heat treatment time (Figure 2), indicating that basil was suffering from severe oxidative stress. It is worth pointing out that in *Ocimum basilicum* L. cv. 'Genovese', the content of H_2O_2 did not change significantly and the MDA content decreased under high temperature at elevated CO_2 [9]. The difference in oxidative stress between the two varieties under high temperature may be caused by the difference in treatment conditions and genetic background, which needs further study. Despite suffering from a reduced photosynthetic capacity and severe oxidative stress, basil could survive for five days under heat stress (Figure 1d), and might live longer if watered again. Overall, basil (*Ocimum basilicum* Linn. var. *pilosum* (Willd.) Benth.) exhibits a certain ability to resist heat stress.

4.2. Involvement of Key Metabolites in Response to Heat Stress

Secondary metabolites play a variety of functions in plant physiological processes such as the response to environmental changes, innate immunity generation, growth, and development [39,40]. Metabolomics has been widely used in abiotic stress research due to its high-throughput detection of chemical substances, and has been proposed as an effective means of enhancing heat resistance [41]. In soybean, for example, an increase in sugar and nitrogen metabolism provided an obvious help in dealing with heat stress [42], and metabolomics analysis of wheat grain at the filling stage also showed that the content of carbohydrates decreased significantly under heat stress, while the contents of amino acid and biosynthetic starch increased [43]. In this study, 29 significantly DAMs were found between CK and H1, while 37 DAMs found between CK and H3 (Table S2). Venn diagram analysis revealed that 10 DAMs were specifically found in the CK-H1 comparison pair, and 14 DAMs were found in the CK-H3 pair, suggesting that these DAMs were specifically involved in response to early or long-term heat stress (Figure S1). Moreover, 18 common DAMs were observed in CK-H1 and CK-H3 comparison pairs, including 10 up-regulated DAMs such as pecazine, 4-hydroxylphenyllactic acid, beta-sitosterol, conduritol-beta-epoxide, and quinic acid, and 8 DAMs including L-serine, D-glucosone, D-arabinopyranose, and L-alanine. All these key DAMs may play crucial roles in regulating heat resistance, and the detailed mechanism needs further study.

4.3. Molecular Basis of Basil Resistance to Heat Stress

Transcriptome analysis showed that a total of 19,400 DEGs were identified in the three groups (WT-H1, WT-H3, and H1-H3), and 15,066, 15,445, and 192 DEGs were found in WT-H1, WT-H3, and H1-H3, respectively. Venn analysis revealed that 11,166 common DEGs exhibiting similar expression trends were observed in the CK-H1 and CK-H3 comparison group, accounting for about 58% of all DEGs, and their expression levels altered immediately after 1 d of heat treatment, being maintained at these levels after 3 d of heat treatment; 3830 DEGs were distinctively observed in WT-H1, 4212 DEGs were observed in WT-H3, and their expressions responded specifically to heat stress time without causing significant differences in H1-H3 (Figure 7). Interestingly, it should be noted that 17 common DEGs were observed in all comparation groups, such as the heat shock protein 70 family (HSP70) and the gibberellin regulated protein and multicopper oxidase family, and their expressions levels were more closely correlated with heat treatment time. In sum, there was a significant difference at the transcriptional level between the heat-stressed groups and the CK, while a slight difference was caused by the time of heat treatment.

All DEGs were highly enriched in 30 KEGG pathways including glyoxylate and dicarboxylate metabolism, starch and sucrose metabolism, and the biosynthesis of other secondary metabolites (Figure 9). According to the Mapman analysis results, the majority of DEGs were significantly down-regulated in the pathways related to the photosynthesis of lipids, ascorbate, glutathione, starch, and sucrose, indicating that these metabolic pathways were suppressed by heat stress; on the contrary, some metabolic pathways were strengthened by heat stress, because most DEGs involved in terpenes, phenylopropanoids, phenolics, and nucleotides were significantly up-regulated (Figure 10).

When plants are subjected to heat stress, the expression and accumulation of many heat shock proteins (HSPs) are always induced to enhance their adaptability to extreme environments [44]. In the current study, a total of 37 HSPs were found in all DEGs (Table S7). Among them, 35 HSPs were significantly down-regulated after heat treatment of 1 d and 3 d, probably because the expressions of HSPs are always activated immediately after a short-term heat treatment and would decrease at a later time [45]. In comparison with the CK, however, a small heat shock protein HSP18.2 (Cluster-9139.239298), respectively, increased by 3.36-fold and 5.72-fold after 1 d and 3 d of heat treatment, and the mitochondrial localized HSP60-3A (Cluster-9139.246542) also increased by 1.96-and 3.51-fold after 1 d and 3 d of heat stress, respectively, implying that the two HSPs could be involved in the regulation of long-term heat stress. In addition, some research studies have demonstrated that transcription factors could play crucial roles in heat stress. For example, a heat stress transcription factor *LlHsfA4* enhanced thermotolerance via regulating ROS metabolism in Lilies [46], and *AtMYB30* regulated the responses of *Arabidopsis* to oxidative and heat stress through calcium signaling, partially mediated by ANNEXIN (ANN) genes [47]. Li et al. (2011) reported that the constitutive expression of *WRKY25*, *WRKY26*, or *WRKY33* enhanced the heat stress resistance of *Arabidopsis* [48]. In this work, a total of 206 differentially expressed transcription factors (TFs) were identified in all DEGs, and they belonged to various families (Table S6). Among them, 58 TFs were significant differentially expressed after 1 d of heat stress, and 59 TFs were significant differentially expressed after 3 d of heat stress, indicating that these TFs were specifically involved in the early-stage or long-term response to high temperature. The other common 128 differentially expressed TFs were observed in the CK-H1 and CK-H3 comparison group, but not in the H1-H3 group, meaning that these TFs may be rapidly induced by heat stress and maintained at a certain level of expression to regulate other genes expression. Overall, all these DEGs could contribute to heat resistance, and their potential regulatory mechanisms should be explored in the future.

5. Conclusions

In conclusion, the potential mechanisms of basil (*Ocimum basilicum* Linn. var. *pilosum* (Willd.) Benth.) in response to heat stress were explored by combining physiological changes with multi-omics analysis. Heat stress disturbed the development of basil by causing severe oxidative damage and inhibiting photosynthesis. Metabolome and transcriptome analyses identified a total of 51 DAMs and 19,400 DEGs in basil after heat stress, and they could be enriched in diverse metabolic pathways such as glyoxylate and dicarboxylate metabolism, glycolysis or gluconeogenesis, and aminoacyl-tRNA biosynthesis. This paper has provided meaningful information to further understand the mechanism of the plant response to heat stress, which may also be useful in crop breeding.

Supplementary Materials: The following supporting information can be downloaded at: https://www.mdpi.com/article/10.3390/agronomy12061434/s1, Figure S1: Venn diagram analysis of DAMs amongst the comparison groups; Figure S2: K-means cluster analysis of all DEGs; Table S1: Primer sequences in this study; Table S2: All metabolites identified in samples: Table S3: DAMs in each comparison group: Table S4: DEGs in each comparison group; Table S5: DEGs in alanine, aspartate, and glutamate metabolism; Table S6: Differentially expressed transcription factors under heat stress; Table S7: Differentially expressed HSPs under heat stress.

Author Contributions: Q.W., A.Q. and L.Y. conceived the ideas and designed the methodology; L.Q., C.L. and J.W. collected the data; L.Q., C.L. and D.L. analyzed the data; Q.W. and A.Q. led the writing of the manuscript. All authors have read and agreed to the published version of the manuscript.

Funding: This research was funded by the Department of Science and Technology of Ningbo (2021IJCGY020198), and the Natural Science Foundation of Zhejiang (LQ19C140002).

Institutional Review Board Statement: Not applicable.

Informed Consent Statement: Not applicable.

Data Availability Statement: Any data and codes used in this study are available upon request from the corresponding author.

Acknowledgments: We are especially grateful to Yingxian Zhao for his guidance on the paper.

Conflicts of Interest: The authors declare no conflict of interest.

References

1. Varga, F.; Carović-Stanko, K.; Ristić, M.; Grdiša, M.; Liber, Z.; Šatović, Z. Morphological and Biochemical Intraspecific Characterization of *Ocimum basilicum*, L. *Ind. Crop. Prod.* **2017**, *109*, 611–618. [CrossRef]
2. Carović-Stanko, K.; Orlić, S.; Politeo, O.; Strikić, F.; Kolak, I.; Milos, M.; Satovic, Z. Composition and Antibacterial Activities of Essential Oils of Seven Ocimum Taxa. *Food Chem.* **2010**, *119*, 196–201. [CrossRef]
3. Majdi, C.; Pereira, C.; Dias, M.I.; Calhelha, R.C.; Alves, M.J.; Rhourri-Frih, B.; Charrouf, Z.; Barros, L.; Amaral, J.S.; Ferreira, I.C.F.R. Phytochemical Characterization and Bioactive Properties of Cinnamon Basil (*Ocimum basilicum* cv. '*Cinnamon*') and Lemon Basil (*Ocimum* × *Citriodorum*). *Antioxidants* **2020**, *9*, 369. [CrossRef] [PubMed]
4. Politeo, O.; Jukic, M.; Milos, M. Chemical Composition and Antioxidant Capacity of Free Volatile Aglycones from Basil (*Ocimum basilicum*, L.) Compared with Its Essential Oil. *Food Chem.* **2007**, *101*, 379–385. [CrossRef]
5. Sestili, P.; Ismail, T.; Calcabrini, C.; Guescini, M.; Catanzaro, E.; Turrini, E.; Layla, A.; Akhtar, S.; Fimognari, C. The Potential Effects of *Ocimum basilicum* on Health: A Review of Pharmacological and Toxicological Studies. *Expert Opin. Drug Metab. Toxicol.* **2018**, *14*, 679–692. [CrossRef] [PubMed]
6. Benner, J.P. Pesticidal Compounds from Higher Plants. *Pestic. Sci.* **1993**, *39*, 95–102. [CrossRef]
7. Zhang, J.W.; Li, S.K.; Wu, W.J. The Main Chemical Composition and In Vitro Antifungal Activity of the Essential Oils of *Ocimum basilicum* Linn. var. *Pilosum* (Willd.) Benth. *Molecules* **2009**, *14*, 273–278. [CrossRef]
8. Haider, S.; Iqbal, J.; Naseer, S.; Yaseen, T.; Shaukat, M.; Bibi, H.; Ahmad, Y.; Daud, H.; Abbasi, N.L.; Mahmood, T. Molecular Mechanisms of Plant Tolerance to Heat Stress: Current Landscape and Future Perspectives. *Plant Cell Rep.* **2021**, *40*, 2247–2271. [CrossRef]
9. Barickman, T.C.; Olorunwa, O.J.; Sehgal, A.; Walne, C.H.; Reddy, K.R.; Gao, W. Yield, Physiological Performance, and Phytochemistry of Basil (*Ocimum basilicum*, L.) under Temperature Stress and Elevated CO(2) Concentrations. *Plants* **2021**, *10*, 1072. [CrossRef]
10. Kumari, P.; Rastogi, A.; Yadav, S. Effects of Heat Stress and Molecular Mitigation Approaches in Orphan Legume, Chickpea. *Mol. Biol. Rep.* **2020**, *47*, 4659–4670. [CrossRef]
11. Hasanuzzaman, M.; Nahar, K.; Alam, M.M.; Roychowdhury, R.; Fujita, M. Physiological, Biochemical, and Molecular Mechanisms of Heat Stress Tolerance in Plants. *Int. J. Mol. Sci.* **2013**, *14*, 9643–9684. [CrossRef] [PubMed]
12. Perera, R.S.; Cullen, B.R.; Eckard, R.J. Growth and Physiological Responses of Temperate Pasture Species to Consecutive Heat and Drought Stresses. *Plants* **2019**, *8*, 227. [CrossRef] [PubMed]
13. Zhao, J.; Lu, Z.; Wang, L.; Jin, B. Plant Responses to Heat Stress: Physiology, Transcription, Noncoding RNAs, and Epigenetics. *Int. J. Mol. Sci.* **2020**, *22*, 117. [CrossRef] [PubMed]
14. Mathur, S.; Agrawal, D.; Jajoo, A. Photosynthesis: Response to High Temperature Stress. *J. Photochem. Photobiol. B Biol.* **2014**, *137*, 116–126. [CrossRef]
15. Allakhverdiev, S.I.; Kreslavski, V.D.; Klimov, V.V.; Los, D.A.; Carpentier, R.; Mohanty, P. Heat Stress: An Overview of Molecular Responses in Photosynthesis. *Photosynth. Res.* **2008**, *98*, 541–550. [CrossRef]

16. Wassie, M.; Zhang, W.; Zhang, Q.; Ji, K.; Cao, L.; Chen, L. Exogenous Salicylic Acid Ameliorates Heat Stress-Induced Damages and Improves Growth and Photosynthetic Efficiency in Alfalfa (*Medicago sativa*, L.). *Ecotoxicol. Environ. Saf.* **2020**, *191*, 110206. [CrossRef]
17. Song, Y.; Chen, Q.; Ci, D.; Shao, X.; Zhang, D. Effects of High Temperature on Photosynthesis and Related Gene Expression in Poplar. *BMC Plant Biol.* **2014**, *14*, 111. [CrossRef]
18. Ma, G.; Zhang, M.; Xu, J.; Zhou, W.; Cao, L. Transcriptomic Analysis of Short-Term Heat Stress Response in Pinellia Ternata Provided Novel Insights into the Improved Thermotolerance by Spermidine and Melatonin. *Ecotoxicol. Environ. Saf.* **2020**, *202*, 110877. [CrossRef]
19. Alam, M.N.; Zhang, L.; Yang, L.; Islam, M.R.; Liu, Y.; Luo, H.; Yang, P.; Wang, Q.; Chan, Z. Transcriptomic Profiling of Tall Fescue in Response to Heat Stress and Improved Thermotolerance by Melatonin and 24-Epibrassinolide. *BMC Genom.* **2018**, *19*, 224. [CrossRef]
20. Geng, X.-M.; Liu, X.; Ji, M.; Hoffmann, W.A.; Grunden, A.; Xiang, Q.-Y.J. Enhancing Heat Tolerance of the Little Dogwood *Cornus canadensis*, L. f. with Introduction of a Superoxide Reductase Gene from the Hyperthermophilic Archaeon Pyrococcus Furiosus. *Front. Plant Sci.* **2016**, *7*, 26. [CrossRef]
21. Wang, L.; Ma, K.-B.; Lu, Z.-G.; Ren, S.-X.; Jiang, H.-R.; Cui, J.-W.; Chen, G.; Teng, N.-J.; Lam, H.-M.; Jin, B. Differential Physiological, Transcriptomic and Metabolomic Responses of Arabidopsis Leaves under Prolonged Warming and Heat Shock. *BMC Plant Biol.* **2020**, *20*, 86. [CrossRef] [PubMed]
22. Sita, K.; Sehgal, A.; Kumar, J.; Kumar, S.; Singh, S.; Siddique, K.H.M.; Nayyar, H. Identification of High-Temperature Tolerant Lentil (Lens Culinaris Medik.) Genotypes through Leaf and Pollen Traits. *Front. Plant Sci.* **2017**, *8*, 744. [CrossRef] [PubMed]
23. Devireddy, A.R.; Tschaplinski, T.J.; Tuskan, G.A.; Muchero, W.; Chen, J.-G. Role of Reactive Oxygen Species and Hormones in Plant Responses to Temperature Changes. *Int. J. Mol. Sci.* **2021**, *22*, 8843. [CrossRef] [PubMed]
24. Ozga, J.A.; Kaur, H.; Savada, R.P.; Reinecke, D.M. Hormonal Regulation of Reproductive Growth under Normal and Heat-Stress Conditions in Legume and Other Model Crop Species. *J. Exp. Bot.* **2017**, *68*, 1885–1894. [CrossRef]
25. Muñoz-Espinoza, V.A.; López-Climent, M.F.; Casaretto, J.A.; Gómez-Cadenas, A. Water Stress Responses of Tomato Mutants Impaired in Hormone Biosynthesis Reveal Abscisic Acid, Jasmonic Acid and Salicylic Acid Interactions. *Front. Plant Sci.* **2015**, *6*, 997. [CrossRef]
26. Ohama, N.; Sato, H.; Shinozaki, K.; Yamaguchi-Shinozaki, K. Transcriptional Regulatory Network of Plant Heat Stress Response. *Trends Plant Sci.* **2017**, *22*, 53–65. [CrossRef]
27. Dai, X.; Wang, Y.; Chen, Y.; Li, H.; Xu, S.; Yang, T.; Zhang, X.; Su, X.; Xia, Z. Overexpression of NtDOG1L-T Improves Heat Stress Tolerance by Modulation of Antioxidant Capability and Defense-, Heat-, and ABA-Related Gene Expression in Tobacco. *Front. Plant Sci.* **2020**, *11*, 568489. [CrossRef]
28. Jakovljević, D.; Momčilović, J.; Bojović, B.; Stanković, M. The Short-Term Metabolic Modulation of Basil (*Ocimum basilicum*, L. Cv. '*Genovese*') after Exposure to Cold or Heat. *Plants* **2021**, *10*, 590. [CrossRef]
29. Fiehn, O.; Wohlgemuth, G.; Scholz, M.; Kind, T.; Lee, D.Y.; Lu, Y.; Moon, S.; Nikolau, B. Quality Control for Plant Metabolomics: Reporting MSI-Compliant Studies. *Plant J.* **2008**, *53*, 691–704. [CrossRef]
30. Chong, J.; Xia, J. MetaboAnalystR: An R Package for Flexible and Reproducible Analysis of Metabolomics Data. *Bioinformatics* **2018**, *34*, 4313–4314. [CrossRef]
31. Wang, L.; Si, Y.; Dedow, L.K.; Shao, Y.; Liu, P.; Brutnell, T.P. A Low-Cost Library Construction Protocol and Data Analysis Pipeline for Illumina-Based Strand-Specific Multiplex RNA-Seq. *PLoS ONE* **2011**, *6*, e26426. [CrossRef]
32. Chen, C.; Chen, H.; Zhang, Y.; Thomas, H.R.; Frank, M.H.; He, Y.; Xia, R. TBtools: An Integrative Toolkit Developed for Interactive Analyses of Big Biological Data. *Mol. Plant* **2020**, *13*, 1194–1202. [CrossRef] [PubMed]
33. Bari, R.; Jones, J.D.G. Role of Plant Hormones in Plant Defence Responses. *Plant Mol. Biol.* **2009**, *69*, 473–488. [CrossRef] [PubMed]
34. Shi, J.; Chen, Y.; Xu, Y.; Ji, D.; Chen, C.; Xie, C. Differential Proteomic Analysis by ITRAQ Reveals the Mechanism of Pyropia Haitanensis Responding to High Temperature Stress. *Sci. Rep.* **2017**, *7*, 44734. [CrossRef]
35. Centritto, M.; Brilli, F.; Fodale, R.; Loreto, F. Different Sensitivity of Isoprene Emission, Respiration and Photosynthesis to High Growth Temperature Coupled with Drought Stress in Black Poplar (*Populus nigra*) Saplings. *Tree Physiol.* **2011**, *31*, 275–286. [CrossRef]
36. Wang, J.; Liang, C.; Yang, S.; Song, J.; Li, X.; Dai, X.; Wang, F.; Juntawong, N.; Tan, F.; Zhang, X.; et al. ITRAQ-Based Quantitative Proteomic Analysis of Heat Stress-Induced Mechanisms in Pepper Seedlings. *PeerJ* **2021**, *9*, e11509. [CrossRef]
37. Tan, W.; Meng, Q.W.; Brestic, M.; Olsovska, K.; Yang, X. Photosynthesis Is Improved by Exogenous Calcium in Heat-Stressed Tobacco Plants. *J. Plant Physiol.* **2011**, *168*, 2063–2071. [CrossRef]
38. Djanaguiraman, M.; Prasad, P.V.V.; Al-Khatib, K. Ethylene Perception Inhibitor 1-MCP Decreases Oxidative Damage of Leaves through Enhanced Antioxidant Defense Mechanisms in Soybean Plants Grown under High Temperature Stress. *Environ. Exp. Bot.* **2011**, *71*, 215–223. [CrossRef]
39. Berini, J.L.; Brockman, S.A.; Hegeman, A.D.; Reich, P.B.; Muthukrishnan, R.; Montgomery, R.A.; Forester, J.D. Combinations of Abiotic Factors Differentially Alter Production of Plant Secondary Metabolites in Five Woody Plant Species in the Boreal-Temperate Transition Zone. *Front. Plant Sci.* **2018**, *9*, 1257. [CrossRef]
40. Kroymann, J. Natural Diversity and Adaptation in Plant Secondary Metabolism. *Curr. Opin. Plant Biol.* **2011**, *14*, 246–251. [CrossRef]

41. Raza, A. Metabolomics: A Systems Biology Approach for Enhancing Heat Stress Tolerance in Plants. *Plant Cell Rep.* **2022**, *41*, 741–763. [CrossRef] [PubMed]
42. Das, A.; Rushton, P.J.; Rohila, J.S. Metabolomic Profiling of Soybeans (*Glycine max*, L.) Reveals the Importance of Sugar and Nitrogen Metabolism under Drought and Heat Stress. *Plants* **2017**, *6*, 21. [CrossRef] [PubMed]
43. Wang, X.; Hou, L.; Lu, Y.; Wu, B.; Gong, X.; Liu, M.; Wang, J.; Sun, Q.; Vierling, E.; Xu, S. Metabolic Adaptation of Wheat Grain Contributes to a Stable Filling Rate under Heat Stress. *J. Exp. Bot.* **2018**, *69*, 5531–5545. [CrossRef] [PubMed]
44. Wu, X.; Wang, J.; Wu, X.; Hong, Y.; Li, Q.Q. Heat Shock Responsive Gene Expression Modulated by MRNA Poly(A) Tail Length. *Front. Plant Sci.* **2020**, *11*, 1255. [CrossRef]
45. Chen, L.; Chen, Q.; Kong, L.; Xia, F.; Yan, H.; Zhu, Y.; Mao, P. Proteomic and Physiological Analysis of the Response of Oat (*Avena sativa*) Seeds to Heat Stress under Different Moisture Conditions. *Front. Plant Sci.* **2016**, *7*, 896. [CrossRef]
46. Wang, C.; Zhou, Y.; Yang, X.; Zhang, B.; Xu, F.; Wang, Y.; Song, C.; Yi, M.; Ma, N.; Zhou, X.; et al. The Heat Stress Transcription Factor LlHsfA4 Enhanced Basic Thermotolerance through Regulating ROS Metabolism in Lilies (*Lilium longiflorum*). *Int. J. Mol. Sci.* **2022**, *23*, 572. [CrossRef]
47. Liao, C.; Zheng, Y.; Guo, Y. MYB30 Transcription Factor Regulates Oxidative and Heat Stress Responses through ANNEXIN-Mediated Cytosolic Calcium Signaling in Arabidopsis. *New Phytol.* **2017**, *216*, 163–177. [CrossRef]
48. Li, S.; Fu, Q.; Chen, L.; Huang, W.; Yu, D. Arabidopsis Thaliana WRKY25, WRKY26, and WRKY33 Coordinate Induction of Plant Thermotolerance. *Planta* **2011**, *233*, 1237–1252. [CrossRef]

agronomy

MDPI

Article

Combined Effect of Potassium Permanganate and Ultraviolet Light as Ethylene Scavengers on Post-Harvest Quality of Peach at Optimal and Stressful Temperatures

Ramiro Alonso-Salinas [1,†], José Ramón Acosta-Motos [2,3,*,†], Estrella Núñez-Delicado [1,2], José Antonio Gabaldón [1] and Santiago López-Miranda [1]

1 Molecular Recognition and Encapsulation Research Group (REM), Campus de los Jerónimos 135, Universidad Católica de Murcia (UCAM), Guadalupe, 30107 Murcia, Spain; ralonso4@alu.ucam.edu (R.A.-S.); enunez@ucam.edu (E.N.-D.); jagabaldon@ucam.edu (J.A.G.); slmiranda@ucam.edu (S.L.-M.)

2 Cátedra de Emprendimiento en el Ámbito Agroalimentario UCAM-Santander, Campus de los Jerónimos, Universidad Católica San Antonio de Murcia (UCAM), 30107 Murcia, Spain

3 Group of Fruit Tree Biotechnology, Department of Plant Breeding, CEBAS-CSIC, Campus Universitario de Espinardo, 3A, 30100 Murcia, Spain

* Correspondence: jracosta@ucam.edu; Tel.: +34-968-278756

† These authors contributed equally to this work.

Abstract: The aim of this study was to evaluate the combination of two ethylene removal methods and temperature on the post-harvest quality of peaches. For this purpose, filters with potassium permanganate ($KMnO_4$) and lamps emitting ultraviolet light (UV) were mounted on machines which enabled air movement in the conservation chambers, facilitating the removal of ethylene by $KMnO_4$ and photocatalysis simultaneously. This system was used at two temperatures, 1 °C and 25 °C, simulating an ideal storage temperature in industry and extreme temperature to observe faster ripening, respectively. The results obtained showed that this combination of ethylene scavengers favoured the efficient elimination of this gas. Consequently, the use of this innovative technique made possible a better preservation of fruit firmness, colour, soluble solids content, pH, total acidity, and maturity index. Moreover, using this method in peaches subjected to 25 °C increased their survival by seven days more than those without this system, indicating the effectiveness of ethylene scavengers even under these extreme temperatures.

Keywords: climacteric fruit; ethylene scavengers; fruit quality; potassium permanganate; *Prunus persica*; UV radiation

Citation: Alonso-Salinas, R.; Acosta-Motos, J.R.; Núñez-Delicado, E.; Gabaldón, J.A.; López-Miranda, S. Combined Effect of Potassium Permanganate and Ultraviolet Light as Ethylene Scavengers on Post-Harvest Quality of Peach at Optimal and Stressful Temperatures. *Agronomy* **2022**, *12*, 616. https://doi.org/10.3390/agronomy12030616

Academic Editor: Noam Alkan

Received: 22 January 2022
Accepted: 27 February 2022
Published: 1 March 2022

1. Introduction

The nutritional and organoleptic quality, and the shelf-life characteristics of peaches [*Prunus persica* (L.) Batsch] are affected by the interaction of multiple factors. Fruit quality cannot be improved during post-harvest, but it can be maintained through the application of innovative conservation techniques [1–3]. Apart from the crucial control of the optimal storage temperature, ethylene concentration is one of the most important parameters during post-harvest conservation of climacteric fruit. Ethylene (C_2H_4) is a phytohormone that, even in low concentrations, can produce undesirable changes in physical and chemical parameters in fruits, such as changes in firmness, colour, pH, or maturity index [4–7]. Therefore, limiting its presence has proven to be an economically and commercially key process for avoiding post-harvest losses and food wastage. Moreover, ethylene removal technology could guarantee the safety and maintenance of fruit qualities for an increasingly demanding consumer market [8–11].

Many methods exist for the elimination of this phytohormone, with the non-intrusive being more common, characterized by not coming into direct contact with the fruit [12–14]. The two most effective non-intrusive preservation methods include the use of palladium and potassium permanganate ($KMnO_4$) as ethylene oxidants [9]. The first has the disadvantage of being more expensive; therefore, the oxidation of ethylene using $KMnO_4$ is the most suitable method in terms of cost-effectiveness [12]. This process of ethylene removal is based on an oxidation-reduction process, as $KMnO_4$ is a strong oxidizing agent that promotes the rapid dissociation of ethylene into carbon dioxide, manganese dioxide, and potassium hydroxide [15]. This process is activated by ethylene released by the climacteric fruit as a result of its natural metabolism. $KMnO_4$ undergoes a colour change from violet to dark brown when it is saturated, confirming the elimination of ethylene during the reaction [7,13]. To support this process, this hyperoxidant molecule is introduced into porous materials with high adsorption power, such as zeolite, sepiolite, vermiculite, alumina, or activated carbon [16]. These materials are widely used in sachets placed in boxes during the transport of fruit [17].

Another non-invasive method for ethylene removal is photocatalysis using ultraviolet light (UV), a low-cost and environmentally friendly technology that can be used to degrade a variety of aqueous and gas-phase pollutants [5,18]. Ethylene photo-degradation induced by radiation of UV light generates oxidizing agents, including hydroxyl radical (OH•) as reactive oxygen species (ROS), which are highly reactive and convert ethylene into CO_2 and H_2O [12,19] supporting $KMnO_4$ action. Although the mechanism of photocatalytic oxidation has been elucidated previously [5,13], the exact reaction mechanism still remains under debate owing to the presence of various reaction intermediates which have not been clarified completely [9].

The aim of this study was to evaluate the combined effect of two different ethylene removal methods, such as oxidation by potassium permanganate and photocatalysis by UV light, on the post-harvest quality of peaches stored at two different temperatures (1 °C and 25 °C). The study covered a period of 24 days of storage with measurements of physical parameters, such as weight, size, firmness, or colour, and measurements of biochemical parameters related to maturity, such as soluble solids content, pH, titratable acidity, maturity index, and microbiological incidence.

2. Materials and Methods

2.1. Plant Material

Forty kilograms of peaches of the yellow flesh Rojo de Rito variety were supplied by Thader Cieza, S.C.L. (Cieza, Murcia, Spain) through the intermediation of the agrarian cooperative FECOAM ("Federación de Cooperativas Agrarias de Murcia"). The fruit were harvested manually and maintained in refrigerated conditions (1 °C) for two days until the start of the trial. All peaches selected had a homogenous weight, size and colour. The harvesting indices forecast by the supplying company were as follows:

- Weight: 179.5 ± 2.2 g
- Size: 72.0 ± 2 mm
- Firmness: 30.2 ± 1.6 N
- Soluble solids content (SSC): 10.9%
- Total acidity (TA): 3.8%
- Ratio SSC/TA (MI): 2.86

2.2. Experimental Design

A total of 250 peaches were randomly distributed into four conservation chambers (CCs) that were 150 litres in volume (Eurofred Cool Head RCG200, Eurofred S.A., Barcelona, Catalonia, Spain).

As a first factor, the filters for ethylene removal were composed of $KMnO_4$ anchored into the active centre of sepiolite. The composition of the filters in terms of granulometry and other adsorbent substances was patented in Spain by the company "Nuevas Tecnologías Agroalimentarias (KEEPCOOL)" (Molina de Segura, Murcia, Spain), patent No. 2548787 (2016). The adsorbent material was covered by a semi-permeable paper, which allowed the entry of air rich in ethylene and the output of air clean of this phytohormone, while also avoiding the entry of water or other particles into the filter. Potassium permanganate filters were installed inside an air-flow-forcing machine (M-CAM 50, KEEPCOOL, Molina de Segura, Murcia, Spain) to ensure appropriate movement of the air inside the CCs and to support ethylene removal. The machine incorporated a photocatalytic ultraviolet light system (TUV 254 nm, Philips, Amsterdam, Netherland) to aid the potassium permanganate filters in the removal of ethylene. The ultraviolet light was focused on the air coming out of the filters, not on the fruit. Throughout the article, the machine, UV light and filter combination will be referred to as the ethylene scavenger (ES).

As a second factor, two different temperatures were analysed: refrigeration temperature set at 1 °C and room temperature of approximately 25 °C for non-refrigerated treatments. Taking into account the combination of the two factors, the following four treatments were established:

- NoES-R (control): No Ethylene Scavenger + Refrigeration temperature.
- ES-R: Ethylene Scavenger + Refrigeration temperature.
- NoES-NoR: No Ethylene Scavenger + No Refrigeration temperature.
- ES-NoR: Ethylene Scavenger + No Refrigeration temperature.

2.3. Conservation Chambers Atmosphere

The CCs' temperature and relative humidity (RH) were registered with a Testo 184 H1 Data Logger, (Titisee-Neustadt, Baden-Württemberg, Germany).

The ethylene (C_2H_4), carbon dioxide (CO_2), and oxygen (O_2) concentration of the four CCs was monitored daily using a gas analyzer (Felix Three F-950, Felix Instruments, Camas, WA, USA). To carry out the gas measurements, the CCs were equipped with a hermetically sealed probe in order not to disturb the atmosphere, through which the necessary amount of air used by the gas analyzer mentioned above could be extracted. C_2H_4 and CO_2 were finally expressed as $mL \times kg^{-1} \times h^{-1}$, to compare the concentrations of the two gases. The O_2 was expressed as a percentage. The initial measurement at day 0 was taken 6 h after receipt of the product.

2.4. Physical Parameters

The weight was measured using a precision balance (Gram RZ, Gram Group, L'Hospitalet de Llobregat, Catalonia, Spain), expressed in grams. The size of the peaches, according to their equatorial diameter, was measured with a caliper (Wurth vernier caliper, Baden-Württemberg, Germany) and expressed in millimetres.

Pulp colour was determined in the layer of flesh immediately under the skin (2 mm) with the CIELAB system using a colorimeter (Hunterlab Colorflex EZ, Hunterlab Reston, VA, USA), with the measurements performed in two different places on the non-blushed side. Coordinate (a*) in the CIELAB system indicates the colour's position between green and red (negative values indicate green, positive values indicate red), and coordinate (b*) indicates the colour's position between blue and yellow (negative values indicate blue, positive values indicate yellow)

The firmness of the peaches was measured in the equatorial zone, away from the suture, with a CT3 texturometer (AMETEK Brookfield, Middleboro, MA, USA) equipped with a cylindrical probe measuring 35 mm in height and 6 mm in diameter, which penetrated 10 mm at a speed of $0.5 mm \times s^{-1}$. Peach firmness was considered as the maximum force (N) measured during probe penetration.

All physical parameters were measured at the following storage days: 0, 3, 7, 10, 14, 17, 22 and 24. Seven peaches per treatment were used for each of above-mentioned days of analysis.

2.5. Maturity Parameters

Soluble solid content (SSC), pH, and titratable acidity (TA) were measured on fruit samples using the method adapted from [20].

Twenty grams of peach were taken and added to 20 mL of distilled water, then homogenized with a mixer (Ultra turrax T25, LabWare Wilmington, DE, USA) for 30 s. The homogenate was centrifuged at 3600× g for 10 min in a centrifuge (Eppendorf Centrifuge 5810, Hamburg, Germany) and the supernatant was used to obtain SSC, pH, and TA.

The SSC was determined by a manual refractometer (Atago Manual master-α, Atago Tokyo, Japan) at 20 °C and expressed as a percentage (sugar equivalents in g × 100 g^{-1}). The pH was determined with a pH-meter (HI 2221, Hanna Instruments Eibar, Gipuzkoa, Spain).

The determination of TA was carried out by adapting the method described by [20]. For this, 20 g of fresh peach were weighed and brought to a volume of 100 mL with deionized water; the resulting mixture was titrated to a pH of 8.1 ± 0.1 with NaOH 0.1 N and constant stirring. The percentage of acid in the sample was calculated and expressed as a percentage according to [4].

The maturity index (MI) was determined by dividing SSC (%) by TA (%). The expression of this parameter is dimensionless.

All maturity parameters were measured at the following storage days: 0, 3, 7, 10, 14, 17, 22 and 24. Seven peaches per treatment were used for each of above-mentioned days of analysis.

2.6. *Microbiological Incidence*

Microbiological incidence was estimated by visual inspection on each fruit according to [4]. Any peaches showing any fungal or bacterial growth were considered infected and discarded. Data were expressed as the percentage of peaches affected.

2.7. *Statistical Analysis*

The descriptive statistics (mean and standard error of the mean [SEM]) and the different tests described below were performed using the StatGraphics Centurion XV software package (StatPoint Technologies, Inc. Warrenton, VA, USA). The Shapiro–Wilk test was performed to check the normality of the data. In addition, to check the homogeneity of the variance, Bartlett's test was applied. The data were analysed using an analysis of variance (two-way ANOVA), as four independent treatments and two factors were available (3 and 7 days). Next, the data were processed using an analysis of variance (one-way ANOVA) when the three independent treatments were available (10 and 14 days) and for all figures. Next, a t-test was performed when only two independent treatments were available (17, 22 and 24 days). Finally, Tukey's Multiple Range Test was utilized to separate means and detect significant differences (p-value < 0.05).

3. Results and Discussion

Due to the temperature conditions and ethylene concentration, not all the treatments could be analysed towards the end of the trial. As will be observed in the following tables and figures, no analyses were carried out in NoES-NoR from day 7 of the trial and in ES-NoR from day 14 until the end of the study. This was due to the losses in these treatments caused by ripening, rotting or microbiological damage.

3.1. *Changes in the Conservation Chambers Atmosphere*

The peach is a climacteric fruit, which means that its ripening process continues once harvested, being highly affected by the presence or absence of ethylene. Climacteric fruits increase the production of ethylene during post-harvest ripening, being the gas responsible for the coordination of the ripening process. The increase in ethylene production is promoted by the same gas during ripening and is associated with an increase in the respiration rate of the peach. It is therefore important to keep the ethylene rate at a low level in order to better preserve the quality of the fruit.

The temperature was set at 1 ± 1 °C for treatments kept refrigerated (NoES-R and ES-R), and at 25 ± 1 °C for treatments kept at room temperature (NoES-NoR and ES-NoR). Regarding relative humidity (RH), both refrigerated treatments showed values close to

80%, and in the non-refrigerated treatments RH values were close to 100%. The evolution of the concentration of the gases throughout the storage time is shown in Figure 1.

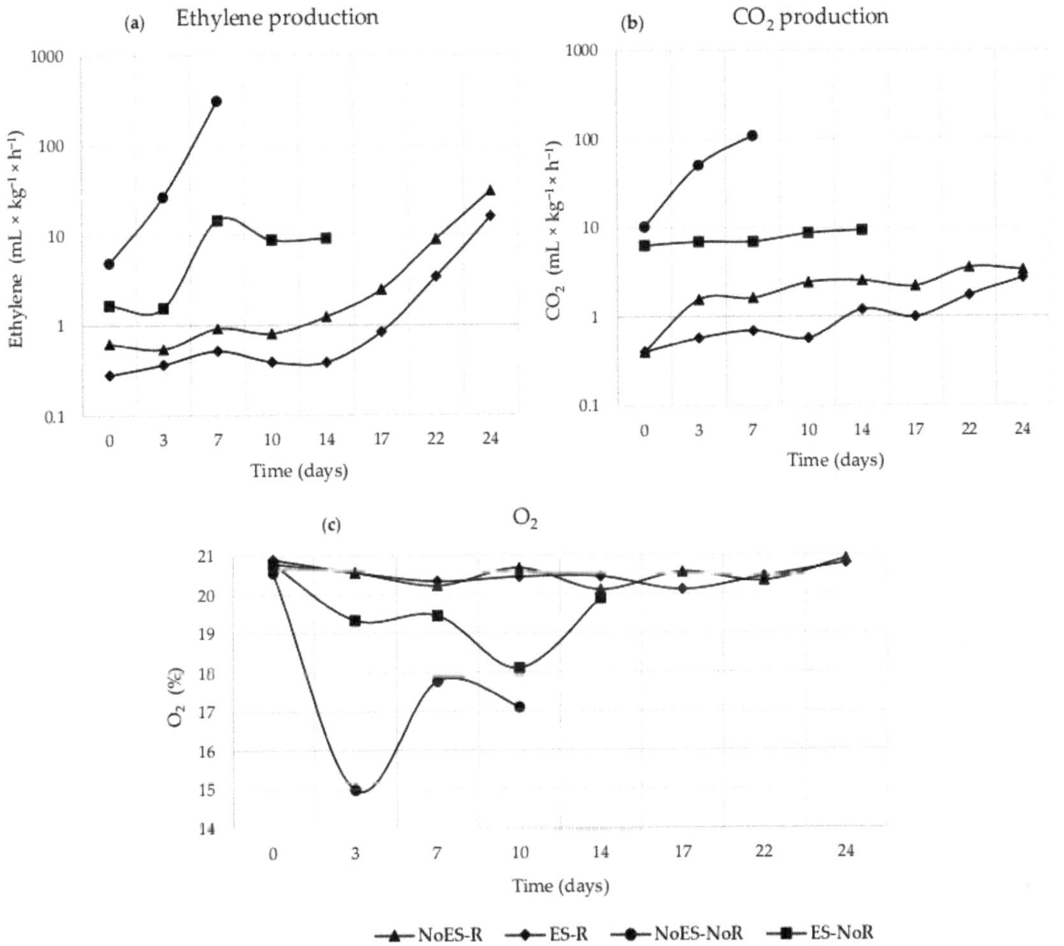

Figure 1. Ethylene (**a**) and CO_2 (**b**) production expressed as mL × kg^{-1} × h^{-1} and O_2 percentage (**c**) over the storage time in peaches subjected to the different treatments (NoES-R, ES-R, NoES-NoR and ES-NoR). The *y*-axis in Figure 1a,b corresponding to ethylene and CO_2 production, respectively, are displayed on a base 10 logarithmic scale.

3.1.1. Ethylene

When comparing the evolution of the ethylene production rate of the four treatments over time, the following conclusions can be drawn: for refrigerated conditions, ethylene scavengers were able to remove, between day 0 and 24, a mean of 52% of the total ethylene produced in ES-R (3.11 mL × kg^{-1} × h^{-1}), compared to NoES-R (6.47 mL × kg^{-1} × h^{-1}). The differences found between NoES-R and ES-R suggest a key role of the ethylene scavengers, which markedly decreased the rates of this phytohormone in ES-R. For the non-refrigerated conditions, ethylene scavengers were able to remove 95%, by mean between days 0 and 24, of the total ethylene production in ES-NoR (8.53 mL × kg^{-1} × h^{-1}), compared to NoES-NoR (168.24 mL × kg^{-1} × h^{-1}). It is worth mentioning that the ethylene levels of the samples preserved with scavengers shown in this study are lower than in other studies, due to the combination of ethylene removal systems (potassium permanganate and photocatalysis).

3.1.2. Carbon Dioxide

Comparing the evolution in the CO_2 production rate among different treatments, the following findings can be highlighted: ES-R, with 1.20 mL $\times$ kg^{-1} $\times$ h^{-1}, showed an average decrease of 51% in CO_2 production between days 0 and 24, compared to NoES-R with 2.44 mL $\times$ kg^{-1} $\times$ h^{-1}; ES-NoR, with 7.96 mL $\times$ kg^{-1} $\times$ h^{-1}, had an average decrease of 90% in CO_2 production between days 0 and 24, compared to NoES-NoR with 79.21 mL $\times$ kg^{-1} $\times$ h^{-1} (Figure 1b). Therefore, it can be concluded that the use of ethylene scavengers decreased the CO_2 production rate (Figure 1b), thereby delaying the ripening of peaches. Similar to the results described above, some authors showed how the respiration rate in melons could be delayed by limiting ethylene levels in the first days of the trial, which in turn decreased CO_2 concentration [21].

3.1.3. Oxygen

The oxygen consumption of the fruit also proved to be relevant in the analysis of the activity of peaches (Figure 1c). The absence of an optimal oxygen percentage in combination with a high temperature and an elevated ethylene concentration led to the activation of a hypothetical internal fermentation metabolic process in NoES-NoR, where oxygen concentrations close to 16% were observed (Figure 1c). Although no hypothetical fermentation processes were observed in ES-NoR, oxygen concentrations were below 20%. Oxygen concentrations in the refrigerated treatments (NoES-R and ES-R) were close to 21%, which is the optimal concentration value in the atmosphere (Figure 1c).

3.1.4. Relationship between Ethylene and CO_2

Considering Figure 1a,b, a relationship between both gases can be observed. A reduction in CO_2 production was observed in the treatments where ethylene scavenging techniques were used (ES-R and ES-NoR). This is in agreement with existing literature [9,19,21]. The authors of [8] also reported the effect of ethylene scavengers on the preservation of apple fruit, where a reduction in CO_2 production was observed in the samples with low ethylene concentration.

More specifically, there was a positive correlation between them, as shown by a sudden increase in ethylene concentration preceding a sudden rise in CO_2 production [4]. These changes in both gases contributed to an increase in the respiration rate, which led to an autocatalytic effect that accelerated the maturation of the product [7,12,22]. This was particularly evident in the NoES-NoR treatment, where the peach preservation conditions were stressful, favouring an increase in ethylene levels due to the absence of ethylene scavengers, and in CO_2 concentrations, due to a higher respiratory rate associated with high temperatures.

3.2. Physical Parameters

Fruit weight is an important parameter for fruit producers, especially from an economic point of view, and therefore controlling fruit weight loss is crucial. This loss is associated with an excessive loss of water due to transpiration, related to a low RH. Water loss after harvesting is an unavoidable phenomenon, the effects of which are loss of weight, decrease in size, wilting, abnormal textures, and decrease in quality. Wilting becomes visible when the peach has lost 5% of its initial weight. The dehydration of peaches can be prevented by maintaining a high RH (90–95%) in the environment, while maintaining control of the air speed and protection with physical or chemical barriers [16,23]. Moreover, many authors have described the weight-conserving effect of ethylene removal on different fruits, such as peaches, kiwifruits, apricots, melons or tomatoes [9–11,13,21,22].

Two different behaviours can be clearly distinguished in our weight loss data (Table 1 and Figure 2a). On the one hand, the NoES-R, ES-R, and ES-NoR treatments showed a continuous weight loss observed from day 3 of measurement until day 14 in ES-NoR, and until day 24 in NoES-R and ES-R. It should be noted that weight loss in the refrigerated treatments (NoES-R and ES-R) was associated with low RH (around 80%). However, the weight loss observed in ES-NoR was not associated with the RH (it was close to 100%), but was due to the high temperatures and to a certain availability of O_2, which favours fruit respiration. In contrast, the NoES-NoR treatment did not show weight loss until day 7, the last day of measurement. In NoES-NoR, the individual effect of the factors, and their interaction ($p < 0.001$), implies that the progressive accumulation of ethylene and the high temperatures led to an accelerated respiration in the peaches. This resulted in a higher CO_2 accumulation which, combined with ethylene, caused an excessive decrease in available O_2 in the CC, blocking respiration and favouring fermentation processes that did not affect weight loss.

Table 1. Evolution during storage time of the physical parameters in peaches subjected to the different treatments (NoES-R, ES-R, NoES-NoR and ES-NoR). The parameters measured were weight expressed in grams; size expressed in millimetres; firmness measured in newtons; and two-colour parameters a* and b*. The mean ± standard error of the means (SEM) is shown. Different letters for each treatment represent statistically significant differences according to Tukey's test.

Storage Days	Treatment	Weight (g)	Size (mm)	Firmness (N)	Colour (a*)	Colour (b*)
0		170.4 ± 2.1	70 ± 2	25.9 ± 2.6	19.2 ± 0.7	59.7 ± 0.9
3	NoES-R	163.7 ± 2.4 a	64 ± 2 b	22.6 ± 1.4 ab	17.9 ± 1.5 ab	60.5 ± 1.4 a
	ES-R	162.2 ± 1.6 a	67 ± 2 ab	27.9 ± 1.6 a	15.8 ± 1.1 b	60.7 ± 0.7 a
	NoES-NoR	167.9 ± 6.9 a	70 ± 1 a	13.7 ± 1.1 c	21.3 ± 1.3 a	56.2 ± 1.5 b
	ES-NoR	158.1 ± 3.4 a	64 ± 1 b	17.6 ± 1.4 bc	19.3 ± 0.8 ab	57.9 ± 1.1 ab
	Ethylene Scavengers (ES)	n.s.	n.s.	***	n.s.	n.s.
	Temperature (T)	n.s.	n.s.	***	**	**
	ES × T	n.s.	*	n.s.	n.s.	n.s.
7	NoES-R	144.8 ± 3.4 b	64 ± 2 a	22.2 ± 2.2 ab	20.5 ± 0.8 b	59.1 ± 1.0 ab
	ES-R	150.7 ± 0.6 b	66 ± 2 a	26.0 ± 0.6 a	19.8 ± 0.9 bc	60.9 ± 2.1 a
	NoES-NoR	172.6 ± 0.0 a	68 ± 1 a	11.9 ± 0.5 c	26.7 ± 0.5 a	53.6 ± 0.8 c
	ES-NoR	145.5 ± 1.7 b	65 ± 1 a	16.8 ± 2.1 bc	27.2 ± 1.4 a	54.2 ± 1.1 bc
	Ethylene Scavengers (ES)	***	n.s.	***	***	n.s.
	Temperature (T)	***	n.s.	**	n.s.	***
	ES × T	***	n.s.	n.s.	***	n.s.
10	NoES-R	142.0 ± 2.2 a	62 ± 2 a	23.9 ± 2.2 a	19.1 ± 0.9 b	61.0 ± 1.3 a
	ES-R	131.1 ± 2.7 b	66 ± 1 a	21.3 ± 1.1 a	18.4 ± 1.1 b	62.8 ± 0.8 a
	NoES-NoR	-	-	-	-	-
	ES-NoR	135.0 ± 3.1 ab	66 ± 1 a	18.3 ± 1.6 a	32.0 ± 0.8 a	47.8 ± 0.3 b
	One-way ANOVA	*	n.s.	n.s.	***	***
14	NoES-R	131.1 ± 3.7 a	66 ± 2 a	17.3 ± 1.5 a	20.8 ± 1.2 b	55.6 ± 2.4 a
	ES-R	115.8 ± 2.8 b	65 ± 2 a	15.7 ± 1.3 a	18.1 ± 1.0 b	58.6 ± 3.0 a
	NoES-NoR	-	-	-	-	-
	ES-NoR	120.8 ± 0.0 ab	63 ± 1 a	14.3 ± 1.2 a	32.7 ± 0.7 a	44.3 ± 1.3 b
	One-way ANOVA	*	n.s.	n.s.	***	***
17	NoES-R	114.7 ± 2.3 a	61 ± 1 a	20.3 ± 2.3 a	21.4 ± 1.2 a	50.0 ± 2.9 b
	ES-R	105.7 ± 1.6 b	61 ± 1 a	15.0 ± 2.9 a	18.3 ± 0.8 b	59.8 ± 1.5 a
	NoES-NoR	-	-	-	-	-
	ES-NoR	-	-	-	-	-
	t-test	*	n.s.	n.s.	*	**
22	NoES-R	96.7 ± 3.7 a	57 ± 1 a	12.1 ± 2.0 a	21.1 ± 0.5 a	53.0 ± 1.7 a
	ES-R	90.8 ± 3.1 a	60 ± 2 a	14.5 ± 1.8 a	19.1 ± 0.5 b	51.5 ± 2.7 a
	NoES-NoR	-	-	-	-	-
	ES-NoR	-	-	-	-	-
	t-test	n.s.	n.s.	n.s.	**	n.s.
24	NoES-R	84.3 ± 2.6 a	55 ± 2 a	11.8 ± 0.93 a	18.0 ± 0.3 a	47.4 ± 2.0 a
	ES-R	82.2 ± 2.7 a	58 ± 2 a	12.4 ± 2.1 a	19.7 ± 1.0 a	52.1 ± 2.9 a
	NoES-NoR	-	-	-	-	-
	ES-NoR	-	-	-	-	-
	t-test	n.s.	n.s.	n.s.	n.s.	n.s.

Levels of statistical significance are: * $p < 0.05$, ** $p < 0.01$ and *** $p < 0.001$. n.s.: no significant differences. Different letters for each treatment represent statistically significant differences ($p < 0.05$) according to Tukey's test.

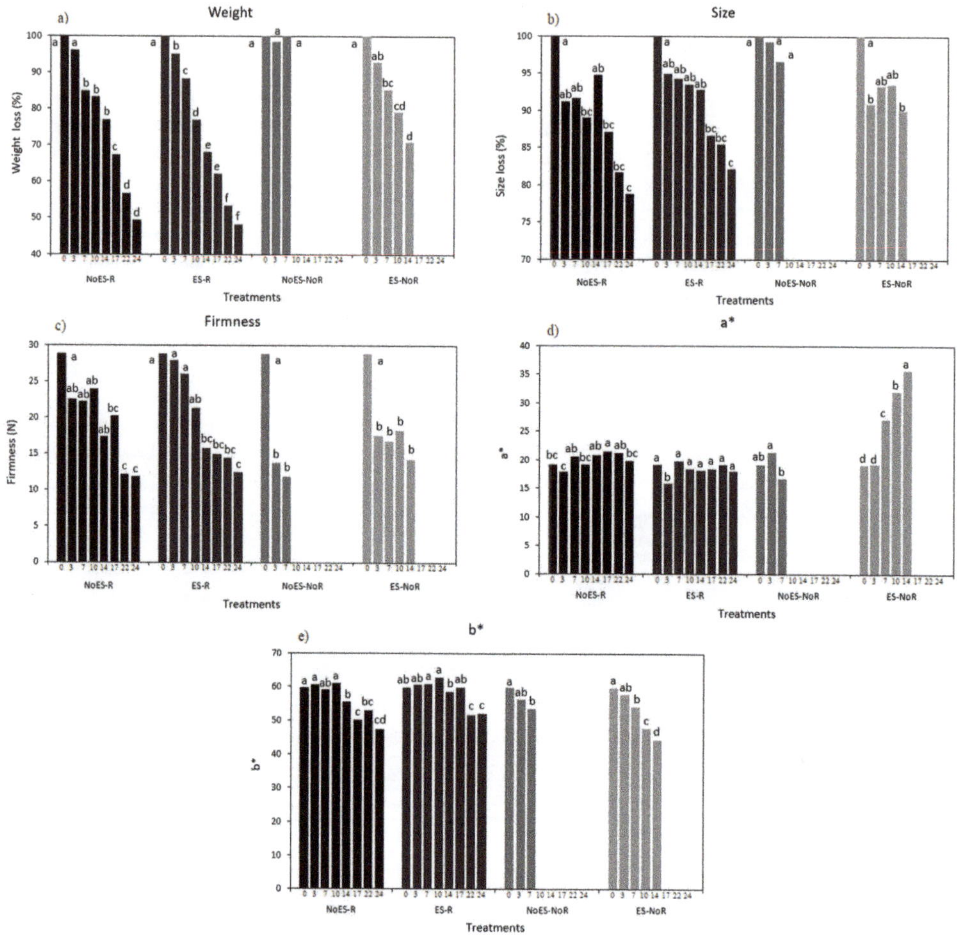

Figure 2. Loss of weight (**a**) and size (**b**) expressed as percentage and compared to initial day; firmness expressed in newtons (**c**); colour expressed as a* (**d**) and colour expressed as b* (**e**) coordinates in CIELAB system in peaches subjected to the different treatments (NoES-R, ES-R, NoES-NoR andES-NoR). Different letters for every day in on treatment represent statistically significant differences according to Tukey's test with the aim of assessing the evolution of each parameter for every treatment.

Fruit size showed a similar trend as fruit weight (Table 1 and Figure 2b). On the one hand, NoES-R, ES-R and ES-NoR treatments showed a continuous decrease in fruit size. In these three treatments, an important reduction in diameter was observed in the first day of conservation (day 3), followed by a stabilization in peach diameter until day 14, to a final, significant decrease until day 24 in NoES-R and ES-R (Figure 2b). In contrast, the NoES-NoR treatment showed a slightly smaller diameter, with significant differences observed on day 3, but not on day 7 with respect to the other three treatments (Table 1 and Figure 2b). This behaviour can be attributable to a possible internal fermentation, which is supported by the low O_2 concentration in the chamber. During anaerobiosis, the fruits ferment, replacing respiration as an energy-producing process. As a result, weight loss is lower than in the other treatments, and therefore the fruit size is preserved [13,21,22].

The decrease in the diameter of peaches, compared with the loss of weight, was less sensitive to the different factors applied. Except in day 3, no statistically significant differences were observed for this parameter (Table 1).

Firmness is one of the main quality attributes that determine the acceptance of the product by the consumer. Different changes in firmness were observed depending on the application of the combined ethylene removal technology and storage temperature (Figure 2c). The ES-R treatment was able to maintain high firmness values during the first week of storage, with values of 26.0 N, while the rest of the treatments showed a reduction of firmness on day 3 and 7, with this result being of significant relevance in the NoES-NoR treatment. This suggests that the use of ethylene scavengers maintained the firmness intact in the short term (days 3 and 7) in the refrigerated treatment (ES-R). In the medium and long term, the firmness values of the ES-R, NoES-R and ES-NoR decreased, with no differences between treatments (Table 1). The NoES-NoR treatment suffered the greatest decreases, losing 47% and 55% of firmness on day 3 and 7, with values of 13.7 N and 11.9 N. These tendencies in NoES-NoR coincided with the highest levels of ethylene and CO_2, and the lowest values of O_2 in this treatment (Figure 1b,c). This is in agreement with the existing literature. For example, [23] showed that the effect of ethylene scavengers on apricots resulted in significantly higher firmness compared to control fruit. Ethylene significantly affects fruit firmness by triggering cell wall hydrolysis, which leads to fruit softening [24,25]. In addition, the expression of polygalacturonase-related genes is associated with ethylene production [26]. The action of this enzyme is considered key in the softening of fruit in general, and stone fruit in particular, increasing its activity in those treatments where the exposure to ethylene was higher [27].

Colour is a parameter that changes significantly during the fruit ripening process, and can be associated with the optimal time of consumption of the fruit [28]. For the variety used in this study, "Rojo de Rito", the ideal values for (a*) and (b*) parameters are 20 and 55, respectively, corresponding to an optimum orange-yellow colour (Table 1).

The (a*) parameter was generally stable throughout the storage time, with a value close to 20 for the refrigerated treatments, although it can be observed that the presence of the ethylene scavengers in ES-R kept the values slightly low compared with the NoES-R treatment, being significant at 17 and 22 days (Table 1 and Figure 2d). However, this changed at the end of the experiment after 24 days, with both treatments showing similar values. In contrast, the NoES-NoR and ES-NoR treatments showed higher a* values than the refrigerated treatments, with these values being statistically significant in NoES-NoR at 3 days, and for two treatments at 7 days with values close to 27. In the medium term, the parameter a* continuously increased in ES-NoR for up to 14 days, reaching values of 33, which indicated a greater presence of red tones in the pulp of the peaches (Figure 2d). Furthermore, important significant differences with respect to the refrigerated treatments were observed in ES-NoR ($p < 0.001$).

With respect to the b* parameter, an overall decreasing trend was observed throughout the storage time, with a different intensity observed depending on the effect of ethylene scavengers and conservation temperature (Table 1, Figure 2e). For the refrigerated treatments, the yellow colour of the peaches was maintained, and even increased, during the first 10 days of conservation, with values of 62, although these values finally decreased at the end of the experiment, with values close to 50 (day 24). The b* value of the non-refrigerated treatments continuously decreased throughout the conservation period, with NoES-NoR showing the lowest values of 53.6 at day 7, but without differences when compared with ES-NoR. In the medium term, the b* parameter decreased up to 14 days, reaching values of 44, with important significant differences with respect to the refrigerated treatments observed in ES-NoR ($p < 0.001$).

By analysing the (a*) and (b*) parameters, it can be stated that the delay in peach ripening caused by ethylene scavengers promoted better colour preservation in the ES-R treatment compared to the control (NoES-R). The fruits of the ES-R treatment had a more stable, greener (a*) and yellower (b*) colour than the control treatment during the storage time. Fruits from the control treatment had a darker, and therefore riper, pulp than fruits from the ES-R treatment. In the 25 °C storage treatments (NoES-NoR and ES-NoR), the ethylene scavengers were able to delay colour change in the short term (days 3 and 7), in the ES-NoR treatment compared to NoES-NoR. However, the NoES-NoR treatment did not last beyond day 7 of storage, due to adverse temperature and ethylene conditions. On days 10 and 14, the ES-NoR treatment showed a marked increase in parameter a* and a decrease in parameter b* compared to the control treatment (Table 1, Figure 2d,e).

The authors of [29] observed the relationship between colour preservation and antioxidant capacity, the concentration of phenolic compounds and chlorophylls in peaches. This study showed that colour preservation is a highly desirable trait for breeding programmes aimed at improving the consumption of peaches selected for their nutraceutical properties. The longer the fruit retains its colour, the higher the quantity of beneficial compounds present in the fruit.

3.3. Maturity Parameters

Soluble solid content (SSC) is used to determine the total ratio of sugars dissolved in a liquid (peach juice in this study). During post-harvest ripening of climacteric fruits, such as peaches, sugars displace acids by certain metabolic processes, increasing SSC and giving the fruit a sweet taste. Therefore, SSC is a key indicator of the ripening stage of the fruit (Table 2) [20,30,31].

Table 2. Evolution during storage time of the maturity parameters in peaches subjected to different treatments (NoES-R, ES-R, NoES-NoR and ES-NoR). The parameters measured were SSC expressed as percentage, pH, TA expressed as percentage, MI as the SSC/TA ratio.

Storage Days	Treatment	SSC (%)	pH	TA (%)	MI
0		11.2 ± 0.4	3.7 ± 0.03	3.7 ± 0.21	3.04 ± 0.16
3	NoES-R	13.0 ± 0.6 a	3.7 ± 0.05 c	3.4 ± 0.03 b	3.82 ± 0.20 ab
	ES-R	11.0 ± 1.0 a	3.6 ± 0.03 c	3.9 ± 0.07 a	2.84 ± 0.30 b
	NoES-NoR	11.7 ± 0.9 a	4.1 ± 0.05 a	2.4 ± 0.04 d	4.91 ± 0.41 a
	ES-NoR	11.0 ± 0.6 a	3.9 ± 0.04 b	2.8 ± 0.07 c	3.94 ± 0.19 ab
	Ethylene Scavengers (ES)	n.s.	**	***	**
	Temperature (T)	n.s.	***	***	**
	ES × T	n.s.	n.s.	n.s.	n.s.
7	NoES-R	11.7 ± 0.9 ab	3.8 ± 0.02 c	3.8 ± 0.30 a	3.16 ± 0.38 b
	ES-R	14.0 ± 1.0 a	3.7 ± 0.01 c	4.3 ± 0.10 a	3.30 ± 0.31 b
	NoES-NoR	8.3 ± 1.2 c	4.5 ± 0.05 a	1.5 ± 0.16 c	5.88 ± 1.46 a
	ES-NoR	10.7 ± 0.3 bc	4.1 ± 0.05 b	2.5 ± 0.25 b	4.39 ± 0.52 ab
	Ethylene Scavengers (ES)	*	***	**	n.s.
	Temperature (T)	**	***	***	*
	ES × T	n.s.	**	n.s.	n.s.
10	NoES-R	13.6 ± 0.7 a	3.8 ± 0.03 b	3.7 ± 0.10 b	3.73 ± 0.27 ab
	ES-R	12.0 ± 0.7 a	3.7 ± 0.03 b	4.0 ± 0.09 a	2.99 ± 0.22 b
	NoES-NoR	-	-	-	-
	ES-NoR	12.0 ± 1.0 a	4.1 ± 0.02 a	2.1 ± 0.06 c	4.55 ± 0.47 a
	One-way ANOVA	n.s.	***	***	***
14	NoES-R	12.4 ± 1.2 a	4.0 ± 0.03 b	4.1 ± 0.16 a	3.48 ± 0.26 ab
	ES-R	13.8 ± 0.9 a	3.9 ± 0.07 b	4.0 ± 0.12 a	3.00 ± 0.19 b
	NoES-NoR	-	-	-	-
	ES-NoR	11.6 ± 0.7 a	4.4 ± 0.05 a	2.6 ± 0.10 b	5.79 ± 0.51 a
	One-way ANOVA	n.s.	***	***	*
17	NoES-R	13.4 ± 1.4 a	4.1 ± 0.04 a	3.2 ± 0.11 b	4.27 ± 0.49 a
	ES-R	16.4 ± 2.9 a	4.0 ± 0.04 a	3.9 ± 0.15 a	4.24 ± 0.76 a
	NoES-NoR	-	-	-	-
	ES-NoR	-	-	-	-
	t-test	n.s.	n.s.	**	n.s.
22	NoES-R	16.8 ± 0.9 a	4.2 ± 0.05 a	3.2 ± 0.10 b	5.26 ± 0.33 a
	ES-R	16.7 ± 0.4 a	4.0 ± 0.11 a	4.0 ± 0.30 a	4.27 ± 0.33 b
	NoES-NoR	-	-	-	-
	ES-NoR	-	-	-	-
	t-test	n.s.	n.s.	*	*
24	NoES-R	21.3 ± 1.1 a	4.3 ± 0.06 a	2.9 ± 0.26 b	7.40 ± 6.4 a
	ES-R	23.3 ± 2.4 a	4.1 ± 0.04 b	3.7 ± 0.22 a	6.49 ± 8.6 a
	NoES-NoR	-	-	-	-
	ES-NoR	-	-	-	-
	t-test	n.s.	*	*	n.s.

Levels of statistical significance are: * $p < 0.05$, ** $p < 0.01$ and *** $p < 0.001$. n.s.: no significant differences. Different letters for each treatment represent statistically significant differences ($p < 0.05$) according to Tukey's test.

Comparing the data obtained, a logical evolution of SSC in the NoES-R, ES-R treatments was observed, as shown by a progressive increase, with values close to 13% on day 3 and close to 22% on day 24. The ES-NoR treatment, despite having adverse storage conditions, did not showed significant differences compared to the refrigerated treatments from day 10 of the experiment until 14, with values of 12%. On the other hand, the NoES-NoR treatment followed an opposite response, decreasing to a value of 8.33% at day 7. This behaviour suggests the hypothesis that the combination of high ethylene and CO_2 production and a decrease in O_2 (see Figure 1) caused internal fermentation (Figure 3a).

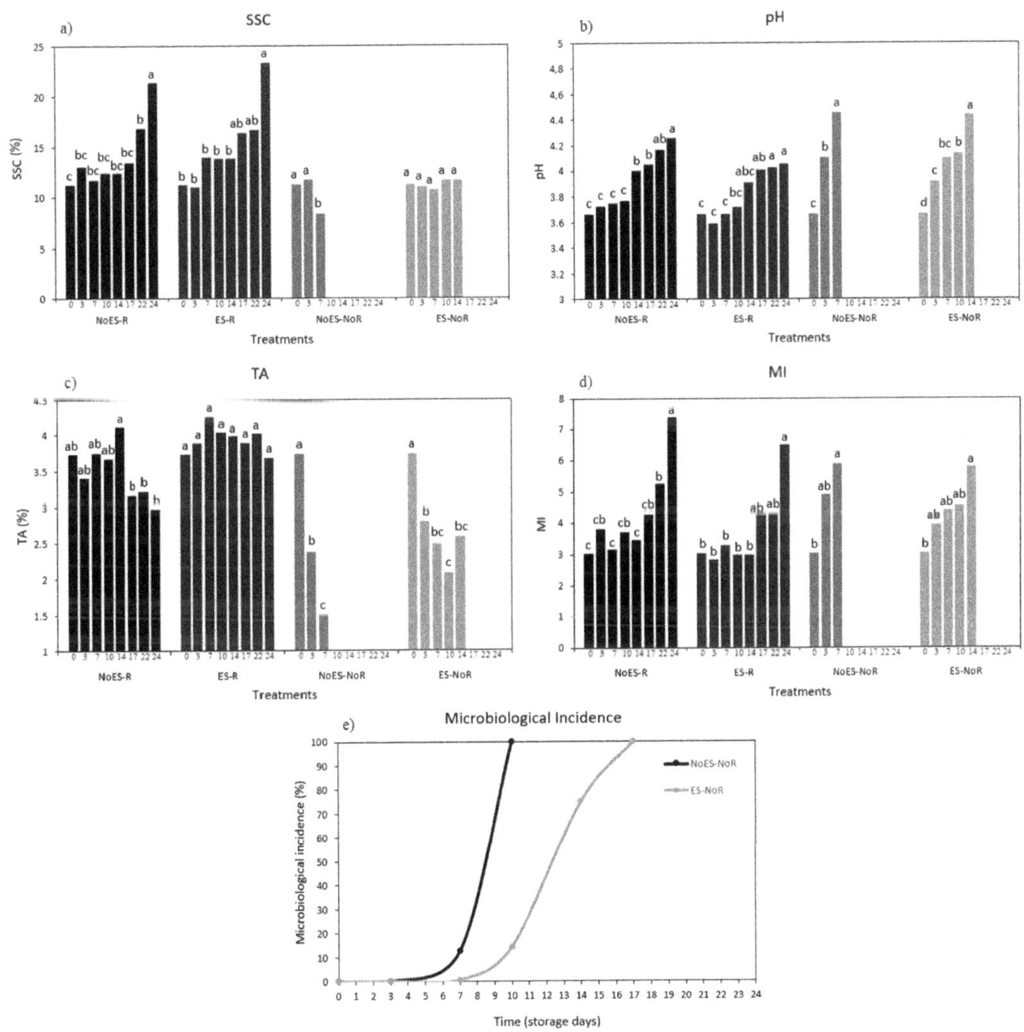

Figure 3. Soluble solid content (SSC) expressed in percentage (**a**), hydrogen potential expressed in pH units (**b**), titratable acidity (TA) calculated as malic acid equivalents and expressed as percentage (**c**), maturity index (MI) calculated as SSC/TA ratio (**d**), and microbiological incidence expressed as percentage (**e**) over the storage time (24 days) in peaches subjected to the different treatments (NoES-R, ES-R, NoES-NoR and ES-NoR). Different letters for each day in one treatment represent statistically significant differences according to Tukey's test with the aim of assessing the evolution of each parameter for every treatment.

Other studies have reported similar results, indicating a slight effect of ethylene scavengers on peach fruit conserved for 36 days at 0 °C [8,32]. The same effect was also observed in baby bananas, where SSC values tended to increase, depending on the doses of $KMnO_4$ and clay provided as ethylene scavengers [33]. In addition, apricots stored at 15 °C showed changes in SSC which were significantly affected by ethylene scavengers [4].

pH is perhaps the most important potentiometric measurement used in the agriculture and food industry and serves to quantify the concentration of H_3O^+ in the juice obtained from the liquefied fruit, which is determined as active acidity. This can be associated with the content of acids present and the capacity of microbial proliferation in conservation, as the acids will act on the fruit as a natural physiological barrier against microbial action (Table 2 and Figure 3b) [9,10].

The pH tended to increase during storage. Already from day 3, the first significant differences could be observed with lower values in NoES-R (3.7) and ES-R (3.6) indicating a key role of temperature ($p < 0.001$). Moreover, on the same day, the ES-NoR treatment, with a pH value of 3.9, showed a lower value than NoES-NoR with a pH of 4.1, indicating a relevant role of the ethylene scavengers utilized ($p < 0.01$) (Table 2). The same separation in treatments was observed on day 7, with both factors being highly relevant ($p < 0.001$), as well as the interaction between them ($p < 0.01$). On days 10 and 14, a clear separation between the refrigerated and ES-NoR treatments was observed ($p < 0.001$). In the long term, on days 17, 20, and 24, significant differences were only observed between the two refrigerated treatments on the last day, indicating the importance of ethylene removal in maintaining juice acidity at lower pH values (Table 2).

From an overall point of view, the ethylene scavengers were able to decrease the pH by 2.54%. In addition, the refrigeration temperatures decreased pH values by 6.74% (Figure 3b).

Some authors have shown an indirect effect of ethylene on pH values in other stone fruit, such as apricots [4]. The elimination of ethylene decreases metabolic processes related to fruit ripening, which minimises sugar production and preserves pH levels [25]. Other authors have shown data for a delayed pH rise; [10] observed, in a review, a slowing down of pH rise in "Golden Delicious" apples after the use of $KMnO_4$-based C_2H_4 scavengers [34]. This effect was also observed in "Kolikutu" [35] and "Williams" [36] bananas, in kiwifruit of the "Hayward" variety [37], and "Karuthacolomban" [38] and "Haden 2H" [39] mangoes. In peaches, [32] observed a significant delay in pH increase in ripening after 36 days of storage in those fruits treated with ethylene scavengers based on $KMnO_4$.

Titratable acidity (TA) represents the total amount of acids in the fruit, and it is expressed as a percentage. TA is inversely proportional to pH and SSC. Acids influence food taste (roughness), colour, microbial stability, and quality (Table 2 and Figure 3c).

The TA values tended to decrease throughout the study period, especially in non-refrigerated treatments and in the NoES-R (control) treatment compared to ES-R. These responses indicate that both factors were relevant when observing significant differences in TA, particularly on day 3 ($p < 0.001$). In the medium term, on days 10 and 14 of the trial, the differences observed in ES-NoR with respect to NoES-R and ES-R were due to the temperature factor ($p < 0.001$). In the long term, significant differences were observed between the two refrigerated treatments associated with ethylene scavengers on days 17 and 20, but not on day 24 of the trial.

From a general point of view, the use of ethylene scavengers was able to avoid 15.9% of TA losses. Similarly, the refrigeration temperature utilized was able to avoid 40.6% of TA losses (Figure 3c).

Other authors have already observed that the elimination of ethylene caused a maintenance of pH levels, delaying the acid degradation process. A natural increase in pH values implies a decrease in TA [32]. This effect was also observed using sachets of $KMnO_4$ in mangoes [40]. In conclusion, the data presented in this study suggest that the use of ethylene scavengers positively affected acid metabolism, with a resulting delay in sugar production and higher acid accumulation.

The maturity index (MI) depends on the total acidity and the soluble solids content and tends to increase during fruit ripening (Table 2 and Figure 3d) [31,41].

The MI values generally increased with the ripening process in climacteric fruits such as peaches, being highly affected by the use of ethylene scavengers and storage temperature. In the values recorded on day 3, significant differences were only observed between the ES-R (2.84) and NoES-NoR (4.91) treatments (Table 2). The responses observed on day 7 were very similar to those observed on day 3, with a high value of 5.88 observed in NoES-NoR. In the medium term, at day 10, significant differences were observed between ES-NoR and ES-R, but not between NoES-R and ES-NoR, indicating a positive role of the ethylene scavengers in delaying maturity in ES-NoR. At 14 days, ES-NoR showed significant differences compared with the refrigerated treatments, reaching a value of 5.79 that was very similar to the one reached by NoES-NoR at day 7, with such values of MI exceeding the optimum quality limits for peaches. In the long term, there were no differences between the refrigerated treatments, and the MI values simply increased and became critical on the last day of the trial at day 24.

From a general point of view, the ethylene scavengers were able to prevent increase in MI values by 9%. In addition, the refrigerated temperature prevented increase in MI values by 15% (Figure 3d).

3.4. Microbiological Incidence

During the entire conservation period, loss of fruits in the ES-R and NoES-R treatments did not occur because of the maintenance of constant optimum refrigeration conditions (Figure 3e).

On the other hand, in the ES-NoR and NoES-NoR treatments, differences were observed from day 7 onwards. On day 7, the NoES-NoR treatment suffered losses of 50% due to microbiological damage, while the ES-NoR treatment had a loss rate of 12%. On day 10, damage in the NoES-NoR treatment reached 100% of the fruit, making further analysis impossible. However, the ES-NoR treatment increased its fruit loss, to a total of 14.28%. On day 14, losses in the NoES-NoR treatment increased to 75%. Finally, on day 17, losses due to microbiological incidence in the ES-NoR treatment reached 100% (Figure 3e).

The application of ultraviolet radiation is known to have a spore-killing effect [36,37]. Therefore, the above data suggest that the use of ultraviolet light helped in both ethylene degradation and spore removal. As the machine forces air through the UV light, these spores could have been affected by photocatalysis [42].

The factors that most affect post-harvest losses of stone fruits are those associated with physiological damage or diseases. Among them, one of the most important is the effect of microorganisms, such as *Monilinia* spp. [43,44]. Several studies have shown that ethylene has an effect on the development of post-harvest diseases depending on the host-pathogen system and fruit [4,27]. In some studies, fungi from the genus *Monilinia* spp. have been inoculated on peach petals, with a conservative effect of ethylene removal observed on their browning process [45]. Research on tomatoes conserved at 11 °C and 22 °C for 28 days has shown that the use of ethylene scavengers supported with thymol led to the highest fungal inhibition (≥91%) in comparison to the control, with this study also concluding that the C_2H_4-scavengers were helpful in controlling post-harvest fungal diseases while preserving fruit quality [46].

In the present study, the decrease in microbiological damage observed in the ES-NoR treatment suggests that the use of ethylene scavengers ($KMnO_4$ and UV radiation) in peaches subjected to a storage temperature of 25 °C also prevented the proliferation of pathogens, thereby extending their survival by 7 days compared to the NoES-NoR treatment.

4. Conclusions

The results obtained provide clear evidence that the combined effect of the photocatalytic action of UV radiation and potassium permanganate favoured the preservation of the post-harvest quality of the fruits stored at 1 °C and 25 °C. This was especially important in the ES-NoR treatment, where metabolic processes would have been more active due to the higher temperatures, although these were slowed down by the ethylene scavengers, extending fruit survival by 7 days compared to the NoES-NoR treatment. In addition, among the refrigerated treatments, a better ES-R performance was also observed due to the effect of the ethylene scavengers on SSC and firmness parameters in the short term (7 days), on MI and colour parameters a* and b* in the medium term (14 days), and on pH and TA in the long term (beyond day 14).

Author Contributions: Conceptualization, R.A.-S., J.R.A.-M. and S.L.-M.; methodology, R.A.-S., J.R.A.-M. and S.L.-M.; software, R.A.-S. and J.R.A.-M.; validation, E.N.-D. and J.A.G.; formal analysis, J.R.A.-M. and S.L.-M.; investigation, R.A.-S., J.R.A.-M. and S.L.-M.; resources, J.R.A.-M. and S.L.-M.; data curation, R.A.-S. and J.R.A.-M.; writing—original draft preparation, R.A.-S., J.R.A.-M. and S.L.-M.; writing—review and editing, R.A.-S., J.R.A.-M. and S.L.-M.; visualization, E.N.-D., J.R.A.-M. and S.L.-M.; supervision, J.R.A.-M., S.L.-M. and J.A.G.; project administration, S.L.-M. and E.N.-D.; funding acquisition, E.N.-D. and J.R.A.-M. All authors have read and agreed to the published version of the manuscript.

Funding: This research was funded by "Nuevas Tecnologías Agroalimentarias (KEEPCOOL)" and "Cátedra Emprendimiento en el Ámbito Agroalimentario" from Universidad Católica San Antonio de Murcia (UCAM) through project CFE/KE/76-19.

Institutional Review Board Statement: Not applicable.

Informed Consent Statement: Not applicable.

Acknowledgments: The authors gratefully acknowledge Antonio Cerdá Cerdá, Thader Cieza, S.C.L. and FECOAM ("Federación de Cooperativas Agrarias de Murcia"), especially Antonio Sanz and Lola Mondéjar for supplying peaches and Mario Fon for his help with English editing.

Conflicts of Interest: The doctoral thesis (PhD) of Ramiro Alonso Salinas, co-author of this work, is being co-financed by the company Nuevas Tecnologías Agroalimentarias (KEEPCOOL), within the UCAM Universidad Católica de Murcia Industrial Doctorate Program. The remaining authors of the article declare that their contribution to this research was carried out in the absence of commercial or financial relationships that could be construed as a possible conflict of interest.

References

1. Minas, I.S.; Tanou, G.; Molassiotis, A. Environmental and orchard bases of peach fruit quality. *Sci. Hortic.* **2018**, *235*, 307–322. [CrossRef]
2. Crisosto, C.H.; Johnson, R.S.; DeJong, T.; Day, K.R. Orchard Factors Affecting Postharvest Stone Fruit Quality. *HortScience* **1997**, *32*, 820–823. [CrossRef]
3. Iqbal, N.; Nazar, R.; Khan MI, R.; Khan, N.A. Variation in photosynthesis and growth of mustard cultivars: Role of ethylene sensitivity. *Sci. Hortic.* **2012**, *135*, 1–6. [CrossRef]
4. Álvarez-Hernández, M.H.; Martínez-Hernández, G.B.; Avalos-Belmontes, F.; Molina, F.D.M.; Artés-Hernández, F. Postharvest quality retention of apricots by using a novel sepiolite–loaded potassium permanganate ethylene scavenger. *Postharvest Biol. Technol.* **2020**, *160*, 111061. [CrossRef]
5. Pathak, N.; Caleb, O.J.; Geyer, M.; Herppich, W.B.; Rauh, C.; Mahajan, P.V. Photocatalytic and Photochemical Oxidation of Ethylene: Potential for Storage of Fresh Produce—A Review. *Food Bioprocess Technol.* **2017**, *10*, 982–1001. [CrossRef]
6. Sakizci, M. Effect of salt modification and acid activation on ethylene adsorption properties of sepiolite. *Adsorption* **2013**, *19*, 1083–1091. [CrossRef]
7. Janjarasskul, T.; Suppakul, P. Active and intelligent packaging: The indication of quality and safety. *Crit. Rev. Food Sci. Nutr.* **2018**, *58*, 808–831. [CrossRef]
8. Gaikwad, K.K.; Singh, S.; Negi, Y.S. Ethylene scavengers for active packaging of fresh food produce. *Environ. Chem. Lett.* **2020**, *18*, 269–284. [CrossRef]
9. Wei, H.; Seidi, F.; Zhang, T.; Jin, Y.; Xiao, H. Ethylene scavengers for the preservation of fruits and vegetables: A review. *Food Chem.* **2021**, *337*, 127750. [CrossRef]
10. Álvarez-Hernández, M.H.; Martínez-Hernández, G.B.; Avalos-Belmontes, F.; Castillo-Campohermoso, M.A.; Contreras-Esquivel, J.C.; Artés-Hernández, F. Potassium Permanganate-Based Ethylene Scavengers for Fresh Horticultural Produce as an Active Packaging. *Food Eng. Rev.* **2019**, *11*, 159–183. [CrossRef]

11. Park, Y.S.; Jung, S.T.; Gorinstein, S. Ethylene treatment of 'Hayward' kiwifruits (Actinidia deliciosa) during ripening and its influence on ethylene biosynthesis and antioxidant activity. *Sci. Hortic.* **2006**, *108*, 22–28. [CrossRef]
12. Kim, S.; Jeong, G.H.; Kim, S.-W. Ethylene Gas Decomposition Using ZSM-5/WO3-Pt-Nanorod Composites for Fruit Freshness. *ACS Sustain. Chem. Eng.* **2019**, *7*, 11250–11257. [CrossRef]
13. Pathak, N. Photocatalysis and Vacuum Ultraviolet Light Photolysis as Ethylene Removal Techniques for Potential Application in Fruit Storage. Ph.D. Thesis, Technische Universität Berlin, Berlin, Germany, 2019. [CrossRef]
14. Yildirim, S.; Röcker, B.; Pettersen, M.K.; Nilsen-Nygaard, J.; Ayhan, Z.; Rutkaite, R.; Radusin, T.; Suminska, P.; Marcos, B.; Coma, V. Active Packaging Applications for Food. *Compr. Rev. Food Sci. Food Saf.* **2018**, *17*, 165–199. [CrossRef] [PubMed]
15. Wills, R.; Warton, M. Efficacy of Potassium Permanganate Impregnated into Alumina Beads to Reduce Atmospheric Ethylene. *J. Am. Soc. Hortic. Sci.* **2004**, *129*, 433–438. [CrossRef]
16. Alvarez-Hernández, M.H.; Artés-Hernández, F.; Ávalos-Belmontes, F.; Castillo-Campohermoso, M.A.; Contreras-Esquivel, J.C.; Ventura-Sobrevilla, J.; Martínez-Hernández, G.B. Current Scenario of Adsorbent Materials Used in Ethylene Scavenging Systems to Extend Fruit and Vegetable Postharvest Life. *Food Bioprocess Technol.* **2018**, *11*, 511–525. [CrossRef]
17. Martínez-Romero, D.; Zapata, P.J.; Guillén, F.; Paladines, D.; Castillo, S.; Valero, D.; Serrano, M. The addition of rosehip oil to Aloe gels improves their properties as postharvest coatings for maintaining quality in plum. *Food Chem.* **2017**, *217*, 585–592. [CrossRef]
18. Ibhadon, A.; Fitzpatrick, P. Heterogeneous Photocatalysis: Recent Advances and Applications. *Catalysts* **2013**, *3*, 189–218. [CrossRef]
19. Kaewklin, P.; Siripatrawan, U.; Suwanagul, A.; Lee, Y.S. Active packaging from chitosan-titanium dioxide nanocomposite film for prolonging storage life of tomato fruit. *Int. J. Biol. Macromol.* **2018**, *112*, 523–529. [CrossRef]
20. Zhang, B.; Peng, B.; Zhang, C.; Song, Z.; Ma, R. Determination of fruit maturity and its prediction model based on the pericarp index of absorbance difference (IAD) for peaches. *PLoS ONE* **2017**, *12*, e0177511. [CrossRef]
21. Le Nguyen, L.P.; Zsom, T.; Sao Dam, M.; Baranyai, L.; Hitka, G. Evaluation of the 1-MCP microbubbles treatment for shelf-life extension for melons. *Postharvest Biol. Technol.* **2019**, *150*, 89–94. [CrossRef]
22. Sammi, S.; Masud, T. Effect of different packaging systems on the quality of tomato (Lycopersicon esculentum var. Rio Grande) fruits during storage. *Int. J. Food Sci. Technol.* **2009**, *44*, 918–926. [CrossRef]
23. Wu, B.; Guo, Q.; Wang, G.-X.; Peng, X.-Y.; Wang, J.-D.; Che, F.-B. Effects of different postharvest treatments on the physiology and quality of 'Xiaobai' apricots at room temperature. *J. Food Sci. Technol.* **2014**, *52*, 2247–2255. [CrossRef]
24. Zhou, H.-W.; Dong, L.; Ben-Arie, R.; Lurie, S. The role of ethylene in the prevention of chilling injury in nectarines. *J. Plant Physiol.* **2001**, *158*, 55–61. [CrossRef]
25. Fan, X.; Shu, C.; Zhao, K.; Wang, X.; Cao, J.; Jiang, W. Regulation of apricot ripening and softening process during shelf life by post-storage treatments of exogenous ethylene and 1-methylcyclopropene. *Sci. Hortic.* **2018**, *232*, 63–70. [CrossRef]
26. Hayama, H.; Shimada, T.; Fujii, H.; Ito, A.; Kashimura, Y. Ethylene-regulation of fruit softening and softening-related genes in peach. *J. Exp. Bot.* **2006**, *57*, 4071–4077. [CrossRef]
27. Palou, L.; Crisosto, C.H.; Garner, D.; Basinal, L.M. Effect of continuous exposure to exogenous ethylene during cold storage on postharvest decay development and quality attributes of stone fruits and table grapes. *Postharvest Biol. Technol.* **2003**, *27*, 243–254. [CrossRef]
28. Iglesias, I.; Echeverría, G. Differential effect of cultivar and harvest date on nectarine colour, quality and consumer acceptance. *Sci. Hortic.* **2009**, *120*, 41–50. [CrossRef]
29. Serra, S.; Anthony, B.; Masia, A.; Giovannini, D.; Musacchi, S. Determination of Biochemical Composition in Peach (*Prunus persica* L. Batsch) Accessions Characterized by Different Flesh Color and Textural Typologies. *Foods* **2020**, *9*, 1452. [CrossRef]
30. Crisosto, C.; Gugliuzza, G.; Garner, D.; Palou, L. Understanding the role of ethylene in peach cold storage life. *Acta Hortic.* **2001**, 287–288. [CrossRef]
31. Zhang, P.; Shao, X.; Wei, Y.; Xu, F.; Wang, H. At-harvest fruit maturity affects sucrose metabolism during cold storage and is related to chilling injury in peach. *J. Food Sci. Technol.* **2020**, *57*, 2000–2009. [CrossRef]
32. Emadpour, M.; Ghareyazie, B.; Kalaj, Y.R.; Entesari, M.; Bouzari, N. Effect of the potassium permanganate coated zeolite nanoparticles on the quality characteristic and shelf life of peach and nectarine. *J. Agric. Technol.* **2015**, *11*, 1263–1273.
33. García, J.C.; Balaguera-López, H.E.; Herrera, A.O. Conservación del fruto de banano bocadillo (Musa AA Simmonds) con la aplicación de permanganato de potasio ($KMnO_4$). *Rev. Colomb. Cienc. Hortíc.* **2012**, *6*, 161–171. [CrossRef]
34. Sardabi, F.; Mohtadinia, J.; Shavakhi, F.; Jafari, A.A. The effects of 1-methylcyclopropen (1-MCP) and potassium permanganate coated zeolite nanoparticles on shelf-life extension and quality lossof golden delicious apples. *J. Food Process. Preserv.* **2014**, *38*, 2176–2182. [CrossRef]
35. Chamara, D.; Illeperuma, K.; Galappatty, T.; Sarananda, K. Modified atmosphere packaging of 'Kolikuttu' bananas at low temperature. *J. Hortic. Sci. Biotechnol.* **2000**, *75*, 92–96. [CrossRef]
36. Tourky, M.; Tarabih, M.; El-Eryan, E. Physiological studies on the marketability of Williams banana fruits. *Am. J. Plant Physiol.* **2014**, *9*, 1–15. [CrossRef]
37. Ramin, A.; Rezaei, A.; Shams, M. Potassium permanganates and short term hypobaric enhances shelf-life of kiwifruits. *Acta Hortic.* **2010**, 849–852. [CrossRef]
38. Illeperuma, C.K.; Jayasuriya, P. Prolonged storage of 'Karuthacolomban' mango by modified atmosphere packaging at low temperature. *J. Hortic. Sci. Biotechnol.* **2002**, *77*, 153–157. [CrossRef]

39. Castro, J.; Conte, R.; Carvalho, C.; Rossetto, C. Effects of postharvest treatments and film packaging on quality of 'Haden 2H' mangoes. *Acta Hortic.* **2010**, *864*, 295–298. [CrossRef]
40. Jeronimo, E.M.; Brunini, M.A.; Arruda, M.C.D.; Cruz, J.C.S.; Gava, G.J.D.C.; Silva, M.D.A. Qualidade de mangas 'Tommy Atkins' armazenadas sob atmosfera modificada. *Ciênc. Agrotecnol.* **2007**, *31*, 1122–1130. [CrossRef]
41. Álvarez-Hernández, M.H.; Martínez-Hernández, G.B.; Avalos-Belmontes, F.; Rodríguez-Hernández, A.M.; Castillo-Campohermoso, M.A.; Artés-Hernández, F. An Innovative Ethylene Scrubber Made of Potassium Permanganate Loaded on a Protonated Montmorillonite: A Case Study on Blueberries. *Food Bioprocess Technol.* **2019**, *12*, 524–538. [CrossRef]
42. Katara, G.; Hemvani, N.; Chitnis, S.; Chitnis, V.; Chitnis, D.S. Surface disinfection by exposure to germicidal UV light. *Indian J. Med. Microbiol.* **2008**, *26*, 241–242. [CrossRef]
43. Forges, M.; Bardin, M.; Urban, L.; Aarrouf, J.; Charles, F. Impact of UV-C Radiation Applied during Plant Growth on Pre- and Postharvest Disease Sensitivity and Fruit Quality of Strawberry. *Plant Dis.* **2020**, *104*, 3239–3247. [CrossRef] [PubMed]
44. Muzzaffar, S.; Bhat, M.M.; Wani, T.A.; Wani, I.A.; Masoodi, F.A. Postharvest Biology and Technology of Apricot. In *Postharvest Biology and Technology of Temperate Fruits*; Mir, S.A., Shah, M.A., Mir, M.M., Eds.; Springer International Publishing: Cham, Switzerland, 2018; pp. 201–222. [CrossRef]
45. Vall-Llaura, N.; Giné-Bordonaba, J.; Usall, J.; Larrigaudière, C.; Teixidó, N.; Torres, R. Ethylene biosynthesis and response factors are differentially modulated during the interaction of peach petals with Monilinia laxa or Monilinia fructicola. *Plant Sci.* **2020**, *299*, 110599. [CrossRef] [PubMed]
46. Álvarez-Hernández, M.H.; Martínez-Hernández, G.B.; Castillejo, N.; Martínez, J.A.; Artés-Hernández, F. Development of an antifungal active packaging containing thymol and an ethylene scavenger. Validation during storage of cherry tomatoes. *Food Packag. Shelf Life* **2021**, *29*, 100734. [CrossRef]

agronomy

MDPI

Article

Control of Substrate Water Availability Using Soil Sensors and Effects of Water Deficit on the Morphology and Physiology of Potted *Hebe andersonii*

Daniel Bañón [1,*], Beatriz Lorente [1], Sebastián Bañón [2], María Fernanda Ortuño [1], María Jesús Sánchez-Blanco [1] and Juan José Alarcón [1]

1 Department of Irrigation, Centro de Edafología y Biología Aplicada del Segura (CEBAS-CSIC), 30100 Murcia, Spain; blorente@cebas.csic.es (B.L.); mfortuno@cebas.csic.es (M.F.O.); quechu@cebas.csic.es (M.J.S.-B.); jalarcon@cebas.csic.es (J.J.A.)

2 Department of Agricultural Engineering, UPCT—Technical University of Cartagena, 30203 Cartagena, Spain; sebastian.arias@upct.es

* Correspondence: dbanon@cebas.csic.es; Tel.: +34-968-396-200

Abstract: Many plant producers tend to overwater crops to prevent water stress and salt-induced damage. These practices waste irrigation water and cause leaching that harms the environment and increases production costs. In order to optimize water consumption and minimize the environmental impact of plant production, this study aimed to determine the physiological and morphological responses of *Hebe andersonii* to three substrate volumetric water contents (49%, 39%, and 32%). The experiment was conducted in a greenhouse with an irrigation protocol that consisted of adding small volumes of water to avoid leaching while monitoring substrate moisture with dielectric soil sensors. The results showed that moderately low substrate moisture improved the water-use efficiency, while growth was significantly reduced under more severe water deficit conditions (but without leaf chlorosis or abscission). The photosynthetic activity of Hebe was primarily controlled by the stomatal aperture, which was co-determined by the substrate moisture and seasonal temperature. Hebe leaves promoted non-photochemical quenching when carbon assimilation was limited by a water deficit, and accumulated solutes through an osmotic adjustment process (especially Cl^-, Na^+, and K^+) to maintain their water status. Overall, *Hebe andersoni* cv. Variegata could successfully grow and improve its water-use efficiency in low substrate moisture and under a non-draining irrigation regime.

Keywords: deficit irrigation; plant physiology; ornamental plants; water relations; water-use efficiency; abiotic stress

Citation: Bañón, D.; Lorente, B.; Bañón, S.; Ortuño, M.F.; Sánchez-Blanco, M.J.; Alarcón, J.J. Control of Substrate Water Availability Using Soil Sensors and Effects of Water Deficit on the Morphology and Physiology of Potted *Hebe andersonii*. *Agronomy* **2022**, *12*, 206. https://doi.org/10.3390/agronomy12010206

Academic Editor: Giovanni Caruso

Received: 30 November 2021
Accepted: 12 January 2022
Published: 15 January 2022

1. Introduction

Managing the irrigation of potted crops is a complex task because substrates can easily dry out due to their low water retention capacity and the small volume of the containers [1]. These conditions lead many ornamental plant producers to overwater their crops in order to avoid drought, while others irrigate the potting substrate to full container capacity to maximize growth, regardless of the plant's water requirements. These practices waste irrigation water and cause leachate, which harms the environment and increases production costs [2,3]. Several studies have suggested that applying more water than required to cover evapotranspiration is undesirable for economic and social reasons [3]. Additionally, it is well known that high substrate moisture can increase plant susceptibility to fungi [4].

Another factor that leads to excessive water use in pot cultures is salt flushing, which is a technique used to leach salts out of the substrate and prevent them from accumulating in the root system. Salt flushing presents the disadvantage of draining away fertilizers, leaving them unavailable to the plant and diminishing vegetative development [3,5]. Another drawback is that the discharge of pesticides and other chemical pollutants commonly found in agricultural drainage is likely to violate environmental policies [6]. Such arguments, added to increasing consumer demands, are promoting plant production based on environmentally friendly practices.

One indicator that measures sustainability in crop production is water-use efficiency (WUE) [7]. In short, the goal of optimizing WUE consists of minimizing water consumption without compromising crop yield. In this sense, closed-loop irrigation systems have been developed to minimize water drainage from greenhouses, but their commercial applications are still limited [8] because of high investment costs and challenges in recycling nutrient solutions. An alternative approach to improving irrigation efficiency would be to maintain the substrate moisture constant via the recurrent addition of small volumes of water [3,9]. In the case of ornamental plants, Nemali and van Iersel [2] reduced water demand and improved plant quality by implementing this strategy. Such irrigation systems can reduce or eliminate leaching, which should result in substantial water savings in commercial-scale production [10]. However, little research has been done to identify the effects of low substrate moisture on the quality and production of ornamental potted crops [11].

Despite the benefits of high water-use efficiencies, low substrate moisture levels may expose plants to severe water stress, and negatively affect plant metabolic and physiological processes [12]. However, plant tolerance to a water deficit depends on the species and the magnitude and duration of the water deficit [13]. Therefore, it is necessary to know the adaptive capacity to drought and the physiological mechanisms of tolerance for each cultivar of interest. Understanding the adaptation mechanisms to drought would allow for identifying the most resistant varieties and help to decide on optimal irrigation conditions [13–15].

Adaptations in plant growth and morphology in response to a water deficit have been widely described [16,17]. Morphological variations due to a water deficit help to reverse metabolic imbalances and improve plant water status [18]. Some examples of such variations are the size and thickness of leaves [3] or the root-to-shoot ratio [19]. Stomatal adjustment is another adaptive process used by plants in the face of water stress [15,20]. The stomatal aperture affects transpiration, CO_2 uptake, and growth. In general, gas exchange measurements are considered efficient indicators of plant fitness to water stress [3]. Similarly, changes in CO_2 assimilation can be studied by analyzing some parameters related to chlorophyll fluorescence [21].

Soil–plant water relations can also influence nutrient availability [22]. Given that reduced transpiration hinders the translocation of some nutrients to the aerial part of plants [23], changes in nutrient composition can be used as a tool to diagnose the level of water stress of a plant [24]. For instance, an accumulation of solutes can be important for lowering TLP and hence maintaining turgor, even with a declining water potential, and, in turn, maintaining open stomata and CO_2 assimilation at levels that would otherwise be inhibited [25]. Similarly, numerous nutrients can be used as indicators for plant stress since many of them regulate plant metabolism, and act as enzyme activators under stressful conditions [26].

Hebe is a genus of shrubs native to Oceania that holds great interest in ornamental horticulture worldwide. These versatile plants are excellent for borders, containers, and mass plantings. It has gained distinct relevance in some Western countries, such as the United Kingdom [27]. British nurseries have produced numerous hybrid cultivars of Hebe that have become popular for home decoration and landscaping. *Hebe x andersonii* is one of these hybrids and the subject of this paper. Although not native to the Mediterranean region, this hybrid is widely used for ornamental purposes in Mediterranean climates given its excellent adaptability. The fleshy and waxy leaves of Hebe help it to retain moisture and tolerate dry spells. Its grayish-green leaves with yellow-cream spots combine ornamentally with its purple-violet flowers that bloom in clusters at the apex of the stems [28].

Although Hebe is frequently produced in nurseries of Oceania, America, and Europe, there is little research on the impact of irrigation on its growth and physiology. Information on substrate moisture conditions to optimize water use is essential if over-irrigation is to be avoided and crop health ensured. Given the scarce information on the response of Hebe to a water deficit and the global need for sustainable practices in nursery production, the objectives of this experiment were as follows: (i) to ascertain and evaluate the morphological and physiological responses of Hebe to water availability, and (ii) to identify irrigation strategies to improve water-use efficiency without substantial loss of ornamental quality. These objectives were addressed by assessing the effects of substrate moisture on the growth, ornamental quality, water relations, gas exchange, photochemical behavior, plant mineral composition, and water-use efficiency of Hebe.

2. Materials and Methods

2.1. Plants and Culture Conditions

Seedlings of the ornamental shrub *Hebe x andersonii* cv. Variegata (7–8 cm high) were obtained from a commercial grower (Viveros Bermejo S.L., Totana, Spain). Round PVC pots (15.5 cm diameter and 2.3 L volume) were used to transplant the seedlings. Pots were filled with a commercial soilless substrate composed of peat, coconut fiber, and perlite (67/30/3, *v*/*v*/*v*) (Fertiberia S.A., Madrid, Spain). The water release curve for the substrate was determined in the laboratory following De Boodt's method [29]. Ten different suctions were applied to the sample: 10, 20, 30, 40, 50, 60, 70, 80, 90, and 100 hPa. The moisture release properties of the substrate were: 63.2% maximum water holding capacity, 28.1% easily available water (EAW), and 5.8% water buffer capacity (Figure 1). Plant available water, 33.9% for our substrate, is defined as the amount of water held between the maximum water holding capacity and the wilting point. Usually, water available to plants in a substrate is defined as the VWC between −1 and −10 kPa of the soil water potential (Ψ_w) [30]. This includes the EAW (−1 to −5 kPa) and the water buffer capacity (−5 to −10 kPa) [29]. The critical Ψ_w at which many crops cultivated in a substrate undergo water stress is around −10 kPa. The pots were placed in openings (16 cm× 16 cm) in a metal grid made of corrugated bars (8 mm ø) 80 cm off the ground. The whole setup was placed in a polycarbonate greenhouse (15 m × 8 m × 6 m) at the Agricultural Experimental Station of the Technical University of Cartagena, Spain (lat. 37°35′ N, long. 0°59′ W). The experiment ran from February to September 2020.

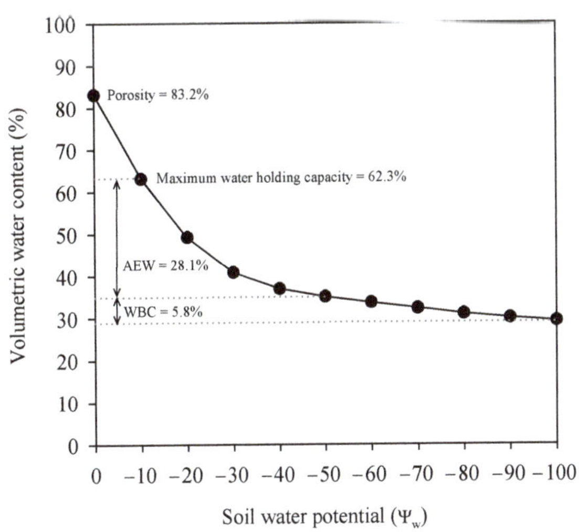

Figure 1. Water retention curve of the substrate (peat, coconut fiber, perlite (67/30/3, *v*/*v*/*v*)) following De Boodt's method. The value of VWC at saturation (Ψ_w = 0) was estimated from the measurement of porosity (83.2%). EAW: easily available water, WBC: water buffer capacity.

2.2. Irrigation System and Treatments

Pressure-compensating and anti-drain emitters were used to irrigate the plants (Netafim Ltd., Corporate Headquarters, Israel). The emitters were tested for flow rate consistency before the experiment began. An irrigation pipe (50 cm × 4 mm) was used to supply water to the emitters, with their flow rates ranging between 1.44 and 1.56 L h^{-1}. The emitters were connected to a straight-arrow dripper inserted into the substrate of the pots. Two emitters per pot and 30 plants per treatment were employed. Centrifugal electric pumps (Mod. Prisma15 3M, ESPA, Banyoles, Spain) were used to drive the water through the emitters. Latching solenoid valves (Aquanet Plus, Netafim, Fresno, CA, USA) were used to open or close the water flow. A separate tank, pump, and irrigation lines were provided for each treatment. The irrigation solution (stored in black 1000 L polypropylene tanks) was prepared with commercial complex fertilizer with a nutrient balance of 4-1.7-4.5-4-1.4 (N-P_2O_5-K_2O-CaO-MgO). Nitric acid was added to adjust the pH of the irrigation solution. The final EC and pH of the irrigation solution were 1.77 dS m^{-1} and 7.45, respectively. The concentration of the main ions in the nutrient solution is listed in Table 1.

Table 1. Composition of major ions in the irrigation solution used in the experiment.

Ion	Concentration (mg L^{-1})
NO_3^-	280.23
NH_4^+	1.89
$H_2PO_4^-$	17.09
K^+	56.06
Ca^{2+}	125.72
Mg^{2+}	57.18
SO_4^{2-}	258.99
Cl^-	245.28
Na^+	161.37
HCO_3^-	105.11
B^{3+}	0.61
Mn^{2+}	0.45
Fe^{3+}	0.07
Zn^{2+}	0.07
Cu^{2+}	0.09

A data logger for temperature and humidity (LOG 32 TH, Dostmann electronic GmbH, Germany) was used to measure and record air temperature and humidity in the greenhouse. Data were collected every minute and daily averages were calculated. Vapor pressure deficit (VPD) values were computed using the equation described by Steiner et al. [31]. Daily average temperature and VPD data are shown in Figure 2.

The experiment consisted of three treatments of different substrate moisture levels under a non-leaching irrigation regime. The average VWC percentages were 49.28, 38.44, and 32.16% for each treatment, which, in this work, are referred to as the control, VWC39%, and VWC32%, respectively. The Ψ_w values that correspond to these water content percentages can be drawn from Figure 1. The irrigation was controlled using soil moisture sensors (EC5; METER Group, Inc., Pullman, WA, USA) connected to a CR1000 datalogger with a 16-channel relay controller (SMD-CD16D; Campbell Scientific Inc., Logan, UT, USA) operating solenoid valves on each of the tanks that contained the irrigation solutions. EC5 sensors determined the VWC by measuring the dielectric constant of the medium (voltage) using capacitance frequency domain technology [32]. For more information, see Decagon-METER Group's EC5 manual.

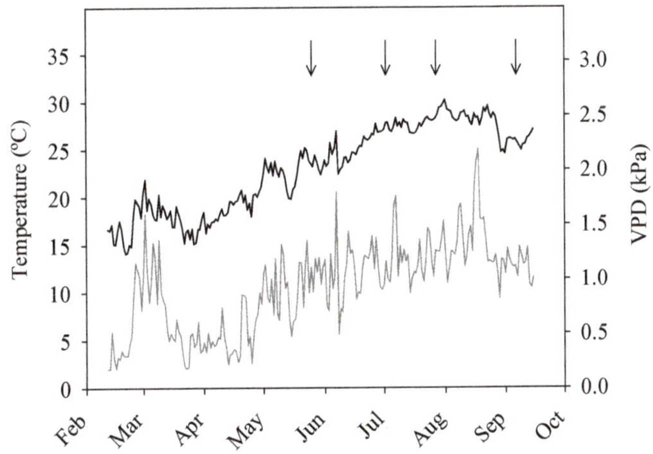

Figure 2. Daily average temperature (black line) and vapor pressure deficit (VPD) (gray line) during the experiment. The arrows indicate the days on which physiological parameters were measured.

Soil sensors were fully inserted vertically in the east-facing quadrant of the root ball, between two emitters. Three probes per treatment were installed in random pots, with one sensor per pot. The volumetric water content (VWC) was calculated from the voltage output of the EC5 sensors (mV) using a substrate-specific calibration equation. The calibration line was obtained following the procedure described by Valdes et al. [33]: VWC = 4.942 mV 0.676, $R^2 = 0.98$. The CR1000 data-logger was programmed using Loggernet 3 (Campbell Scientific Inc.) to log the sensor output every 1 h during the experiment. The outputs were immediately converted to VWC by the CR1000 using the calibration equation. An example of the evolution of substrate moisture over time is given in Figure 3.

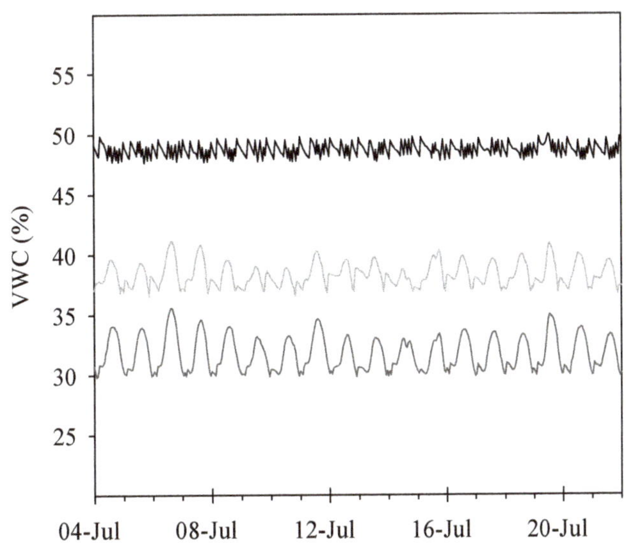

Figure 3. Fluctuations in the substrate moisture when irrigating with small volumes of water. Lines represent three different volumetric water contents (VWCs): 49.28% VWC (control, black line), VWC 38.44% (VWC39%, light grey line), and VWC 32.16% (VWC32%, dark grey line).

From 9 a.m. to 2 p.m., the CR1000 compared the VWC every hour with a specific irrigation set-point for each treatment (see Figure 3). Irrigation was activated automatically and independently in each treatment when the set-point was reached. The CR1000 was programmed to count every irrigation event automatically. For all treatments, the water applied at each irrigation event was the same, around 50 mL pot^{-1} (3 L h^{-1} flow rate in a 60 s irrigation event). Water consumption was the result of multiplying the total number of irrigation events by the volume of each irrigation.

Easily available water (EAW) is the percentage of water that is immediately available for plant uptake. Plants can take up this water with little capillary force. As plants continuously extract water from the substrate, the available water decreases, and plants must exert more effort to extract water from the substrate [34]. The VWC set-points selected for the treatments in this experiment provided different percentages of depletion of EAW. The control treatment (49% VWC) triggered the irrigation when half of the EAW was depleted; therefore, the plant used little energy to take up the water through its roots. The VWC39% treatment started irrigating when 88% of the EAW was depleted; the plants had to exert some effort to absorb water. Meanwhile, VWC32% plants had to absorb the water buffer since the EAW was completely depleted, which entailed drought stress.

The water applied to each treatment corresponded to water consumption since there was no leaching. Evapotranspiration was calculated by dividing the total volume of water applied by the number of days of the experiment (152 days). WUE was calculated as the ratio between the dry mass produced and the water transpired between the beginning and end of the experimental period.

2.3. Growth, SPAD, and Color

Plant height and width were measured in ten plants per treatment at the end of the experiment. Two width measurements were taken: one at its widest section, and the other perpendicular to it. Plant height was measured from the substrate surface to the most distal shoot. The number of shoots (with and without flowers) per plant was also determined.

The dry weights of the roots, stems, and leaves were determined in the same plants. The different plant tissues were dried until constant weight in an isothermal oven at 65 °C (Mod. 210, PSelecta, Barcelona, Spain). Then, the dry mass was weighed with an analytical balance (Mod. TE2145, Sartorius Weighing Technology, GmBH, Goettingen, Germany). The root to shoot ratio was calculated using the dry weight values.

The average leaf area was measured using Easy Leaf Area software [35], which analyses the area of the leaves in digital photographs by counting pixels and referring them to the image of a red square of a known area. Fifteen leaves per plant and 10 plants per treatment were randomly chosen for this purpose. The specific leaf area (SLA) was obtained by dividing the leaf area by the dry weight.

SPAD (relative chlorophyll content) values were measured using a chlorophyll meter (SPAD-502, Minolta Camera Co., Ltd., Osaka, Japan). Leaf color was analyzed with a Minolta CR10 colorimeter (Konica Minolta Sensing, Inc., Osaka, Japan), which calculated the color coordinates CIELab (lightness, hue angle, and chrome). Six plants per treatment were sampled for SPAD and color measurements. In each plant, two leaves located at mid-height and facing south were selected for each existing shoot. Additionally, the number of total stems and flowering stems per plant was measured.

2.4. Substrate EC and Plant Mineral Content

The electrical conductivity of the 1:2 extract v/v ($EC_{1:2}$) was tested at the end of the experiment, following the procedure described by Sonneveld and van den Ende [36], whereby one volume of the substrate was combined with two volumes of distilled water. Then, the mixture was stirred by hand and left to stand for half an hour. Finally, the samples were filtered through filter paper and the EC of the leachate was measured with a hand-held EC tester (ECTestr11, Eutech Instruments Pte Ltd., Singapore).

Plant mineral content was determined in six pots per treatment. For this, oven-dried leaves, stems, and roots were ground to a fine dry powder. Inorganic elements (Na^+, K^+, P, Ca^{2+}, and Mg^{2+}) were determined using inductively coupled plasma emission spectrophotometry (IRIS Intrepid II XDL ICP-OES, Thermo Fischer Scientific, Waltham, MA, USA). Plant tissues were extracted by mixing 100 mg of dry powder with 40 mL of deionized water. The mixture was stirred for 30 min in a rotary shaker at 30 rpm and passed through a 0.45 µm PTFE syringe filter. Chloride concentration was analyzed in the aqueous extract using a chloride analyzer (Mod. 926, Sherwood Scientific, Cambridge, UK). Total nitrogen was measured with a nitrogen analyzer (Mod. Flash EA 1112, Thermo Fischer Scientific, Waltham, MA, USA).

2.5. Plant Water Status

Leaf water potential (Ψ_l), leaf osmotic potential (Ψ_s), leaf turgor potential (Ψ_t), and leaf osmotic potential at full turgor (Ψ_{100s}) were determined in six plants per treatment during the central hours of illumination at four stages of growth. As an additional metric for determining plant drought tolerance, water potential at the turgor loss point (Ψ_{tlp}) was calculated for each treatment using Equations (6) and (7) proposed by [37]. Ψ_l was measured with a pressure chamber (Mod. 3000; Soil Moisture Equipment Co., Santa Barbara, CA, USA) in mature leaves according to Scholander et al. [38]. The leaves were wrapped in a plastic bag, cut from the main stem with a sharp blade, and quickly placed in the chamber, being careful to leave enough petiole to be inserted into the nozzle of the pressure chamber. Compressed nitrogen gas was used to increase the pressure in the chamber at a rate of 0.03 MPa s^{-1} until the suppuration of a sap drop was visible. The same leaves used for Ψ_l were frozen in liquid nitrogen (−196 °C) and stored at −30 °C. After thawing, the osmotic potential (Ψ_s) was measured in the extracted sap using a WESCOR 5520 vapor pressure osmometer (Wescor Inc., Logan, UT, USA). The leaf turgor potential was estimated as the difference between Ψ_l and Ψ_s. The leaf osmotic potential at full turgor was measured using the same procedure for Ψ_s, but on leaves that had been previously placed in distilled water until full saturation was reached.

2.6. Gas Exchange and Chlorophyll Fluorescence

The gas and photochemical parameters considered in this work were as follows: (i) stomatal conductance (g_s), (ii) net photosynthetic rate (P_n), (iii) intrinsic water-use efficiency (P_n/g_s), (iv) apparent electron transport rate (ETR), and (v) the ETR/P_n ratio. These parameters were quantified for Hebe leaves using a portable photosynthesis system (Mod. LI-6400; LI-COR Inc., Lincoln, NE, USA). The CO_2 concentration was set at 400 ppm, the flow rate was set at 500 µmol s^{-1}, and the photosynthetically active radiation (PAR) was set at 1000 µmol m^{-2} s^{-1} [39]. The effective photochemical quantum yield of photosystem II (ePSII) was used to estimate the electron transport rate (ETR) according to the equation of Krall and Edwards (1992): ETR = ePSII × PFD × 0.84 × 0.5, where PFD represents the photon flux density incident on the leaf, 0.84 represents the leaf absorbance, and 0.5 represents a factor implying an equal distribution of energy between photosystems II and I. The intrinsic water-use efficiency was estimated as the ratio between P_n and g_s (P_n/g_s). Gas exchange, water potential, and chlorophyll fluorescence values were measured simultaneously around noon to minimize the influence of temperature variations.

At the end of the experiment, chlorophyll fluorescence parameters were measured with a pulse-modulated fluorometer (Mod. FMS-2, Hansatech Instruments, Norfolk, UK). The parameters determined were as follows: (i) maximum quantum yield of PSII (Fv/Fm), (ii) effective quantum yield of PSII (ePSII), and (iii) non-photochemical quenching (NPQ). Each leaf was dark-adapted with a shutter-plate leaf clip 30 min before each measurement, following the method described by Sheng et al. [40]. Once in darkness, the minimum fluorescence (F_o) was measured, and a light irradiation pulse of 5000 µmol m^{-2} s^{-1} was administered for 0.7 to measure the maximum fluorescence of the dark-adapted leaf (F_m). The sample was then irradiated with actinic light of 400 µmol m^{-2} s^{-1} for 150 s to measure fluorescence under stationary light (Fs). Next, radiation of 5000 µmol m^{-2} s^{-1} was administered for 0.7 s to determine the maximum fluorescence of the light-adapted leaf ($F_{m'}$). Finally, the actinic radiation was turned off and far-red radiation was turned on for a period of 5 s to reoxidize the centers of PSII. Then, the minimum fluorescence of the light-adapted leaf ($F_{o'}$) was measured. Fv/Fm, ePSII, and NPQ were estimated from F_o, F_m, F_s, $F_{m'}$, and $F_{o'}$, according to the equations described in Brestic and Zivcak [41].

2.7. Leaf Cross-Sectional Anatomy

For the analysis at the end of the experiment, six healthy mature leaves per treatment were selected from the fifth fully expanded leaf starting from the tip. Leaf cross-sections were cut from the middle portion of each leaf, avoiding the central nerve. Four sections per leaf of approximately 10 µm thickness were cut with a hand microtome (Mod. 501, Nahita, Navarra, Spain). The sections were stained with 0.5% toluidine blue and transferred to a glass slide for microscopic observation. Measurements were taken with an optical microscope at ×40 magnification (Mod. BX51, Olympus, Tokyo, Japan). The thickness of the total blade, palisade parenchyma (PP), spongy parenchyma (SP), epidermis (E), and cuticle (C) were measured with a micrometric eyepiece.

2.8. Experimental Design and Statistical Analysis

Plants were arranged on crop benches in a randomized block design. Each of the three treatments (control, VWC39%, and VWC32%) was divided into three blocks, and each block was randomly populated with ten plants. Differences were assessed via one-way analysis of variance (ANOVA) using Statgraphics Centurion (v.XVI, StatPoint Technologies, Inc., Warrenton, VA, USA). When the ANOVA indicated significant effects, means were separated by the Least Significant Difference (LSD) test. Plotting was performed with the SigmaPlot program (v.14.5, Systat Software Inc., San Jose, CA, USA).

3. Results and Discussion

3.1. Water Consumption and Water-Use Efficiency

Hebe consumed less water when the substrate moisture decreased (Table 2), a pattern that has been reported in many other species exposed to a water deficit [1,10,42,43]. In this experiment, evapotranspiration equaled the water consumption per day and per plant since no leaching was produced during irrigation and, therefore, water consumption was entirely due to evapotranspiration. Consequently, evapotranspiration varied in the same proportion as water consumption (Table 2). The reduction in water consumption can be explained considering that evaporation losses occur mainly in the outermost layer of the substrate [44]. As humidity decreases, the outermost layer of the substrate dries out and the rate of evaporative water loss is reduced. Kool et al. [45] and Navarro et al. [46] indicated that water loss through evaporation represented 30% or more of the evapotranspiration in other ornamental shrubs. Additionally, transpiration was also diminished by other factors, such as growth reduction, stomatal regulation, and morphological changes [47].

Table 2. Water consumption, average daily evapotranspiration (ET), and water-use efficiency (WUE) of Hebe plants after 152 days of cultivation at three different volumetric water contents (VWCs): 49.28% VWC (control), VWC 38.44% (VWC39%), and VWC 32.16% (VWC32%).

	Control	VWC39%	VWC32%
Water consumption (L pot^{-1}) *	37.95	27.70	16.40
ET (mL pot^{-1} day^{-1}) *	249.67	182.24	107.89
WUE (g L^{-1})	1.70 a	1.83 b	2.47 c

Different letters in the same row indicate statistically significant differences between means at $p < 0.05$ according to the least significant difference (LSD) test. * No statistical analysis was performed.

In addition to reducing water consumption, low substrate moisture improved the water-use efficiency of Hebe by 8% in VWC39% and 45% in VWC32%. The water-use efficiency increased because water consumption was reduced more than plant growth, and Hebe required less water to produce one unit of dry weight. In our experiment, the water-use efficiency was generally high since there was no leaching, and water losses were entirely due to evapotranspiration. The lack of leaching also implied a minimal environmental impact since no water was wasted and no chemical fertilizers were leached.

3.2. Growth and Development

The dry weights of the Hebe plants decreased significantly with lower substrate moisture (Table 3). Growth decline is one of the first adaptive responses to a water deficit [48], which is usually followed by a decline in transpiration [49,50]. Although this reduction affected all plant organs (leaves, stems, and roots), the leaves were the least affected compared with the roots and stems (Table 3). Nevertheless, all dry weight values were significantly lower in VWC32% than in VWC39% because the former suffered greater water stress. In particular, the reductions in dry weight in VWC39% were 13%, 19%, and 36% for leaves, stems, and roots, respectively, while in VWC32%, these reductions were 24%, 45%, and 47%, respectively, compared with the control. The difference in the development of root dry weight compared with the aerial dry weight (leaves and stems) resulted in the reduction of the root-to-shoot ratio in the treatments with restricted water availability (Table 3).

Table 3. Growth and development parameters of Hebe plants under three different volumetric water contents (VWCs) over 152 days: control (49.28% VWC), VWC39% (VWC 38.44%), and VWC32% (VWC 32.16%).

	Control	VWC39%	VWC32%
Plant dry weight (g)	64.42 c	50.6 b	40.5 a
Leaf dry weight (g)	26.72 c	23.3 b	20.3 a
Stem dry weight (g)	18.27 c	14.8 b	9.97 a
Root dry weight (g)	19.43 c	12.49 b	10.22 a
Root-to-shoot ratio	0.423 b	0.328 a	0.338 a
Leaf area (dm^2)	28.32 c	22.45 b	13.5 a
SLA * (cm^2 g^{-1})	105.99 c	96.35 b	66.5 a
SPAD **	52.11 b	47.78 ab	43.43 a
Plant height (cm)	30.83 b	31.43 b	26.33 a
Plant width (cm)	39.83 c	37.54 b	33.58 a
No. of stems	10.97 b	10.03 ab	9.67 a
No. of flowerings stems	3.33 b	3.01 b	1.23 a

Different letters in the same row indicate statistically significant differences between means at $p < 0.05$ according to the least significant difference (LSD) test. * SLA (specific leaf area). ** SPAD (relative chlorophyll content in leaves).

The leaf area of Hebe was reduced with decreasing substrate moisture (Table 3). Leaf growth is considered one of the parameters most sensitive to a water deficit [51,52]. This reduction is attributed to an adaptive mechanism that helps to limit water loss through transpiration [53]. Low water availability reduced the specific leaf area of VWC32% plants (Table 3), which is a behavior that has been observed in other species under a water deficit [54–56]. Lower SLA corresponds to thicker and/or denser leaf blades, resulting in improved drought resistance due to lower transpiration rates [13]. Microscopic measurements confirmed the increase in leaf blade thickness suggested by the decrease in SLA in VWC32%, as is discussed in the following section.

SPAD values in VWC32% were significantly reduced compared with the control (Table 3), which could suggest a loss of leaf chlorophyll. However, this reduction in chlorophyll was not visually apparent. The absence of significant effects in the colorimetry study (data not shown) confirmed the absence of yellowing and necrosis in the leaves; therefore, the impact of the water deficit on chlorophyll content was not large.

Plants in VWC32% were shorter and narrower than control plants, whereas those in VWC39% only decreased in width (Table 3). The differences in width could be related to an overall reduction in the number of shoots, while the reduction in plant height could be driven by a decline in leaf water potential since small reductions in water potential affect cell expansion [57]. However, under less water-limiting conditions (VWC39%), water stress was not sufficient to affect plant height (Table 3).

Only VWC32% plants reduced both the number of stems and the number of flowering stems. Burnett and van Iersel [42] found a directly proportional relationship between the number of shoots and substrate moisture in *Gaura lindheimeri*. In other shrubs, size was negatively correlated with substrate moisture [58,59]. Gradual leaf drop is an adaptation mechanism in response to a water deficit [60], which attempts to maximize photosynthetic activity by concentrating nutrients and resources on fewer leaves. Although leaf abscission was observed in several species under a water deficit [61,62], in our study, none of the three treatments promoted leaf abscission.

The influence of low moisture substrate on flowering capacity is well established in ornamental species, such as carnations, petunias, impatiens, and geraniums [3,13,63,64]. In our study, while VWC39% plants did not modify their number of flowering stems, the same parameter was considerably reduced in VWC32% plants. Zhen and Burnett [1] noted the inhibition of flowering stems in Lavandula when substrate moisture was lower than 20%. In this study, 32% VWC was sufficient to reduce the presence of flowers in Hebe. However, this does not have a major impact from an ornamental point of view since this species has scarce and short-lived flowers.

3.3. Leaf Anatomy Changes

Hebe leaves in the control group presented a cross-sectional structure in which the palisade parenchyma was more developed than the spongy parenchyma, the upper epidermis was thicker than the lower epidermis, and both cuticles had similar thickness (Table 4). Only VWC32% significantly increased the total leaf thickness compared with the control. In addition, VWC32% decreased the proportion of spongy parenchyma in favor of thicker palisade parenchyma and increased the thickness of the lower cuticle and lower epidermis. On the other hand, VWC39% increased the thickness of the lower epidermis and lower cuticle compared with the control (Table 4).

Table 4. Thickness of the cross-sectional layer of Hebe leaves (µm) and percentage of each component in response to three different volumetric water contents (VWCs) after 152 days: control (VWC 49.28%), VWC39% (VWC 38.44%), and VWC32% (VWC 32.16%).

	Control	VWC39%	VWC32%
Total leaf thickness (µm)	474.88 a	486.89 ab	506.2 b
Upper cuticle (%)	1.67 a	1.69 a	1.58 a
Upper epidermis (%)	3.46 a	3.47 a	3.53 a
Palisade parenchyma (%)	51.83 a	52.19 ab	54.37 b
Spongy parenchyma (%)	40.34 b	40.09 b	37.93 a
Lower epidermis (%)	1.17 a	1.44 b	1.47 b
Lower cuticle (%)	1.52 b	1.12 a	1.12 a

Different letters in the same row indicate statistically significant differences between means at $p < 0.05$ according to the least significant difference (LSD) test.

As noted above, morphological adaptations to a water deficit modulated transpiration and photosynthesis. The sensitivity of both processes may not necessarily be identical. For instance, an increase in leaf thickness will decrease transpiration while increasing photosynthesis [65,66]. Transpiration occurs not only through the stomata but also through the cuticle [67]. In this sense, Holmgren et al. [3] reported that, depending on the species, water loss through the cuticle is 1.7 to 28.6% of that through the stomata. The characteristics of Hebe leaves (fleshy and with a waxy cuticle) suggest a good aptitude for restricting non-stomatal water losses. Thick leaves have higher photosynthetic activity per unit weight because they have a greater density of pigments, proteins, and other metabolites [54]. Moreover, thick leaves have a higher SLA and, therefore, less transpiration surface area exposed to the air [68]. The leaves in VWC32% were the thickest and showed more developed palisade parenchyma than spongy parenchyma. Since the palisade parenchyma is richer in chloroplasts than the spongy parenchyma, a reduction in the latter would make it easier for CO_2 to reach the chloroplasts present in the palisade parenchyma [69,70]. This anatomical response would imply an increase in chloroplast density in order to adapt the photosynthetic efficiency of Hebe to the new stress conditions.

3.4. Plant Water Relations

Five months after the start of the experiment (May), all treatments had similar leaf water potentials (Ψ_l's) (Figure 4A); however, in July the treatments with reduced water availability started to show lower Ψ_l values than the control. These differences became more apparent at the end of the experiment (September) when the differences in Ψ_l between all treatments became significant (Figure 4A). The turgor potentials (Ψ_t's) were similar for all treatments until July (Figure 4B) but increased at the end of the experiment in VWC39% and VWC32%. Decreases in osmotic potentials at full turgor (Ψ_{100s}'s) were observed in the last two measurements of the experiment in the VWC32% and VWC39% treatments (Figure 4C), indicating the plants under a water deficit developed an osmotic adjustment.

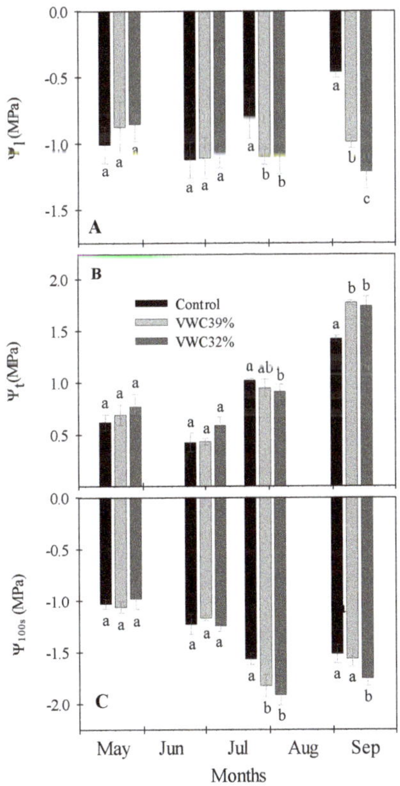

Figure 4. Water–plant relations in Hebe during the experiment: leaf water potential at midday (Ψ_l) (**A**), leaf turgor potential (Ψ_t) (**B**), and leaf osmotic potential at full turgor (Ψ_{100s}) (**C**). Control (black), VWC39% (light gray), and VWC32% (dark gray) indicate Hebe plants cultivated at 49.28%, 38.44%, and 32.16% VWC, respectively. Values are means (n = 5) and vertical bars indicate ± standard error. Different letters indicate statistically significant differences between means at $p < 0.05$ according to the least significant difference (LSD) test.

The measurements of Ψ_l indicated that Hebe took around six months to change its water status (Figure 4A), which highlights its good adaptation to water stress. The evolution of Ψ_{100s} indicated that the plants developed osmotic adjustment to maintain leaf turgor during the warmer and drier summer. When comparing treatments, it can be observed that an osmotic regulation was promoted by decreasing the water availability. This suggested that Hebe increased the concentration of osmoregulatory solutes to prevent turgor loss. However, this osmotic adjustment process itself can damage cell organelles or restrict growth since plant solutes are retained and cannot be used for other cellular processes [20,71]. Although it has been suggested that osmotic adjustments may have a positive effect on the esthetics of plants subjected to a water deficit [72], this did not contribute to the maintenance of Hebe inflorescences in the most severe treatment. However, this osmotic adjustment probably contributed to the maintenance of leaf quality in both the VWC39% and VWC32% treatments. Osmotic adjustments also play a role in reducing the turgor loss point of plants and maintaining turgor, even at low water potentials [73]. The Ψ_{tlp} values of Hebe were -1.62 ± 0.13 MPa for the control, -1.67 ± 0.09 MPa for VWC39%, and -1.89 ± 0.11 MPa for VWC32%. The turgor loss points were lower than the Ψ_l's values (Figure 4A), which indicates that the leaves did not lose their turgor under the amount of water stress applied in this experiment. This decrease suggested that Hebe plants adapted to the water stress by lowering the Ψ_{tlp}.

3.5. Leaf Gas Exchange

The photosynthetic activity of Hebe decreased from the beginning of the experiment, and only at the end of the experiment were there significant differences between the treatments (Figure 5A). Stomatal conductance (g_s) behaved similarly to P_n (Figure 5B) and consequently the evolution of the P_n/g_s ratio did not change during the experiment (Figure 5C). While the ETR decreased throughout the experiment (Figure 5D), the ETR/P_n ratio increased until July (Figure 5E), indicating that the ETR decreased less than P_n. Only at the end of the experiment were there statistically significant differences between all treatments regarding the ETR/P_n ratio (Figure 5E).

The decrease in P_n with increasing summer temperature showed the sensitivity of this process to heat [74]. Given the strong correlation between g_s and P_n, it is likely that the influence of environmental conditions on g_s controlled P_n values via the stomatal aperture. The strong relationship between these two variables was evident from the stability of the P_n/g_s ratio observed during the experiment. Stomatal closure in the face of a water deficit is a common physiological behavior that usually leads to parallel decreases in P_n and g_s [75–77]. The substrate moisture had a smaller effect than the temperature on gas exchange parameters because only a small reduction in P_n and g_s was observed in VWC39% and VWC32% at the end of the experimental period. At this time, the plants that received less water were more sensitive to environmental stress, as reflected by the lower Ψ_l values (Figure 4A).

The ETR gradually fell with temperature in a similar pattern to P_n (Figure 5D). Silim et al. [78] found lower ETR values at 27 °C in cultivated *Populus* sp. than at 19 °C, and other authors have suggested that water stress can damage the photochemical apparatus of plants [79]. For example, Singal et al. [80] found a decrease in the activity of many enzymes of the Calvin cycle in plants subjected to water stress. In the case of Hebe, the evolution of ETR and P_n caused the ETR/P_n ratio to increase with temperature throughout the experiment. Especially at the end of the experiment, there was a clear difference in the ETR/P_n ratio among the three irrigation treatments (Figure 5E).

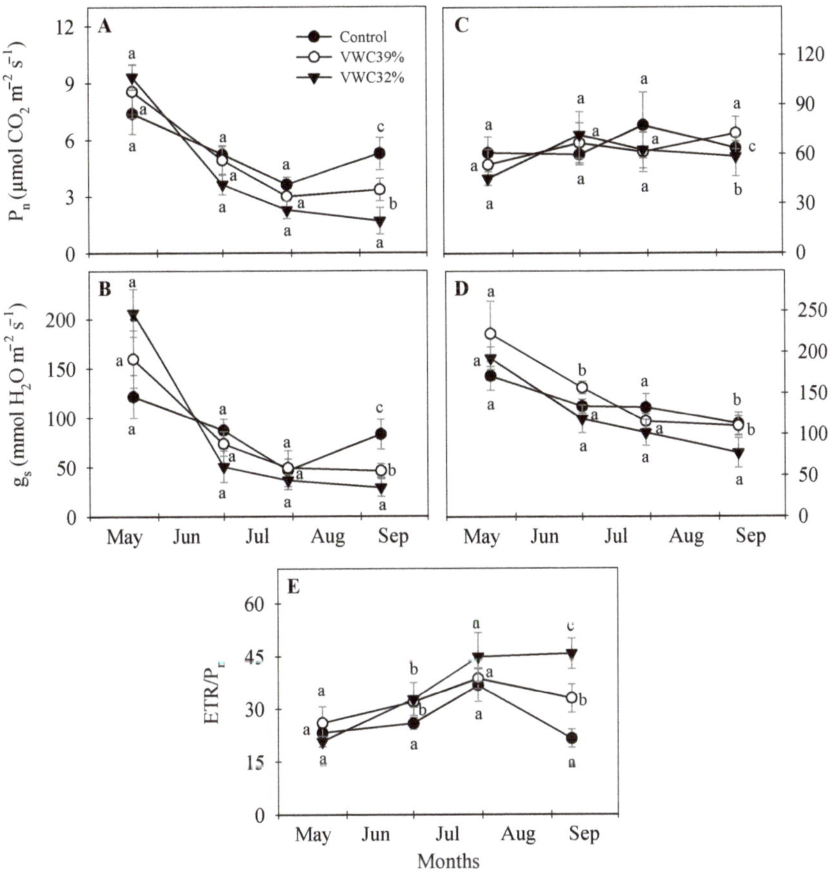

Figure 5. Time evolution of net photosynthesis at midday (P_n) (**A**), stomatal conductance at midday (g_s) (**B**), P_n/g_s ratio (**C**), ETR (**D**), and ETR/P_n ratio (**E**). Control (black), VWC39% (light gray), and VWC32% (dark gray) denote Hebe plants grown at 49.28%, 38.44%, and 32.16% VWC, respectively. Values are means (n = 5) and vertical bars show ± standard error. Different letters indicate statistically significant differences between means at $p < 0.05$ according to the least significant difference (LSD) test.

3.6. Chlorophyll Fluorescence

The maximum quantum yield of the photosystem II (Fv/Fm) and the effective quantum yield (ePSII) were similar in all treatments at the end of the experiment (Figure 6). However, the non-photochemical quenching (NPQ) increased in VWC32%, while the ETR/Pn ratio increased in both VWC39% and VWC32%. An increase in the ETR/P_n ratio is an indication that some of the photochemical excitation energy was used in other non-CO_2 assimilatory processes [81]. Since the stability of the Fv/Fm and ePSII ratios indicated good photochemical functioning [82], this excess energy was probably dissipated as heat. This hypothesis was supported by the increase in NPQ (Figure 6), which is a defense mechanism that dissipates the excess energy and prevents its photochemical apparatus from being damaged under the stress conditions tested in this work [83].

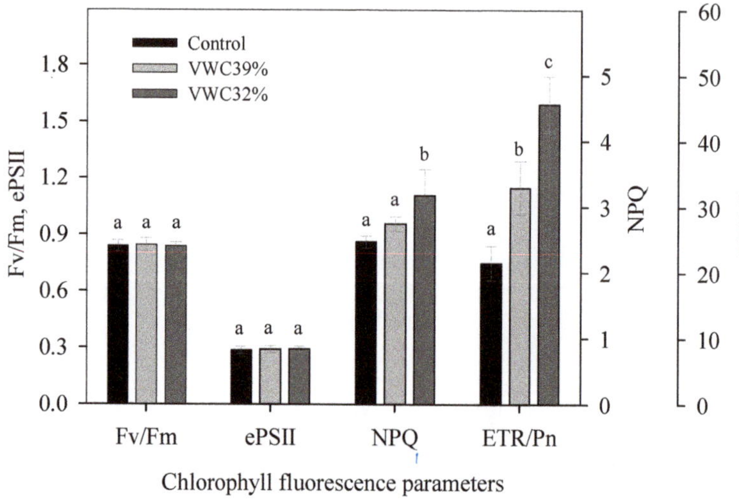

Figure 6. Effect of substrate moisture on several chlorophyll fluorescence parameters of Hebe leaves at the end of the experiment: maximum quantum yield of photosystem II (PSII) (Fv/Fm), effective quantum yield (ePSII), non-photochemical quenching (NPQ), and the ratio of the electron transport rate to the photosynthesis rate (ETR/P_n). Control (black), VWC39% (light gray), and VWC32% (dark gray) denote Hebe plants grown over 152 days at 49.28%, 38.44%, and 32.16% VWC, respectively. Values are means (n = 5) and vertical bars indicate ± standard error. Different letters indicate statistically significant differences between means at $p < 0.05$ according to the least significant difference (LSD) test.

3.7. Substrate EC and Plant Mineral Content

The $EC_{1:2}$ of the substrate at the end of the experimental period was similar between the treatments, with 1.98 ± 0.19, 2.04 ± 0.20, and 2.11 ± 0.22 dS m^{-1} in the control, VWC39%, and VWC32%, respectively. Not leaching plant pots can cause salts to accumulate in the substrate and, depending on the EC of the irrigation solution, salinity may reach harmful values at the end of the experiment that can impair growth and cause leaf damage [84,85]. $EC_{1:2}$ values around 2 dS m^{-1} have been related to reduced growth, wilting, leaf necrosis, and chlorosis in many crops [86]; thus, Hebe may be considered moderately tolerant to salinity.

In general, the plant nutrient content increased with decreasing substrate moisture, except for calcium and magnesium, which practically did not change (Table 5). Reduced substrate moisture often produces varied effects on the mineral content of leaves [87]. In our experiment, lower substrate moisture increased N, K^+, P, Cl^-, and Na^+, especially in VWC32% plants.

The nitrogen content increased in the stems and roots but did not change in the leaves. It was suggested that small changes in tissue nitrogen can decisively affect plant growth [88], and in this sense, the slight increase in nitrogen in VWC32% plants could have had a positive effect on plant growth. A water deficit promoted an accumulation of phosphorus only in the stems (Table 5). The aerial part of Hebe accumulated potassium when the substrate moisture decreased, and only the roots of VWC32% plants increased their potassium content. This increase suggests that Hebe can efficiently incorporate potassium into its tissues under an irrigation regime at low substrate moisture. Potassium has traditionally been associated with good plant quality [89], better resistance to phytosanitary problems [90], and the adjustment of the osmotic potential of plant tissues [24].

Table 5. Plant macronutrient (N, P, K^+) and mineral content (Ca^{2+}, Mg^{2+}, Na^+, Cl^-) in leaves, stems, and roots of Hebe at the end of the experiment. Control, VWC39%, and VWC32% denote Hebe plants grown over 152 days at 49.28%, 38.44%, and 32.16% VWC, respectively.

Element (mg g^{-1})	Plant Organ	Control	VWC39%	VWC32%
N	Leaf	19.19 a	19.74 a	20.68 a
	Stem	13.57 a	18.37 b	19.49 b
	Root	11.8 a	15.23 b	16.2 b
P	Leaf	2.03 a	2.25 a	2.14 a
	Stem	2.19 a	2.78 b	3.43 c
	Root	0.78 a	0.73 a	0.81 a
K^+	Leaf	22.09 a	26.55 b	28.63 c
	Stem	6.77 a	9.09 b	16.6 c
	Root	3.72 a	3.21 a	5.93 b
Ca^{2+}	Leaf	13.45 a	12.73 a	12.27 a
	Stem	3.4 a	3.74 a	4.07 a
	Root	8.02 a	8.71 a	9.43 a
Mg^{2+}	Leaf	3.28 a	3.21 a	3.45 a
	Stem	2.74 a	3.13 a	3.23 a
	Root	3.36 b	2.54 a	2.81 ab
Na^+	Leaf	2.18 a	3.31 b	5.51 c
	Stem	1.92 a	3.81 b	8.49 c
	Root	12.42 b	9.01 a	9.67 a
Cl^-	Leaf	11.01 a	19.11 b	30.41 c
	Stem	4.47 a	8.94 b	18.31 c
	Root	24.69 c	20.46 b	17.19 a

Different letters in the same row indicate statistically significant differences between means at $p < 0.05$ according to the least significant difference (LSD) test.

Sodium and chloride behaved similarly in response to the decreasing substrate moisture, where the content of both ions increased in the aerial part and decreased in the roots (Table 5). This trend suggested that lower substrate moisture promotes the translocation of saline ions from the roots to the shoot [91]. This can be explained by the fact that plants need to osmotically adjust in response to a water deficit and maintain turgor and turgor-dependent processes. It is worth mentioning that the highest Cl^- and Na^+ concentrations in Hebe leaves (30 and 5.5 mg g^{-1}) did not produce visual leaf damage. In this sense, Bañón et al. [82] reported similar salt concentrations in the leaves of *Lantana camara* that also did not cause leaf damage.

4. Conclusions

Hebe showed significant differences in the physiological and morphological responses depending on the substrate moisture. Hebe plants at 39% VWC improved water-use efficiency without substantially reducing growth and esthetics. In contrast, further reduction in water availability negatively affected the flowering and growth of Hebe plants at 32% VWC. Low water availability decreased the photosynthetic activity due to a decrease in stomatal conductance without damaging the photochemical apparatus. The low substrate moisture favored the accumulation of osmoregulatory solutes, especially Cl^-, Na^+, and K^+, which contributed to the maintenance of plant turgor and decreased their leaf turgor loss point. Overall, *Hebe andersoni* cv. Variegata can be considered as a good candidate for improving the water-use efficiency and the sustainability of nursery production, as it can grow successfully without leaching and at constant low substrate moisture.

Author Contributions: Conceptualization, M.J.S.-B., J.J.A. and S.B.; data curation, S.B. and M.F.O.; formal analysis, D.B. and S.B.; funding acquisition, M.J.S.-B. and M.F.O.; investigation, D.B., B.L. and S.B.; methodology, D.B. and B.L.; project administration, M.J.S.-B. and M.F.O.; resources, M.J.S.-B.; supervision, M.F.O. and J.J.A.; visualization, D.B. and B.L.; writing—original draft, D.B.; writing—review and editing, M.J.S.-B., J.J.A. and S.B. All authors have read and agreed to the published version of the manuscript.

Funding: This research was funded by the Ministry of Science, Innovation, and Universities of Spain; the European Regional Development Fund (RTI2018-093997-B-I00).

Institutional Review Board Statement: Not applicable.

Informed Consent Statement: Not applicable.

Data Availability Statement: Data sharing is not applicable.

Conflicts of Interest: The authors declare no conflict of interest. The funders had no role in the design of the study; in the collection, analyses, or interpretation of data; in the writing of the manuscript; or in the decision to publish the results.

References

1. Zhen, S.; Burnett, S.E. Effects of Substrate Volumetric Water Content on English Lavender Morphology and Photosynthesis. *HortScience* **2015**, *50*, 909–915. [CrossRef]
2. Nemali, K.S.; van Iersel, M.W. An Automated System for Controlling Drought Stress and Irrigation in Potted Plants. *Sci. Hortic.* **2006**, *110*, 292–297. [CrossRef]
3. Álvarez, S.; Navarro, A.; Bañón, S.; Sánchez-Blanco, M.J. Regulated Deficit Irrigation in Potted Dianthus Plants: Effects of Severe and Moderate Water Stress on Growth and Physiological Responses. *Sci. Hortic.* **2009**, *122*, 579–585. [CrossRef]
4. Singh, V.K.; Singh, Y.; Kumar, P. Diseases of Ornamental Plants and Their Management. In *Eco-Friendly Innovative Approaches in Plant Disease Management*; International Book Distributors and Publisher: New Delhi, India, 2012; pp. 543–572. ISBN 81-7089-375-5.
5. Álvarez, S.; Gómez-Bellot, M.J.; Castillo, M.; Bañón, S.; Sánchez-Blanco, M.J. Osmotic and Saline Effect on Growth, Water Relations, and Ion Uptake and Translocation in Phlomis Purpurea Plants. *Environ. Exp. Bot.* **2012**, *78*, 138–145. [CrossRef]
6. Mangiafico, S.S.; Newman, J.; Merhaut, D.J.; Gan, J.; Faber, B.; Wu, L. Nutrients and Pesticides in Stormwater Runoff and Soil Water in Production Nurseries and Citrus and Avocado Groves in California. *HortTechnology* **2009**, *19*, 360–367. [CrossRef]
7. Ucar, Y.; Kazaz, S.; Eraslan, F.; Baydar, H. Effects of Different Irrigation Water and Nitrogen Levels on the Water Use, Rose Flower Yield and Oil Yield of Rosa Damascena. *Agric. Water Manag.* **2017**, *182*, 94–102. [CrossRef]
8. Rouphael, Y.; Raimondi, G.; Caputo, R.; Pascale, S.D. Fertigation Strategies for Improving Water Use Efficiency and Limiting Nutrient Loss in Soilless Hippeastrum Production. *HortScience* **2016**, *51*, 684–689. [CrossRef]
9. Belayneh, B.E.; Lea-Cox, J.D.; Lichtenberg, E. Costs and Benefits of Implementing Sensor-Controlled Irrigation in a Commercial Pot-in-Pot Container Nursery. *HortTechnology* **2013**, *23*, 760–769. [CrossRef]
10. Iersel, M.W.V.; Dove, S.; Kang, J.-G.; Burnett, S.E. Growth and Water Use of Petunia as Affected by Substrate Water Content and Daily Light Integral. *HortScience* **2010**, *45*, 277–282. [CrossRef]
11. Montesano, F.F.; van Iersel, M.W.; Boari, F.; Cantore, V.; D'Amato, G.; Parente, A. Sensor-Based Irrigation Management of Soilless Basil Using a New Smart Irrigation System: Effects of Set-Point on Plant Physiological Responses and Crop Performance. *Agric. Water Manag.* **2018**, *203*, 20–29. [CrossRef]
12. Caser, M.; Lovisolo, C.; Scariot, V. The Influence of Water Stress on Growth, Ecophysiology and Ornamental Quality of Potted Primula Vulgaris 'Heidy' Plants. New Insights to Increase Water Use Efficiency in Plant Production. *Plant Growth Regul.* **2017**, *83*, 361–373. [CrossRef]
13. Sánchez-Blanco, M.J.; Ortuño, M.F.; Bañon, S.; Álvarez, S. Deficit Irrigation as a Strategy to Control Growth in Ornamental Plants and Enhance Their Ability to Adapt to Drought Conditions. *J. Hortic. Sci. Biotechnol.* **2019**, *94*, 137–150. [CrossRef]
14. Toscano, S.; Ferrante, A.; Romano, D. Response of Mediterranean Ornamental Plants to Drought Stress. *Horticulturae* **2019**, *5*, 6. [CrossRef]
15. Tribulato, A.; Toscano, S.; Di Lorenzo, V.; Romano, D. Effects of Water Stress on Gas Exchange, Water Relations and Leaf Structure in Two Ornamental Shrubs in the Mediterranean Area. *Agronomy* **2019**, *9*, 381. [CrossRef]
16. Toscano, S.; Ferrante, A.; Tribulato, A.; Romano, D. Leaf Physiological and Anatomical Responses of Lantana and Ligustrum Species under Different Water Availability. *Plant Physiol. Biochem.* **2018**, *127*, 380–392. [CrossRef]
17. Rafi, Z.N.; Kazemi, F.; Tehranifar, A. Morpho-Physiological and Biochemical Responses of Four Ornamental Herbaceous Species to Water Stress. *Acta Physiol. Plant* **2018**, *41*, 7. [CrossRef]
18. Giordano, M.; Petropoulos, S.A.; Cirillo, C.; Rouphael, Y. Biochemical, Physiological, and Molecular Aspects of Ornamental Plants Adaptation to Deficit Irrigation. *Horticulturae* **2021**, *7*, 107. [CrossRef]
19. Fascella, G.; Maggiore, P.; Demma Carà, M.; Zizzo, G.V. Growth and Flowering Response of Euphorbia × Lomi Poysean Cultivars under Two Irrigation Regimes. *Acta Hortic.* **2011**, *893*, 939–943. [CrossRef]

20. Álvarez, S.; Gómez-Bellot, M.J.; Acosta-Motos, J.R.; Sánchez-Blanco, M.J. Application of Deficit Irrigation in Phillyrea Angustifolia for Landscaping Purposes. *Agric. Water Manag.* **2019**, *218*, 193–202. [CrossRef]
21. Castillo-Campohermoso, M.A.; Broetto, F.; Rodríguez-Hernández, A.M.; de Abril Alexandra Soriano-Melgar, L.; Mounzer, O.; Sánchez-Blanco, M.J. Plant-Available Water, Stem Diameter Variations, Chlorophyll Fluorescence, and Ion Content in Pistacia Lentiscus under Salinity Stress. *Terra Latinoam.* **2020**, *38*, 103–111. [CrossRef]
22. Wang, Y.; Sun, Y.; Niu, G.; Deng, C.; Wang, Y.; Gardea-Torresdey, J. Growth, Gas Exchange, and Mineral Nutrients of Ornamental Grasses Irrigated with Saline Water. *HortScience* **2019**, *54*, 1840–1846. [CrossRef]
23. Steudle, E. Water Uptake by Roots: Effects of Water Deficit. *J. Exp. Bot.* **2000**, *51*, 1531–1542. [CrossRef] [PubMed]
24. Marschner, P.; Rengel, Z. Chapter 12—Nutrient Availability in Soils. In *Marschner's Mineral Nutrition of Higher Plants (Third Edition)*; Marschner, P., Ed.; Academic Press: San Diego, CA, USA, 2012; pp. 315–330. ISBN 978-0-12-384905-2.
25. Álvarez, S.; Rodríguez, P.; Broetto, F.; Sánchez-Blanco, M.J. Long Term Responses and Adaptive Strategies of Pistacia Lentiscus under Moderate and Severe Deficit Irrigation and Salinity: Osmotic and Elastic Adjustment, Growth, Ion Uptake and Photosynthetic Activity. *Agric. Water Manag.* **2018**, *202*, 253–262. [CrossRef]
26. Demetriou, G.; Neonaki, C.; Navakoudis, E.; Kotzabasis, K. Salt Stress Impact on the Molecular Structure and Function of the Photosynthetic Apparatus—The Protective Role of Polyamines. *Biochim. Biophys. Acta—Bioenerg.* **2007**, *1767*, 272–280. [CrossRef]
27. Warrington, I.J.; Southward, R.C. Seasonal Frost Tolerance of Hebe Species and Cultivars. *N. Z. J. Crop Hortic. Sci.* **1995**, *23*, 437–445. [CrossRef]
28. Kristensen, L.N. The genus Hebe: A botanical report. *Tidsskr. Planteavls* **1989**, *93*, 369–437.
29. de Boodt, M.; Verdonck, O.; Cappaert, I. Determination and Study of the Water Availability of Substrates for Ornamental Plant Growing. *Acta Hortic.* **1974**, *1*, 51–58. [CrossRef]
30. Arguedas Rodríguez, F.R. Calibrating Capacitance Sensors to Estimate Water Content, Matric Potential, and Electrical Conductivity in Soilless Substrates. Master's Thesis, University of Maryland, College Park, MD, USA, 2009.
31. Steiner, J.L.; Howell, T.A.; Schneider, A.D. Lysimetric Evaluation of Daily Potential Evapotranspiration Models for Grain Sorghum. *Agron. J.* **1991**, *83*, 240–247. [CrossRef]
32. Gaskin, G.J.; Miller, J.D. Measurement of Soil Water Content Using a Simplified Impedance Measuring Technique. *J. Agric. Eng. Res.* **1996**, *63*, 153–159. [CrossRef]
33. Valdés, R.; Miralles, J.; Ochoa, J.; Sánchez Blanco, M.J.; Bañón Arias, S. Saline Reclaimed Wastewater Can Be Used to Produce Potted Weeping Fig (Ficus Benjamina L.) with Minimal Effects on Plant Quality. *Span. J. Agric. Res.* **2012**, *10*, 1167. [CrossRef]
34. Miralles, J.; van Iersel, M.W. A Calibrated Time Domain Transmissometry Soil Moisture Sensor Can Be Used for Precise Automated Irrigation of Container-Grown Plants. *HortScience* **2011**, *46*, 889–894. [CrossRef]
35. Easlon, H.M.; Bloom, A.J. Easy Leaf Area: Automated Digital Image Analysis for Rapid and Accurate Measurement of Leaf Area. *Appl. Plant Sci.* **2014**, *2*, 1400033. [CrossRef]
36. Sonneveld, C.; van den Ende, J. Soil Analysis by Means of a 1:2 Volume Extract. *Plant Soil* **1971**, *35*, 505–516. [CrossRef]
37. Petruzzellis, F.; Savi, T.; Bacaro, G.; Nardini, A. A Simplified Framework for Fast and Reliable Measurement of Leaf Turgor Loss Point. *Plant Physiol. Biochem.* **2019**, *139*, 395–399. [CrossRef]
38. Scholander, P.F.; Bradstreet, E.D.; Hemmingsen, E.A.; Hammel, H.T. Sap Pressure in Vascular Plants. *Science* **1965**, *148*, 339–346. [CrossRef]
39. Gómez-Bellot, M.J.; Nortes, P.A.; Sánchez-Blanco, M.J.; Ortuño, M.F. Sensitivity of Thermal Imaging and Infrared Thermometry to Detect Water Status Changes in Euonymus Japonica Plants Irrigated with Saline Reclaimed Water. *Biosyst. Eng.* **2015**, *133*, 21–32. [CrossRef]
40. Sheng, M.; Tang, M.; Chen, H.; Yang, B.; Zhang, F.; Huang, Y. Influence of Arbuscular Mycorrhizae on Photosynthesis and Water Status of Maize Plants under Salt Stress. *Mycorrhiza* **2008**, *18*, 287–296. [CrossRef]
41. Brestic, M.; Zivcak, M. PSII Fluorescence Techniques for Measurement of Drought and High Temperature Stress Signal in Crop Plants: Protocols and Applications. In *Molecular Stress Physiology of Plants*; Rout, G.R., Das, A.B., Eds.; Springer: Berlin/Heidelberg, Germany, 2013; pp. 87–131. ISBN 978-81-322-0807-5.
42. Burnett, S.E.; van Iersel, M.W. Morphology and Irrigation Efficiency of Gaura Lindheimeri Grown with Capacitance Sensor-Controlled Irrigation. *HortScience* **2008**, *43*, 1555–1560. [CrossRef]
43. Garland, K.F.; Burnett, S.E.; Day, M.E.; van Iersel, M.W. Influence of Substrate Water Content and Daily Light Integral on Photosynthesis, Water Use Efficiency, and Morphology of Heuchera Americana. *J. Am. Soc. Hort. Sci.* **2012**, *137*, 57–67. [CrossRef]
44. Paço, T.A.; Ferreira, M.I.; Rosa, R.D.; Paredes, P.; Rodrigues, G.C.; Conceição, N.; Pacheco, C.A.; Pereira, L.S. The Dual Crop Coefficient Approach Using a Density Factor to Simulate the Evapotranspiration of a Peach Orchard: SIMDualKc Model versus Eddy Covariance Measurements. *Irrig. Sci.* **2012**, *30*, 115–126. [CrossRef]
45. Kool, D.; Agam, N.; Lazarovitch, N.; Heitman, J.L.; Sauer, T.J.; Ben-Gal, A. A Review of Approaches for Evapotranspiration Partitioning. *Agric. For. Meteorol.* **2014**, *184*, 56–70. [CrossRef]
46. Navarro, A.; Bañon, S.; Olmos, E.; Sánchez-Blanco, M.J. Effects of Sodium Chloride on Water Potential Components, Hydraulic Conductivity, Gas Exchange and Leaf Ultrastructure of Arbutus Unedo Plants. *Plant Sci.* **2007**, *172*, 473–480. [CrossRef]
47. Waisel, Y.; Eshel, A.; Kafkafi, U. *Plant Roots: The Hidden Half*, 3rd ed.; Books in soils, plants, and the environment; M. Dekker: New York, NY, USA; Basel, Switzerland, 2002; ISBN 978-0-8247-0631-9.

48. Chaves, M.M.; Oliveira, M.M. Mechanisms Underlying Plant Resilience to Water Deficits: Prospects for Water-Saving Agriculture. *J. Exp. Bot.* **2004**, *55*, 2365–2384. [CrossRef]
49. Wu, Y.; Huang, M.; Warrington, D.N. Growth and Transpiration of Maize and Winter Wheat in Response to Water Deficits in Pots and Plots. *Environ. Exp. Bot.* **2011**, *71*, 65–71. [CrossRef]
50. Lu, Y.; Ma, D.; Chen, X.; Zhang, J. A Simple Method for Estimating Field Crop Evapotranspiration from Pot Experiments. *Water* **2018**, *10*, 1823. [CrossRef]
51. Jones, P.; Jones, J.W.; Allen, L.H., Jr. Seasonal Carbon and Water Balances of Soybeans Grown Under Stress Treatments in Sunlit Chambers. *Trans. ASAE* **1985**, *28*, 2021–2028. [CrossRef]
52. Fernández-García, N.; Olmos, E.; Bardisi, E.; García-De la Garma, J.; López-Berenguer, C.; Rubio-Asensio, J.S. Intrinsic Water Use Efficiency Controls the Adaptation to High Salinity in a Semi-Arid Adapted Plant, Henna (Lawsonia Inermis L.). *J. Plant Physiol.* **2014**, *171*, 64–75. [CrossRef] [PubMed]
53. Álvarez, S.; Sánchez-Blanco, M.J. Changes in Growth Rate, Root Morphology and Water Use Efficiency of Potted Callistemon Citrinus Plants in Response to Different Levels of Water Deficit. *Sci. Hortic.* **2013**, *156*, 54–62. [CrossRef]
54. Liu, F.; Stützel, H. Biomass Partitioning, Specific Leaf Area, and Water Use Efficiency of Vegetable Amaranth (*Amaranthus* spp.) in Response to Drought Stress. *Sci. Hortic.* **2004**, *102*, 15–27. [CrossRef]
55. del Amor, F.M.; Marcelis, L.F.M. Differential Effect of Transpiration and Ca Supply on Growth and Ca Concentration of Tomato Plants. *Sci. Hortic.* **2006**, *111*, 17–23. [CrossRef]
56. Dwyer, J.M.; Hobbs, R.J.; Mayfield, M.M. Specific Leaf Area Responses to Environmental Gradients through Space and Time. *Ecology* **2014**, *95*, 399–410. [CrossRef] [PubMed]
57. Taiz, L.; Zeiger, E. (Eds.) *Plant Physiology*; Sinauer Associates: Sunderland, MA, USA, 2010; ISBN 978-0-87893-565-9.
58. Álvarez, S.; Bañón, S.; Sánchez-Blanco, M.J. Regulated Deficit Irrigation in Different Phenological Stages of Potted Geranium Plants: Water Consumption, Water Relations and Ornamental Quality. *Acta Physiol. Plant.* **2012**, *35*, 1257–1267. [CrossRef]
59. Alem, P.; Thomas, P.A.; van Iersel, M.W. Controlled Water Deficit as an Alternative to Plant Growth Retardants for Regulation of Poinsettia Stem Elongation. *HortScience* **2015**, *50*, 565–569. [CrossRef]
60. Ammar, A.; Ben Aissa, I.; Mars, M.; Gouiaa, M. Comparative Physiological Behavior of Fig (Ficus Carica L.) Cultivars in Response to Water Stress and Recovery. *Sci. Hortic.* **2020**, *260*, 108881. [CrossRef]
61. Giovannelli, A.; Deslauriers, A.; Fragnelli, G.; Scaletti, L.; Castro, G.; Rossi, S.; Crivellaro, A. Evaluation of Drought Response of Two Poplar Clones (Populus x Canadensis Monch "I-214" and P. Deltoides Marsh. 'Dvina') through High Resolution Analysis of Stem Growth. *J. Exp. Bot.* **2007**, *58*, 2673–2683. [CrossRef]
62. Gu, M.; Robbins, J.A.; Rom, C.R. The Role of Ethylene in Water-Deficit Stress Responses in Betula Papyrifera Marsh. *HortScience* **2007**, *42*, 1392–1395. [CrossRef]
63. McMichael, B.L.; Jordan, W.R.; Powell, R.D. Abscission Processes in Cotton: Induction by Plant Water Deficit1. *Agron. J.* **1973**, *65*, 202–204. [CrossRef]
64. Blanusa, T.; Vysini, E.; Cameron, R.W.F. Growth and Flowering of Petunia and Impatiens: Effects of Competition and Reduced Water Content Within a Container. *HortScience* **2009**, *44*, 1302–1307. [CrossRef]
65. Fitter, A.; Hay, R. Introduction. In *Environmental Physiology of Plants*; Elsevier: Amsterdam, The Netherlands, 2002; pp. 1–20.
66. Giuliani, R.; Koteyeva, N.; Voznesenskaya, E.; Evans, M.A.; Cousins, A.B.; Edwards, G.E. Coordination of Leaf Photosynthesis, Transpiration, and Structural Traits in Rice and Wild Relatives (Genus Oryza). *Plant Physiol.* **2013**, *162*, 1632–1651. [CrossRef]
67. Jenks, M.A.; Hasegawa, P.M. (Eds.) *Plant Abiotic Stress*; John Wiley & Sons: Hoboken, NJ, USA, 2008; ISBN 978-0-470-99411-5.
68. Lichtenthaler, H.K.; Ač, A.; Marek, M.V.; Kalina, J.; Urban, O. Differences in Pigment Composition, Photosynthetic Rates and Chlorophyll Fluorescence Images of Sun and Shade Leaves of Four Tree Species. *Plant Physiol. Biochem.* **2007**, *45*, 577–588. [CrossRef]
69. Acosta-Motos, J.-R.; Diaz-Vivancos, P.; Álvarez, S.; Fernández-García, N.; Sanchez-Blanco, M.J.; Hernández, J.A. Physiological and Biochemical Mechanisms of the Ornamental Eugenia Myrtifolia L. Plants for Coping with NaCl Stress and Recovery. *Planta* **2015**, *242*, 829–846. [CrossRef]
70. Yahia, E.M.; Carrillo-López, A.; Barrera, G.M.; Suzán-Azpiri, H.; Bolaños, M.Q. Chapter 3—Photosynthesis. In *Postharvest Physiology and Biochemistry of Fruits and Vegetables*; Yahia, E.M., Ed.; Woodhead Publishing: Thorston, UK, 2019; pp. 47–72. ISBN 978-0-12-813278-4.
71. Munns, R.; Passioura, J.B.; Colmer, T.D.; Byrt, C.S. Osmotic Adjustment and Energy Limitations to Plant Growth in Saline Soil. *New Phytol.* **2020**, *225*, 1091–1096. [CrossRef]
72. Zollinger, N.; Kjelgren, R.; Cerny-Koenig, T.; Kopp, K.; Koenig, R. Drought Responses of Six Ornamental Herbaceous Perennials. *Sci. Hortic.* **2006**, *109*, 267–274. [CrossRef]
73. Bartlett, M.K.; Scoffoni, C.; Sack, L. The Determinants of Leaf Turgor Loss Point and Prediction of Drought Tolerance of Species and Biomes: A Global Meta-Analysis: Drivers of Plant Drought Tolerance. *Ecol. Lett.* **2012**, *15*, 393–405. [CrossRef]
74. Maroco, J.P.; Pereira, J.S.; Chaves, M.M. Stomatal Responses to Leaf-to-Air Vapour Pressure Deficit in Sahelian Species. *Aust. J. Plant Physiol.* **1997**, *24*, 381–387. [CrossRef]
75. Medrano, H. Regulation of Photosynthesis of C3 Plants in Response to Progressive Drought: Stomatal Conductance as a Reference Parameter. *Ann. Bot.* **2002**, *89*, 895–905. [CrossRef] [PubMed]

76. Steduto, P.; Hsiao, T.C.; Fereres, E. On the Conservative Behavior of Biomass Water Productivity. *Irrig. Sci.* **2007**, *25*, 189–207. [CrossRef]
77. Yu, D.J.; Kim, S.J.; Lee, H.J. Stomatal and Non-Stomatal Limitations to Photosynthesis in Field-Grown Grapevine Cultivars. *Biol. Plant* **2009**, *53*, 133–137. [CrossRef]
78. Silim, S.N.; Ryan, N.; Kubien, D.S. Temperature Responses of Photosynthesis and Respiration in Populus Balsamifera L.: Acclimation versus Adaptation. *Photosynth. Res.* **2010**, *104*, 19–30. [CrossRef]
79. Lawlor, D.W.; Tezara, W. Causes of Decreased Photosynthetic Rate and Metabolic Capacity in Water-Deficient Leaf Cells: A Critical Evaluation of Mechanisms and Integration of Processes. *Ann. Bot.* **2009**, *103*, 561–579. [CrossRef]
80. Singal, H.R.; Sheoran, I.S.; Singh, R. Effect of Water Stress on Photosynthesis and in Vitro Activities of the PCR Cycle Enzymes in Pigeonpea (Cajanus Cajan L.). *Photosynth. Res.* **1985**, *7*, 69–76. [CrossRef]
81. D'Ambrosio, N.; Arena, C.; De Santo, A.V. Temperature Response of Photosynthesis, Excitation Energy Dissipation and Alternative Electron Sinks to Carbon Assimilation in Beta Vulgaris L. *Environ. Exp. Bot.* **2006**, *55*, 248–257. [CrossRef]
82. Bañón, S.; Miralles, J.; Ochoa, J.; Franco, J.A.; Sánchez-Blanco, M.J. Effects of Diluted and Undiluted Treated Wastewater on the Growth, Physiological Aspects and Visual Quality of Potted Lantana and Polygala Plants. *Sci. Hortic.* **2011**, *129*, 869–876. [CrossRef]
83. Müller, P.; Li, X.-P.; Niyogi, K.K. Non-Photochemical Quenching. A Response to Excess Light Energy1. *Plant Physiol.* **2001**, *125*, 1558–1566. [CrossRef] [PubMed]
84. Navarro, A.; Bañón, S.; Conejero, W.; Sánchez-Blanco, M.J. Ornamental Characters, Ion Accumulation and Water Status in Arbutus Unedo Seedlings Irrigated with Saline Water and Subsequent Relief and Transplanting. *Environ. Exp. Bot.* **2008**, *62*, 364–370. [CrossRef]
85. Cassaniti, C.; Leonardi, C.; Flowers, T.J. The Effects of Sodium Chloride on Ornamental Shrubs. *Sci. Hortic.* **2009**, *122*, 586–593. [CrossRef]
86. Camberato, D.M.; Lopez, R.G.; Mickelbart, M.V. *PH and Electrical Conductivity Measurements in Soilless Substrates*; Purdue University Libraries: West Lafayette, IN, USA, 2009; pp. 1–7.
87. Alam, S.M. Nutrient Uptake by Plants Under Stress Conditions. In *Handbook of Plant and Crop Stress*; Pessarakli, M., Ed.; CRC Press: Boca Raton, FL, USA, 1999; pp. 285–313.
88. Lawlor, D.W.; Lemaire, G.; Gastal, F. Nitrogen, Plant Growth and Crop Yield. In *Plant Nitrogen*; Lea, P.J., Morot-Gaudry, J.-F., Eds.; Springer: Berlin/Heidelberg, Germany, 2001; pp. 343–367.
89. Lester, G.E.; Jifon, J.L.; Makus, D.J. Impact of Potassium Nutrition on Food Quality of Fruits and Vegetables: A Condensed and Concise Review of the Literature. *Better Crops* **2010**, *94*, 18–21.
90. Yermiyahu, U.; Israeli, L.; David, D.R.; Faingold, I.; Elad, Y. Higher Potassium Concentration in Shoots Reduces Gray Mold in Sweet Basil. *Phytopathology* **2015**, *105*, 1059–1068. [CrossRef]
91. Gómez-Bellot, M.J.; Álvarez, S.; Castillo, M.; Bañón, S.; Ortuño, M.F.; Sánchez-Blanco, M.J. Water Relations, Nutrient Content and Developmental Responses of Euonymus Plants Irrigated with Water of Different Degrees of Salinity and Quality. *J. Plant Res.* **2013**, *126*, 567–576. [CrossRef]

agronomy

MDPI

Review

Mitigation of Calcium-Related Disorders in Soilless Production Systems

Virginia Birlanga [1], José Ramón Acosta-Motos [2,3,*] and José Manuel Pérez-Pérez [4,*]

1 Bayer CropScience, 30319 Miranda, Spain; virginia.birlanga@bayer.com
2 Group of Fruit Tree Biotechnology, CEBAS-CSIC, 30100 Murcia, Spain
3 Cátedra Emprendimiento en el Ámbito Agroalimentario, Campus de los Jerónimos, Universidad Católica San Antonio de Murcia (UCAM), no. 135 Guadalupe, 30107 Murcia, Spain
4 Instituto de Bioingeniería, Universidad Miguel Hernández, 03202 Elche, Spain
* Correspondence: jracosta@ucam.edu (J.R.A.-M.); jmperez@umh.es (J.M.P.-P.); Tel.: +34-968-278-800 (J.R.A.-M.); +34-966-658-958 (J.M.P.-P.)

Abstract: In the current scenario of human-driven climate change, extreme weather events will likely affect agricultural production worldwide. Soilless production systems have recently arisen as a solution to optimize the use of natural resources, such as water and soil, and hence will contribute to reducing the environmental impact of agriculture. However, nutritional imbalance due to adverse environmental factors, such as drought, high temperatures, and salinity, might produce calcium-related physiological disorders during plant growth, such as blossom-end rot (BER) in fruits and tipburn (TB) in leaves, which are a serious problem in crop production. Here, we discuss the different agronomic, physiological, and genetic factors that favor the induction of BER in tomato and TB in lettuce and anticipate the use of an integration of breeding and technological approaches to alleviate nutritional disorders in soilless production systems.

Keywords: climate change; soilless agriculture; blossom-end rot; tipburn; calcium deficiency

Citation: Birlanga, V.; Acosta-Motos, J.R.; Pérez-Pérez, J.M. Mitigation of Calcium-Related Disorders in Soilless Production Systems. *Agronomy* **2022**, *12*, 644. https://doi.org/10.3390/agronomy12030644

Academic Editor: Leo Sabatino

Received: 2 February 2022
Accepted: 3 March 2022
Published: 6 March 2022

1. Climate Change and Agriculture

Scientists now agree that human activities are the main drivers of climate change [1]. Agriculture, forestry, and other land uses contribute approximately 13% of the carbon dioxide (CO_2), 44% of the methane, and 81% of the nitrogen oxide emissions, which together represent 23% of the net greenhouse gas (GHG) emissions [1]. However, only 29% of total anthropogenic CO_2 emissions during the 2007–2016 period were neutralized by Earth's natural responses; hence, it is expected that the global atmospheric CO_2 levels will further increase [2].

Human-driven climate change will enhance extreme weather events that negatively affect terrestrial ecosystems. As a result, there will be an increase in degradation and desertification in many regions of the planet, leading to a reduction in crop yields and consequently food security will be affected globally. Soil is both a source and a sink for GHGs and performs a crucial role in the exchange of energy, water, and aerosols between the soil surface and the atmosphere. Therefore, sustainable soil management can help mitigate the negative impacts of various environmental stressors, especially those dependent on climate change, on ecosystems, and societies [2]. Farmers are now implementing a set of agricultural practices to reduce the effects of climate change, through changes in tillage practices, the selection of crop species and cultivars that grow more efficiently and are better adapted to adverse conditions, as well as through the implementation of a more sustainable use of natural resources. Therefore, a proper balance must be found by considering the contributions of these new practices to produce better yields, increasing farmers' incomes and other environmental indicators [3,4].

The objective of this review is to provide farmers with a compendium of strategies, tools, and solutions to problems directly related to crop quality, such as blossom end rot (BER) on fruits and tipburn (TB) in leaves. In this study we will analyze the possible triggers of these physiological disorders by studying the environmental factors directly related to climate change. In addition, new production and managing strategies will be described for a more efficient use of resources that contribute to reducing the appearance of these symptoms, whether due to environmental or genetic factors or a combination of both.

2. Soilless Production Systems: Challenges and Solutions

A new model of industrial-scale agriculture, known as soilless agriculture, has emerged in recent years as a system that optimizes the use of natural resources, such as water and soil, and that allows for better environmental control due to its implementation indoors. Soilless agriculture contributes to better plant growth thanks to an adequate management of the root zone in terms of a more uniform and precise control of water and fertilizer needs. With this technique, it is possible to produce healthy vegetables of excellent quality [5,6].

Soilless agriculture not only improves the quality of agricultural products, but also contributes to the reduction of their environmental impact by ensuring a more efficient use of water and fertilizers, mainly nitrates and phosphates (NO_3^- and PO_4^{2-}), which can reach rivers and seas due to leaching by torrential rains, causing the contamination of surface waters by eutrophication [7]. The possibilities provided for helping reduce the environmental impact of agricultural systems include the reuse of industrial waste as a growing medium. Soilless cropping systems in which 50% of the drainage was recirculated, reduced NO_3^- and PO_4^{2-} emissions as compared to systems without drainage recovery [8,9]. At present, soilless farming has become consolidated as a suitable tool to optimize intensive crop production and reduce the use of non-renewable resources.

2.1. Soilless Cropping Systems

Soilless cropping systems can be classified into several types based on the use of the nutrient solution or the physical state of the root growth media (Table 1). Consequently, we distinguish between open-loop systems (Figure 1a), if the nutrient solution is discarded after use, or closed-loop systems (Figure 1b), if the nutrient solution is reformulated after use and returned to the system. The nutrient solution consists of water, oxygen (O_2), and all essential plant nutrients [10]. The root system could grow in the air (aeroponic cultivation), on the liquid nutrient solution (hydroponic cultivation), and on a solid substrate with added nutrient solution (substrate cultivation).

Table 1. Classification of soilless cropping systems.

Classification	Categories	Characteristics
Nutrient solution use	Open-loop systems	The used nutrient solution is discarded
	Closed-loop systems	Nutrient solution is reformulated and returned to the system
Physical state of root growth media	Gaseous (aeroponic cultivation)	Spray column system Schwalbach system Aero-Gro system
	Liquid (hydroponic cultivation)	Deep floated technique Nutrient film technique New growing system technique
	Liquid (aquaponic cultivation)	Nutrient solution is derived from waste from fish production
	Solid (substrate cultivation)	Directly in substrate Systems of cultivation in bags or containers Single unit culture systems

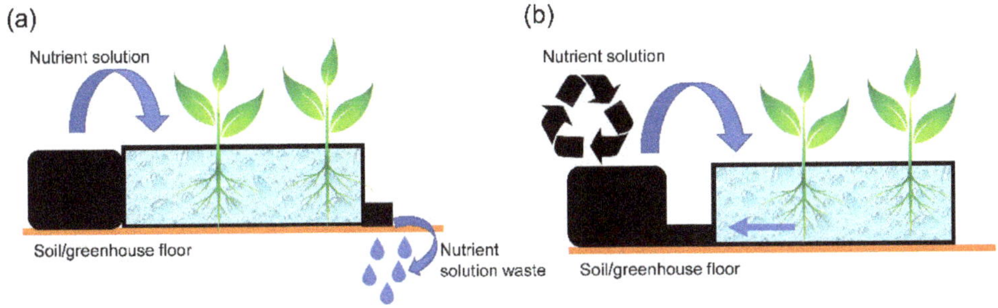

Figure 1. Soilless cropping systems as regards to nutrient solution uses. (**a**) A scheme of an open-loop system in which nutrient solution residues are not recycled, and (**b**) a scheme of a closed-loop system in which nutrient solution residues are reintroduced into the system.

Soilless cropping systems have been improved over time, showing many advantages as compared to conventional systems, because they avoid direct contact with the soil and therefore minimize the problems related to soil diseases [11].

In open-loop systems (Figure 1a), allowing an excess of nutrients and water to the plants compensates for irregular transpiration, prevents salt accumulation, and corrects nutritional imbalances. In these systems, however, a large amount of nutrients and water is drained away, thus increasing production costs and contaminating the surrounding environment [11]. In contrast, in closed-loop systems (Figure 1b) the drainage solution is collected onto a reservoir for additional treatments to reduce the risk of root-borne diseases and to reformulate the nutrient composition, which might then be used for other plots or reintroduced into the system.

2.1.1. Aeroponic Cultivation

For this type of cultivation, the roots are suspended in the air in dark chambers. The nutrient solution is normally sprayed onto the root system at scheduled intervals for optimal aeration [6].

Some of these systems are as follows:

- Spray column system: This consists of a cylindrical platform made of opaque polyvinyl chloride, with lateral perforations through which the plants are introduced. The nutrient solution is sprayed over the upper part of the roots to ensure a permanent contact with the nutrient solution while the lower part of the root is well aerated (Figure 2a).
- Schwalbach system: This consists of a growth chamber in which the roots grow in the air and are kept in complete darkness. The nutrient solution is sprayed at different distribution points located near the leaves to ensure optimal foliar application, after which it drains to the root, where the excess solution is recovered (Figure 2b).
- Aero-Gro system: The nutrient solution is injected onto the roots directly through finely separated droplets at low pressure, avoiding clogging problems in pipes and spray nozzles (Figure 2c).

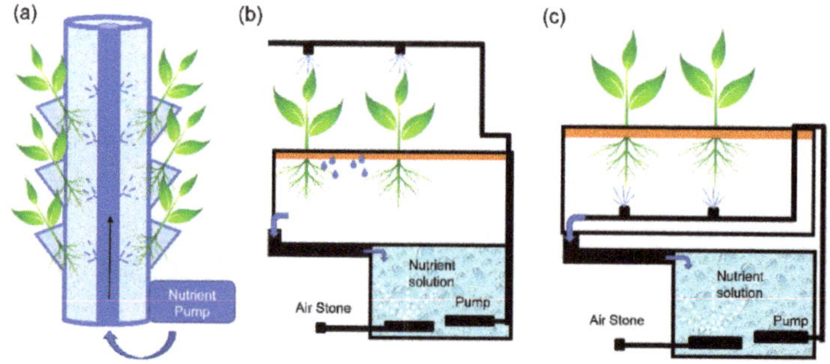

Figure 2. Aeroponic cultivation systems. (**a**) Spray column system, (**b**) Schwalbach system, and (**c**) Aero-Gro system.

2.1.2. Hydroponic Cultivation

As stated above, in hydroponic cultivation the roots are completely submerged in the nutrient solution without any solid substrate. It is very important for light not to reach the nutrient solution to avoid algal blooms, as this would result in low oxygen availability, and this may affect root growth, and consequently, result in reduced plant yield [11].

There are different types of hydroponic systems:

- Deep floating technique (DFT): It incorporates perforated polystyrene sheets as growing units that are placed on top of the tanks filled with the nutrient solution. The aerial part of the plants grows on these sheets with their roots submerged in the tank solution. These systems have an air pump that aerates the nutrient solution (Figure 3a).
- Nutrient film technique (NFT): This system is based on pumping a thin layer of nutrient solution onto the root system through constant flow. This is achieved by placing a small channel with a 1% slope to ensure that the nutritive solution reaches the roots by laminar flow. The excess solution drains into a collecting tank where the conductivity and pH values are restored and the nutrient solution can be pumped back to the top of the channel (Figure 3b).
- New growing system technique (NGST): This system is based on a channel formed by polyethylene bags located internally in three interconnected layers and wrapped by a layer of black polyethylene, which prevents direct contact of light with the root system. The entire system is suspended in the air and leveled to collect drainage at the end of the growing line. The irrigation system is in continuous operation and the drained solution reaches a tank where the nutrient levels are adjusted, heated, and pumped back into the system. The irrigation pipe is located close the root system to facilitate heating of the roots [11] (Figure 3c).

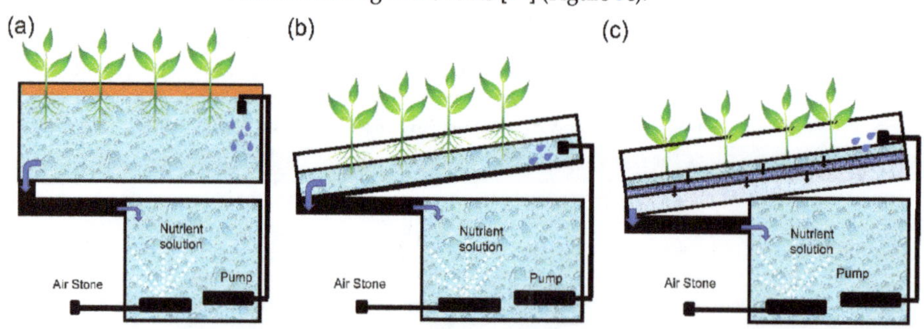

Figure 3. Hydroponic cultivation systems. (**a**) Deep floating technique (DFT) system, (**b**) nutrient film technique (NFT) system, and (**c**) new growing system technique (NGST).

2.1.3. Aquaponic Cultivation

The concept of aquaponics is based on integrating the industrial production of fish (aquaculture) with the cultivation of plants (horticulture), with the aim of establishing a nutritional balance between both species, in such a way that the use of resources (water and nutrients) is shared in the same production system [12,13]. It is based on the use of waste from the aquaculture production, totally or partially, as a nutrient solution for plant growth in a hydroponic cultivation system (Figure 4) [12–14].

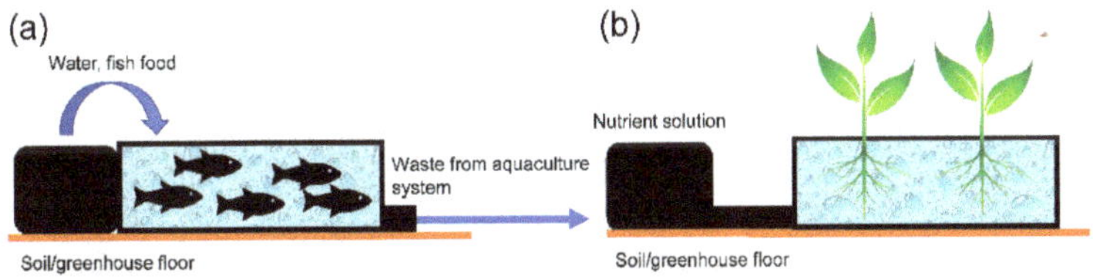

Figure 4. Aquaponic system consisting of an (**a**) aquaculture system for fish production, which is connected to a (**b**) hydroponic system used for crop production.

2.1.4. Cultivation in Organic and Inorganic Substrates

These systems are based on the use of different substrates that provide optimal oxygen and humidity conditions for the correct development of the plant. Organic substrates of natural origin such as peat, or substrates derived from by-products of agricultural activity, such as coconut fiber, cereal straw, or wood shavings, can be used. Inorganic substrates of natural origin with a high porosity, such as sand or volcanic gravel, can also be applied. In addition, inorganic substrates resulting from the industrial transformation such as rock wool, fiberglass, perlite, or vermiculite, are also frequently used [10]. Substrate cultivation systems are characterized by better aeration as compared to water cultivation systems, but at the same time, the flow of water must be continuous to achieve maximum production [11].

Three systems can be distinguished:

- Growing directly on substrate: These systems are delimited by a thick polyethylene mat that prevents the nutrient solution from leaking into the soil. The irrigation system utilized is drip irrigation, and the excess nutrient solution is sent to a tank where the appropriate adjustments will be made for reusing the nutrient solution (Figure 5a,b).
- Growing in bags or containers: The root volume is delimited by elongated two-color polyethylene bags closed at the ends and with two drainage holes filled with substrate. The plants will grow in these bags, the nutritive solution will be dripped in, and the excess solution will be channeled to a tank for further adjustment and reuse [11] (Figure 5c).
- Single unit culture systems: These systems were developed due to the need to control the transmission of fungal diseases in the continuous systems. In this case, the container is the basic cultivation unit and is placed parallel to the drip line. This allows for better control of individual plants, but this system in large-scale production could be prohibitively expensive (Figure 5d,e).

From an environmental and economic point of view, and with the aim of developing cultivation systems that are as sustainable as possible, the implementation of these new types of soilless cropping systems is increasingly widespread. The improvements in these new production systems allow better usage of the nutritive solution and the reuse of the substrates and other supplies, which will lead to a mitigation of the environmental impact of modern agricultural production [10,11].

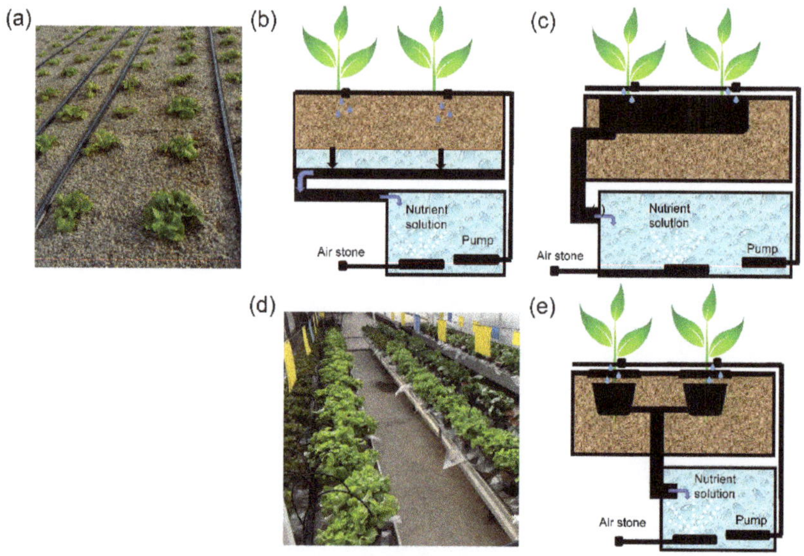

Figure 5. Substrate cultivation systems. (**a**) Lettuce plants growing in sand as an inert substrate, (**b**) a scheme of plants growing directly onto substrate, (**c**) a scheme of plants growing in containers, (**d**) lettuce plants growing in individual containers, and (**e**) a scheme of plants growing in individual containers.

2.2. Physiological Disorders in Soilless Cropping Systems

Various physiological disorders can arise in plants growing in these soilless cropping systems and are mainly caused by nutritional imbalance due to adverse environmental factors and not by the effect of the nutrients themselves [15,16]. Environmental stresses such as those related to temperature, irradiation, or relative humidity, favor the appearance of different physiological disorders [16–19]. The most common of these physiological disorders are BER in fruits such as tomatoes, peppers, squash, cucumbers, melons, etc., and TB, which causes necrosis of the leaf margins in leafy crops such as lettuce or cabbage. BER and TB are generally caused by environmental factors such as soil moisture fluctuations, salinity, and heat stress, among others [18,20]. In these cases, endogenous calcium (Ca^{2+}) content is reduced, and the rapidly growing tissues are mostly affected. Good management of greenhouse environmental conditions, the use of stress-tolerant varieties, and proper handling of the nutrient solution can alleviate these problems [10,21].

3. Physiological Disorders: Blossom-End Rot (BER) and Tipburn (TB)

Climate change is one of the main challenges facing the agricultural sector. Water quantity and quality will be mainly affected by rising temperatures in the coming decades [3,22]. The flexibility of plants to cope with these environmental stresses will depend on their adaptability, and the search for more tolerant genotypes will require the implementation of different strategies to avoid negative effects on plant growth and development.

Physiological disorders such as BER and TB are on the rise due to climate change, are often difficult to predict, and the challenge of controlling the onset of symptoms makes them a serious problem in crop production [15,23,24]. In the early 20th century, BER was believed to be caused by parasitic organisms, chemical toxicity, high transpiration, and lack of soil moisture [25]. However, from the middle of the 20th century, the appearance of BER in tomato, pepper, or watermelon and TB in leafy vegetables, was directly associated to mild Ca^{2+} deficiency in fruits and leaves, respectively [15,18].

Following the new soilless cropping techniques, where the producer provides the necessary nutrient levels that the plant requires at any time, the possibility that mild Ca^{2+}

deficiency alone is not the main cause of these physiological disorders is beginning to be assessed [18,26]. Several authors speculate on the cause–effect relationship of Ca^{2+} deficiency in both disorders, observing that on many occasions, fruits with these symptoms contained equal or higher concentrations of Ca^{2+} in their tissues [26]. Other studies indicated that either low levels or high levels of Ca^{2+} in the nutrient solution led to the appearance of BER in fruits of various species [27]. These results suggest that Ca^{2+} deficiency by itself may not be the causative of BER, but rather that a nutrient imbalance is involved in its appearance. Many authors, in their eagerness to predict and act on time against these problems, have centered their attention on the study of the main triggers of these physiological disorders. These studies distinguish different agronomic, physiological, and genetic factors that favor the induction of BER and TB [15,18,20,26,28]. In the following sections, we will succinctly describe all these factors and their relationships.

3.1. Abiotic Factors Influencing BER and TB

3.1.1. Drought

In agriculture, droughts are generally defined as the periods in which the water losses by transpiration through the leaves, and by evaporation through the soil exceed the amount of water input from precipitation and subsequent water uptake by the roots of the plants. The incidence and intensity of droughts have increased in some regions of the earth, and these are expected to rise in future climate change scenarios [29]. As plants require water for their metabolism, periods of drought can be fatal by reducing crop production to near-unproductive levels (or even causing crop death) or, at best, result in low yields and low quality products. Depending on the decrease in the irrigation levels by droughts, lettuce and carrot yields are expected to be 25–30% lower than usual. Under these conditions, vegetables and fruits such as apples and pears will generally be sweeter, but smaller. For this reason, consumers and markets will likely have to change their expectations [24].

Studies have been carried out on lettuce and tomato with different irrigation regimes, and under field and greenhouse conditions, to assess commercial traits such as growth, crop maturity, and marketability. Several studies have indicated that deeper roots are key drivers of drought tolerance in plants [30,31]. In addition, a higher incidence of BER and TB has been associated with insufficient water uptake by roots [32–35]. In experiments with different cultivars of lettuce, it was found that TB occurred more frequently in iceberg lettuce than in butterhead lettuce, which is also more drought tolerant [34,36,37]. A similar situation was found in tomatoes, where cultivars with larger fruits suffered more BER symptoms than cultivars with smaller tomatoes [19]. Therefore, breeding new varieties with high drought tolerance is essential for developing vegetables that are better adapted to the consequences of projected climate change, such as BER and TB [36,38].

3.1.2. High Temperature

Global warming has led to the increase in the timing, intensity, and duration of heat-related impacts, such as heat waves [2]. In fact, several studies have shown that high temperatures can cause physiological, biochemical, and morphological changes in crops, which leads to inadequate plant development and consequently to yield losses [39–41]. In broccoli, heat can cause malformations, such as uneven heads with large flower buds, bracts on the heads, or soft heads [42]. On the other hand, it has been observed that flavonoid content and glucosinolate composition in the broccoli florets increased with higher temperatures [43].

High light intensities and high temperatures are environmental factors that cause accelerated photosynthesis and growth rates that can trigger BER in tomato and TB in lettuce [42,44–49]. Hence, BER is likely to occur in fruit tissues as the rapid growth rate increases exponentially, and Ca^{2+} supply to other parts of the plant is restricted by mass flow of free Ca^{2+} through the xylem [20]. It has been hypothesized that an increased demand for Ca^{2+} in rapidly growing tissues, such as occurs in fruits during cell growth, might indirectly lead to BER when Ca^{2+} is limited [50]. In addition, it has been proposed that light and temperature influence Ca^{2+} absorption and distribution within the plant, thus limiting Ca^{2+} concentration within the fruit during heat stress [17]. Lettuce grown at high temperatures display an accelerated growth that causes little uniformity in the closure of the head and enhances early flowering of lateral stems, which causes bitterness in the leaves and enhances TB incidence [16,36,37,49,51].

3.1.3. Salinity

Salinity is one of the most recognizable factors that will be enhanced by climate change. In any climate change scenario, salinity occurs because of global warming, which causes ice melting at the poles and thus a rise in sea levels. This further causes coastal waters to come into contact with farmland and contaminate the soil with high levels of salts, which will directly affect the growth, quality, and yield of crops that are grown near the coastline. Saline soils encompass approximately 10% of the land surface, and 50% of irrigated land worldwide [52]. High salt levels in soil lead to aggravated dehydration of plant cells, ion toxicity, and oxidative stress, which can cause growth inhibition, damage at the molecular level, and even plant death [31,53]. In addition, soil salinity prevents nutrient uptake by the plant and alters the permeability of the plasma membrane, causing increased salt accumulation in some plant tissues [17,54]. In fact, the selective uptake of Ca^{2+} over Na^+ is a suitable indicator of salinity stress [55]. Alam and co-authors [56] studied the response of 27 tomato genotypes to various salt treatments to determine their response. They observed that the seedlings from the saline treatments had higher concentrations of Na+ in the leaves, as well as greater root length, fresh and dry weight. It has been observed that in soils with a heterogeneous distribution of salts, the root system absorbs significantly more Ca^{2+} than Na^+, and this could be a critical factor that contributed to greater Ca^{2+}/Na^+ in the fruit and, therefore, to a lower incidence of BER observed in tomato fruits grown in these soils [57]. These authors studied the effect of different saline irrigation regimes under different potential limits of the soil matrix on tomato crop yield and reported BER incidence. The effects of salinity stress on the growth of two types of lettuce under the NFT hydroponic system were analyzed, and it was observed that the amount of fresh and dry matter of the different lettuce types were significantly affected by salinity levels [58].

Water with a low salt content enhanced tomato quality, including fruit density, soluble solids, total acid, vitamin C, and sugar–acid ratio, and had a lower BER incidence than the other more saline treatments. High salinity levels led to a reduction in tomato yield, a decrease in leaf area index and chlorophyll content, together with the appearance of BER symptoms. All this evidence shows that tomato has a moderate salt tolerance index, and mild salinity levels improve osmotic regulation, increase adenosine triphosphatase enzyme activity, and stimulate crop growth [59] Additionally, mild salinity enhanced tomato sensory attributes due to increases in sugar, organic acid, and amino acid contents [59,60]. Inoculation of growth-promoting rhizobacteria in tomato plants has also been shown to improve growth and stress tolerance, resulting in higher crop yields [61–65].

3.2. *Physiological Factors Influencing BER and TB Incidence*

During agricultural production, an appropriate nutrient management is fundamental for the control of BER and TB. It has been observed that when some nutrients, such as K, P, and Mg, are applied above a certain concentration in the nutrient solution (80, 400, and 500 mg L^{-1}), they could decrease Ca^{2+} uptake and increase BER incidence [66]. Indeed, reducing K^+ supply in combination with the use of fertilizers such as $Ca(NO_3)_2$ has been shown to reduce the incidence of BER during soilless tomato production. This occurs directly to the antagonistic effect between the cations in the growing medium, so that by reducing the K^+ concentration, the absorption and mobility of Ca^{2+} can increase [67,68]. It has also been demonstrated that the use of organic fertilizers reduced the incidence of BER [69,70]. The authors found that organic fertilizers not only acted as nutrient sources and increased crop yield, but reduced the effect of BER, probably because they improved

Ca^{2+} absorption and translocation. Ronga and co-authors [70] also suggested that since one of the organic fertilizers they used had milled rice bran with high levels of abscisic acid (ABA), it was possible that the surplus of ABA increased fruit Ca^{2+} uptake directly, as previously reported in tomato fruits [50].

Using pericarp discs from tomato fruits [71], it was shown that exogenously applied Ca^{2+} inhibited BER symptom development in a concentration-dependent manner, but increased symptom severity in tomato fruits when Ca^{2+} was applied to whole plants in the irrigation solution [71]. Unexpectedly, increasing the Ca^{2+} levels of tomato fruits through the expression of the vacuolar H^+/Ca^{2+} antiporter, cation exchanger 1 (CAX1), from *Arabidopsis thaliana*, dramatically increased the occurrence of BER. These latter results suggest that altered Ca^{2+} homeostasis between cytosolic, apoplastic, and vacuolar Ca^{2+} pools might disrupt calcium signaling and lead to localized cell death and enhanced BER incidence [72]. In romaine lettuce cultivars grown in greenhouse conditions, foliar applications of Ca^{2+} resulted in a significant decrease in TB symptoms, which correlated with increased Ca^{2+} concentration in their young leaves as compared with non-treated controls [73]. Several authors have suggested that pectin methylesterases (PME) might be involved in Ca^{2+} transport in tomato plants [74]. They found that silencing PME reduced the concentration of Ca^{2+} bound to the cell wall and improved fruit tolerance to BER [74]. The overexpression of PME was shown to result in Ca^{2+} translocation into cell membranes and, consequently, to Ca^{2+} deficiency in most plant organs, thus enhancing BER incidence in the fruits. Other studies have suggested that the increase in PME synthesis and PME activity overlapped with the critical period for BER development [20]. Taken together, these results indicate that tightly regulated Ca^{2+} homeostasis during periods of rapid growth is required to minimize BER and TB incidence in tomato and lettuce, respectively.

Two stages are involved in fruit growth: cell division influenced by auxin signaling, and cell expansion which is synergistically regulated by auxins and gibberellins (GAs). Fruit ripening occurs when auxin and GA levels decrease with a continuous increase in ABA and ethylene [75]. Phytohormones also regulate a plethora of plant responses to cope with abiotic stress factors [76–78]. Some of these hormones, such as ABA or GAs, have a direct influence on BER [18]. However, a mild level of stress, resulting from one or more interacting environmental factors, does not always result in a certain degree of BER [18]. Rather, it appears that rapid fruit growth promotes a high predisposition to BER and subsequent critical stress is required to trigger cell death [26].

Nevertheless, while it is possible that certain stress conditions may produce hormonal imbalances, it may be likely that hormones involved in cell expansion and fruit development have indirect effects on the incidence of BER. The highest concentration of auxins and GAs in the fruit occurs before cell expansion [79]. The application of auxins and/or GAs is known to increase cell division, rapid fruit growth and BER incidence [80,81]. Therefore, the acceleration of fruit growth and the inability of the plant to supply sufficient Ca^{2+} to the fast-growing fruit could explain the effects of auxins and GAs on BER incidence in most cases.

Although several studies have suggested possible processes by which ABA and GAs regulate BER development in fruit tissue, many of the molecular components involved remain unknown [82]. GAs and ABA can control the expression of genes and gene networks leading to independent and/or antagonistic responses that influence fruit susceptibility to BER [83]. In tomato plants treated with GAs, the expression of genes involved in Ca^{2+} transport and consequently, the concentration of water-soluble Ca^{2+}, was reduced and the incidence of BER concomitantly increased [84]. In turn, the addition of an inhibitor of GA biosynthesis reduced BER in fruits due to increased membrane resistance, thereby decreasing the entry of reactive oxygen (ROS) and other toxic compounds into the fruit [74].

ABA is the main hormone involved in plant stress response. Wang et al. observed that ABA levels negatively correlated with Ca^{2+}, suggesting that ABA plays a regulatory role in response to TB in *Brassica rapa* L. ssp. *pekinensis* [85]. Evidence has been provided indicating an antagonistic interaction between GAs and ABA in the coordination of cation exchange activity (e.g., CAX1) in the tonoplast and thus in the incidence of BER [46,82]. The tomato *procera* (*pro*) mutant, which shows a constitutive GA response, showed a higher BER incidence due to a combined lower Ca^{2+} translocation to the fruit and a reduced delivery of water and nutrients to the fruit, as a result of competition between vegetative organs and fruits for the available Ca^{2+} [86].

Ethylene has also been proposed to be involved in the induction of BER [18]. Ethylene, in addition to its effect on fruit ripening, is known to be involved in the initiation of wound and pathogen responses via Ca^{2+} signals [87]. Early ethylene production, premature ripening, necrosis, and cell death in the apical region of the fruit, have also been found to be symptoms directly related to BER [20,26,88]. However, it is also possible that ethylene and other stress factors that increase ROS production may influence BER, subsequent Ca^{2+} concentration increase, and rapid cell expansion [88]. In persimmon fruits, salinity stress increased ethylene production, which resulted in necrotic lesions in the calyx resembling BER, but the link with endogenous Ca^{2+} levels has not yet been established [89].

3.3. Genetic Factors Influencing BER and TB Incidence

Crop yields are strongly affected by abiotic stress caused by drought, salinity, and high temperatures. Plants respond to these stressors through various biochemical and physiological adaptations, some of which are the result of changes in gene expression [90]. In addition, many studies have emphasized that susceptibility to BER and TB is highly genotype-dependent [19,91]. In tomato, for example, pear tomatoes are more susceptible to BER than round tomatoes, and BER is never observed in cherry tomatoes [19]. In addition, a strong variation in the incidence of TB between different lettuce cultivars has been reported [92,93] that has been used for the development of TB resistant varieties through targeted breeding [92,94,95].

The use of genomic tools has allowed the identification of quantitative trait loci (QTL) for TB incidence in various recombinant inbred line (RIL) populations of lettuce and the subsequent development of linked molecular markers [96]. A major QTL accounts for up to 70% of the phenotypic variance for TB incidence in lettuce. By comparing lines with contrasting haplotypes, the genetic region was narrowed down to a genomic region containing 12 genes, two of which encoded proteins with sequence similarity to Ca^{2+} transporters. These studies will allow the development of molecular markers to introgress the major resistance alleles found into new cultivars of TB-sensitive iceberg genotypes [96,97]. However, more research is needed to identify the underlying candidate genes for these QTL and to assess the effect of their introgression in other lettuce cultivars. Conversely, only a few studies have been conducted on the incidence of TB in hydroponically grown lettuce [98,99].

Ca^{2+} deficiency in maize causes leaf tip rot, which is similar to TB in lettuce. Two maize lines, B73 and Mo17, differed in their Ca^{2+} deficiency symptoms. In a recent study by Wang and coauthors [100], it was suggested that ammonium reduced the seedling's ability to absorb Ca^{2+}, which ultimately caused the observed Ca^{2+} deficiency phenotype in the leaf tip. To identify a QTL associated with Ca^{2+} deficiency in maize leaves, the authors used a RIL mapping population of 276 lines derived from a cross between B73 and Mo17 maize genotypes. Five QTL associated with a variation in the Ca^{2+} deficiency trait were identified, and some candidate genes were selected for further studies [100].

The slow growth rate and the high concentration of Ca^{2+} observed in the fruits of the IL8–3 line, which contain a small chromosome segment of the wild relative *Solanum pennellii* in the tomato cultivar M82, could be related to the low incidence of BER observed in the IL8–3 line [101]. The results of this study suggest that the main factors contributing to the difference in BER incidence between M82 and IL8–3 were fruit growth rate and Ca^{2+} availability (but also other elements, including K^+ and B^+) during the early stages of fruit enlargement [102].

In a recent systematic review published by Kuronuma and Watanabe [84], the authors discussed the latest studies aimed at the identification of genes associated with BER and TB by QTL and transcriptomic analysis. Despite these recent advances, the causative genes for Ca^{2+} deficiency disorders in most crops are not yet known and await further investigation.

4. Solutions to Alleviate Ca-Related Disorders in Soilless Production Systems

In the present section, we will briefly introduce farmers to the tools available to minimize some physiological disorders, such as BER and TB, the incidence of which is likely to increase in the coming decades due to climate change. In intensive production systems, new strategies must be applied to mitigate these Ca-related disorders, in order to synergize crop and environmental factors to achieve efficient production with higher yields [15].

The use of smart management practices could help mitigate these Ca-related disorders but could also be useful in lessening the impact of climate change on crop productivity through better nutrient management [15,103,104]. Continuous monitoring of soilless production systems using low-cost sensors, as well as data-integration management approaches, will be key for establishing criteria and aiding decision-making during crop production [105]. It is now possible to automate a hydroponic growing system using cheap sensors that monitor and control environmental parameters such as light intensity, relative humidity, as well as pH, electrical conductivity, and temperature of the nutrient solution [106–108]. Hasan et al. [109] used drones to detect diseases in tomatoes by analyzing foliar images, which allowed them to adjust the treatment to the most affected regions of the crop. These technologies are based on the need to apply artificial intelligence techniques, such as machine learning, that requires training the initial model with a large amount of data and then using the information gathered from the crops to make predictions [109]. Indeed, the use of sensors that measure physiological processes such as photosynthesis, transpiration, and leaf stomatal conductance, has made it possible to detect and quantify the impact of drought stress in tomato plants [104]. In a recent study, the continuous monitoring of tomatoes grown in an NFT soilless system was performed by combining Netatmo sensors for greenhouse microclimate data collection, with daily fertilizer usage data [110]. Based on these data and on crop yield, the authors concluded that a cost-effective and simplified smart agriculture system allows farmers to apply accurate crop production planning and decision making of cultivation activities, such as maintaining a well-balanced microclimate environment [110]. These tools allow us to remotely or automatically adjust the different abiotic factors that, as mentioned above, can trigger the appearance of BER or TB in crops. It has also been observed that regulation of the size of air bubbles in the hydroponic could increase crop yield [111]. In this sense, it has been shown that the production of microbubbles through specific injectors would facilitate the arrival of oxygen to the finest roots, which is necessary for the effective absorption of essential nutrients and plant growth [111].

New strategies have recently been studied to reduce soil contamination due to the excessive use of agrochemicals. It has been proposed that mitigating excess of plant nutrients by using nanoparticles could lead to more precise nutrition and reduced fertilization in both conventional and hydroponic cropping systems [106]. Nanoparticles have been used as slow-release fertilizers [112,113] or for the elaboration of specific biopesticides [114,115]. In this sense, nanoparticles may provide nutrients in a more soluble and available form to plants [116], and some studies have also found that the use of carbon nanotubes as a soil amendment can double tomato yields and increase agricultural production under certain conditions. Strategies such as the use of nanoparticles for fertilization could help deliver nutrients very precisely, especially at different physiological stages, and thus avoid the effects caused by nutrient imbalances in certain phases of the plant growth cycle, which are more sensitive to the appearance of BER and TB. However, it is not yet clear how the soil ecosystem may be affected by such practices, and therefore, a thorough investigation of the impact and assessment of toxicity at all levels of the ecosystem is required [117].

Plant growth-promoting rhizobacteria (PGPRs) have been used in hydroponic growing systems as biofertilizers and/or biocontrol agents with variable success [106,109,118].

Tomato plants treated with potassium-releasing PGPRs showed a greater reduction in BER levels than untreated plants, which ultimately increased yield in terms of fruit size and weight [119]. In another study, tomato plants treated with *Pseudomonas* sp. LSW25R showed a 61% reduction in BER incidence in a hydroponic system, possibly due to increased Ca^{2+} uptake in their roots [120]. In addition, the exogenous application of ABA to tomato crops has been shown to reduce BER incidence at different Ca^{2+} concentrations in the nutrient solution [23,46]. Additionally, the foliar application of Ca^{2+} in lettuce was found to significantly reduce TB incidence [73] Taken together, the implementation of these strategies could enhance crop production and reduce the excess use of fertilizers [121].

From a genetics point of view, identifying genotypes with a high resilience to nutritional disorders, especially Ca^{2+}, and introgressing the causative genes through breeding, may alleviate physiological disorders such as BER and TB [84]. Targeted breeding combined with the application of precision tools in soilless cultivation will provide us with higher yields, especially in terms of fruit quality in the case of tomato [94], as well as of leaf and head quality in the case of lettuce [92,93].

5. Conclusions

The present review summarizes the factors and mechanisms that trigger TB and BER, and this knowledge can be used for the development of new strategies that could help us mitigate these Ca-related physiological disorders. On the one hand, this evidence can be used to develop new cultivars that are highly tolerant to the factors that cause BER and TB. On the other hand, we propose that soilless cultivation offers many advantages over conventional cultivation, as it allows for the detailed monitoring of physiological processes and nutritional balance of plants using remote sensors. The proposed multidisciplinary strategy to reduce BER and TB levels will bring us higher yields and better quality of the final product.

Author Contributions: Conceptualization, V.B., J.R.A.-M. and J.M.P.-P.; writing—original draft preparation, V.B. and J.R.A.-M.; writing—review and editing, J.M.P.-P.; supervision, J.R.A.-M. and J.M.P.-P. All authors have read and agreed to the published version of the manuscript.

Funding: This research received no external funding.

Institutional Review Board Statement: Not applicable.

Informed Consent Statement: Not applicable.

Acknowledgments: We are grateful to Jorge Benítez Vega (Bayer CropScience) for his support and Mario Fon for his help with English editing.

Conflicts of Interest: The authors declare no conflict of interest.

References

1. An IPCC Special Report on Climate Change, Desertification, Land Degradation, Sustainable Land Management, Food Security and Greenhouse Gas Fluxes in Terrestrial Ecosystems. Available online: https://www.ipcc.ch/srccl/ (accessed on 24 February 2022).
2. Climate Change: Atmospheric Carbon Dioxide. Available online: https://www.climate.gov/news-features/understanding-climate (accessed on 24 February 2022).
3. Reidsma, P.; Ewert, F.; Lansink, A.O.; Leemans, R. Adaptation to Climate Change and Climate Variability in European Agriculture: The Importance of Farm Level Responses. *Eur. J. Agron.* **2010**, *32*, 91–102. [CrossRef]
4. Hamidov, A.; Helming, K.; Bellocchi, G.; Bojar, W.; Dalgaard, T.; Ghaley, B.B.; Hoffmann, C.; Holman, I.; Holzkämper, A.; Krzeminska, D.; et al. Impacts of Climate Change Adaptation Options on Soil Functions: A Review of European Case-Studies. *Land Degrad. Dev.* **2018**, *29*, 2378–2389. [CrossRef] [PubMed]
5. Hussain, A.; Iqbal, K.; Showket, A.; Prasanto, M.; Negi, A.K. A Review On The Science Of Growing Crops Without Soil (Soilless Culture)—A Novel Alternative For Growing Crops. *Int. J. Agric. Crop Sci.* **2014**, *7*, 833–842.
6. Beltrano, J.; Gimenez, D.O. *Cultivo En Hidroponía*, 1st ed.; Edulp integra la Red de Editoriales Universitarias Nacionales (REUN): La Plata, Argentina, 2015.
7. Gruda, N.; Bisbis, M.; Tanny, J. Impacts of Protected Vegetable Cultivation on Climate Change and Adaptation Strategies for Cleaner Production—A Review. *J. Clean. Prod.* **2019**, *225*, 324–339. [CrossRef]

8. Martinez-Mate, M.A.; Martin-Gorriz, B.; Martínez-Alvarez, V.; Soto-García, M.; Maestre-Valero, J.F. Hydroponic System and Desalinated Seawater as an Alternative Farm-Productive Proposal in Water Scarcity Areas: Energy and Greenhouse Gas Emissions Analysis of Lettuce Production in Southeast Spain. *J. Clean. Prod.* **2018**, *172*, 1298–1310. [CrossRef]
9. Urrestarazu, M.; Postigo, A.; Salas, M.; Sánchez, A.; Carrasco, G. Nitrate Accumulation Reduction Using Chloride in the Nutrient Solution on Lettuce Growing by NFT in Semiarid Climate Conditions. *J. Plant Nutr.* **1998**, *21*, 1705–1714. [CrossRef]
10. Baixauli Soria, C.; Aguilar Olivert, J.M. *Cultivo Sin Suelo de Hortalizas: Aspectos Práticos y Experiencias*; Generalitat Valenciana, Conselleria de Agricultura, Pesca y Alimentación: Valencia, Spain, 2002.
11. Maluin, F.N.; Hussein, M.Z.; Nik Ibrahim, N.N.L.; Wayayok, A.; Hashim, N. Some Emerging Opportunities of Nanotechnology Development for Soilless and Microgreen Farming. *Agronomy* **2021**, *11*, 1213. [CrossRef]
12. Goddek, S.; Joyce, A.; Kotzen, B.; Burnell, G.M. *Aquaponics Food Production Systems, Combined Aquaculture and Hydroponic Production Technologies for the Future*; Springer Nature Switzerland AG: Cham, Switzerland, 2020.
13. Maucieri, C.; Nicoletto, C.; Junge, R.; Schmautz, Z. Hydroponic Systems and Water Management in Aquaponics: A Review. *Ital. J. Agron.* **2018**, *13*, 1012. [CrossRef]
14. Lennard, W.; Ward, J. A Comparison of Plant Growth Rates between an NFT Hydroponic System and an NFT Aquaponic System. *Horticulturae* **2019**, *5*, 27. [CrossRef]
15. Hagassou, D.; Francia, E.; Ronga, D.; Buti, M. Blossom End-Rot in Tomato (*Solanum lycopersicum* L.): A Multi-Disciplinary Overview of Inducing Factors and Control Strategies. *Sci. Hortic.* **2019**, *249*, 49–58. [CrossRef]
16. Saure, M.C. Causes of the Tipburn Disorder in Leaves of Vegetables. *Sci. Hortic.* **1998**, *76*, 131–147. [CrossRef]
17. Adams, P.; Ho, L.C. Effects of Environment on the Uptake and Distribution of Calcium in Tomato and on the Incidence of Blossom-End Rot. *Plant Soil* **1993**, *154*, 127–132. [CrossRef]
18. Saure, M. Blossom-End Rot of Tomato (*Lycopersicon esculentum* Mill.)—A Calcium or a Stress-Related Disorder? *Sci. Hortic.* **2001**, *90*, 193–208. [CrossRef]
19. Ho, L.C. The Physiological Basis for Improving Tomato Fruit Quality. *Acta Hortic.* **1999**, *487*, 33–40. [CrossRef]
20. Ho, L.C.; White, P.J. A Cellular Hypothesis for the Induction of Blossom-End Rot in Tomato Fruit. *Ann. Bot.* **2005**, *95*, 571–581. [CrossRef]
21. Camacho Ferre, F.; Cánovas, M.F.; Magán, C.J. *Técnicas de Producción En Cultivos Protegidos-Cultivos Sin Suelo*, 2nd ed.; Caja Rural Intermediterránea, Cajamar.: Barcelona, Spain, 2003.
22. Mendelsohn, R.; Dinar, A. Climate Change, Agriculture and Developing Countries: Does Adaptation Matter? *World Bank Res. Obs.* **1999**, *14*, 277–293. [CrossRef]
23. Casey Barickman, T.; Kopsell, D.A.; Sams, C.E. Foliar Applications of Abscisic Acid Decrease the Incidence of Blossom-End Rot in Tomato Fruit. *Sci. Hortic.* **2014**, *179*, 356–362. [CrossRef]
24. Bisbis, M.B.; Gruda, N.; Blanke, M. Potential Impacts of Climate Change on Vegetable Production and Product Quality—A Review. *J. Clean. Prod.* **2018**, *170*, 1602–1620. [CrossRef]
25. Wedgworth, H.H.; Neal, D.C.; Wallace, J.M.; Ricks, J.R. Wilt and Blossom-End Rot of the Tomato. *Bulletins* **1927**, *247*, 1–18.
26. Saure, M.C. Why Calcium Deficiency Is Not the Cause of Blossom-End Rot in Tomato and Pepper Fruit—A Reappraisal. *Sci. Hortic.* **2014**, *174*, 151–154. [CrossRef]
27. Reitz, N.F.; Shackel, K.A.; Mitcham, E.J. Differential Effects of Excess Calcium Applied to Whole Plants vs. Excised Fruit Tissue on Blossom-End Rot in Tomato. *Sci. Hortic.* **2021**, *290*, 110514. [CrossRef]
28. Sonneveld, C.; van den Ende, J. The Effect of Some Salts on Head Weight and Tipburn of Lettuce and on Fruit Production and Blossom-End Rot of Tomatoes. *Neth. J. Agric. Sci.* **1975**, *23*, 191–201. [CrossRef]
29. Teichmann, C.; Bülow, K.; Otto, J.; Pfeifer, S.; Rechid, D.; Sieck, K.; Jacob, D. Avoiding Extremes: Benefits of Staying below +1.5 °C Compared to +2.0 °C and +3.0 °C Global Warming. *Atmosphere* **2018**, *9*, 115. [CrossRef]
30. Lynch, J.P.; Chimungu, J.G.; Brown, K.M. Root Anatomical Phenes Associated with Water Acquisition from Drying Soil: Targets for Crop Improvement. *J. Exp. Bot.* **2014**, *65*, 6155–6166. [CrossRef]
31. Acosta-Motos, J.R.; Penella, C.; Hernández, J.A.; Díaz-Vivancos, P.; Sánchez-Blanco, M.J.; Navarro, J.M.; Gómez-Bellot, M.J.; Barba-Espín, G. Towards a Sustainable Agriculture: Strategies Involving Phytoprotectants Against Salt Stress. *Agronomy* **2020**, *10*, 194. [CrossRef]
32. Karni, L.; Aloni, B.; Bar-Tal, A.; Moreshet, S.; Keinan, M.; Yao, C. The Effect of Root Restriction on the Incidence of Blossom-End Rot in Bell Pepper (*Capsicum annuum* L.). *J. Hortic. Sci. Biotechnol.* **2000**, *75*, 364–369. [CrossRef]
33. Sun, Y.; Feng, H.; Liu, F. Comparative Effect of Partial Root-Zone Drying and Deficit Irrigation on Incidence of Blossom-End Rot in Tomato under Varied Calcium Rates. *J. Exp. Bot.* **2013**, *64*, 2107–2116. [CrossRef]
34. Périard, Y.; Caron, J.; Lafond, J.A.; Jutras, S. Root Water Uptake by Romaine Lettuce in a Muck Soil: Linking Tip Burn to Hydric Deficit. *Vadose Zone J.* **2015**, *14*, vzj2014.10.0139. [CrossRef]
35. Kuronuma, T.; Ando, M.; Watanabe, H. Tipburn Incidence and Ca Acquisition and Distribution in Lisianthus (*Eustoma grandiflorum* (Raf.) Shinn.) Cultivars under Different Ca Concentrations in Nutrient Solution. *Agronomy* **2020**, *10*, 216. [CrossRef]
36. Ee, R.; Eriksen, L.; Knepper, C.; Cahn, M.D.; Mou, B. Screening of Lettuce Germplasm for Agronomic Traits under Low Water Conditions. *HortScience* **2016**, *51*, 669–679.
37. Barta, D.J.; Tibbitts, T.W. Calcium Localization in Lettuce Leaves with and without Tipburn: Comparison of Controlled-Environment and Field-Grown Plants. *HortScience* **1991**, *116*, 870–875. [CrossRef]

38. Millones-Chanamé, C.E.; de Oliveira, A.M.S.; de Castro, E.M.; Maluf, W.R. Inheritance of Blossom End Rot Resistance Induced by Drought Stress and of Associated Stomatal Densities in Tomatoes. *Euphytica* **2019**, *215*, 120. [CrossRef]
39. Adams, S.R.; Cockshull, K.E.; Cave, C.R.J. Effect of Temperature on the Growth and Development of Tomato Fruits. *Ann. Bot.* **2001**, *88*, 869–877. [CrossRef]
40. Wahid, A.; Gelani, S.; Ashraf, M.; Foolad, M.R. Heat Tolerance in Plants: An Overview. *Environ. Exp. Bot.* **2007**, *61*, 199–223. [CrossRef]
41. Sattar, F.A.; Hamooh, B.T.; Wellman, G.; Ali, M.A.; Shah, S.H.; Anwar, Y.; Mousa, M.A.A. Growth and Biochemical Responses of Potato Cultivars under in Vitro Lithium Chloride and Mannitol Simulated Salinity and Drought Stress. *Plants* **2021**, *10*, 924. [CrossRef]
42. Kałużewicz, A.; Krzesiński, W.; Knaflewski, M. Effect of Temperature on the Yield and Quality of Broccoli Heads. *J. Fruit Ornam. Plant Res.* **2009**, *71*, 51–58. [CrossRef]
43. Mølmann, J.A.B.; Steindal, A.L.H.; Bengtsson, G.B.; Seljåsen, R.; Lea, P.; Skaret, J.; Johansen, T.J. Effects of Temperature and Photoperiod on Sensory Quality and Contents of Glucosinolates, Flavonols and Vitamin C in Broccoli Florets. *Food Chem.* **2015**, *172*, 47–55. [CrossRef]
44. Wiebe, H.-J. The Morphological Development of Cauliflower and Broccoli Cultivars Depending on Temperature. *Sci. Hortic.* **1975**, *3*, 95–101. [CrossRef]
45. Tibbitts, T.W.; Theodore, G.F. Effects of Relative Humidity and Root Temperature on Calcium Concentration and Tipburn Development in Lettuce. *J. Am. Soc. Hortic. Sci.* **1984**, *109*, 128–131.
46. Tonetto de Freitas, S.; McElrone, A.J.; Shackel, K.A.; Mitcham, E.J. Calcium Partitioning and Allocation and Blossom-End Rot Development in Tomato Plants in Response to Whole-Plant and Fruit-Specific Abscisic Acid Treatments. *J. Exp. Bot.* **2014**, *65*, 235–247. [CrossRef]
47. Lee, J.G.; Choi, C.S.; Jang, Y.A.; Jang, S.W.; Lee, S.G.; Um, Y.C. Effects of Air Temperature and Air Flow Rate Control on the Tipburn Occurrence of Leaf Lettuce in a Closed-Type Plant Factory System. *Hortic. Environ. Biotechnol.* **2013**, *54*, 303–310. [CrossRef]
48. Gonzalo, M.J.; Li, Y.-C.; Chen, K.-Y.; Gil, D.; Montoro, T.; Nájera, I.; Baixauli, C.; Granell, A.; Monforte, A.J. Genetic Control of Reproductive Traits in Tomatoes Under High Temperature. *Front. Plant Sci.* **2020**, *11*, 326. [CrossRef] [PubMed]
49. Choi, K.Y.; Paek, K.Y.; Lee, Y.B. Effect of Air Temperature on Tipburn Incidence of Butterhead and Leaf Lettuce in a Plant Factory. In *Transplant Production in the 21st Century*; Springer: Dordrecht, The Netherlands, 2000; pp. 166–171.
50. Tonetto de Freitas, S.; Padda, M.; Wu, Q.; Park, S.; Mitcham, E.J. Dynamic Alternations in Cellular and Molecular Components during Blossom-End Rot Development in Tomatoes Expressing SCAX1, a Constitutively Active Ca2+/H+ Antiporter from Arabidopsis. *Plant Physiol.* **2011**, *156*, 844–855. [CrossRef] [PubMed]
51. Jenni, S.; Truco, M.J.; Michelmore, R.W. Quantitative Trait Loci Associated with Tipburn, Heat Stress-Induced Physiological Disorders, and Maturity Traits in Crisphead Lettuce. *Theor. Appl. Genet.* **2013**, *126*, 3065–3079. [CrossRef]
52. Ruan, C.-J.; Da Silva, J.A.T.; Mopper, S.; Qin, P.; Lutts, S. Halophyte Improvement for a Salinized World. *Crit. Rev. Plant Sci.* **2010**, *29*, 329–359. [CrossRef]
53. Acosta-Motos, J.R.; Ortuño, M.F.; Bernal-Vicente, A.; Diaz-Vivancos, P.; Sanchez-Blanco, M.J.; Hernandez, J.A. Plant Responses to Salt Stress: Adaptive Mechanisms. *Agronomy* **2017**, *7*, 18. [CrossRef]
54. Liu, X.; Baird, W.V. Identification of a Novel Gene, HaABRC5, from *Helianthus annuus* (Asteraceae) That Is Upregulated in Response to Drought, Salinity, and Abscisic Acid. *Am. J. Bot.* **2004**, *91*, 184–191. [CrossRef]
55. Xu, C.; Mou, B. Evaluation of Lettuce Genotypes for Salinity Tolerance. *Hortscience* **2015**, *50*, 1441–1446. [CrossRef]
56. Alam, M.S.; Tester, M.; Fiene, G.; Mousa, M.A.A. Early Growth Stage Characterization and the Biochemical Responses for Salinity Stress in Tomato. *Plants* **2021**, *10*, 712. [CrossRef]
57. Chen, S.; Wang, Z.; Zhang, Z.; Guo, X.; Wu, M.; Rasool, G.; Qiu, R.; Wang, X. Effects of Uneven Vertical Distribution of Soil Salinity on Blossom-End Rot of Tomato Fruit. *HortScience* **2017**, *52*, 958–964. [CrossRef]
58. Al-Maskri, A.; Al-Klharusi, L.; Al-Miqbali, H.; Khran, M.M. Effects of Salinity Stress on Growth of Lettuce (*Lactuca sativa*) under Closed-Recycle Nutrient Film Technique. *Agric. Biol.* **2010**, *12*, 377–380.
59. Zhai, Y.; Yang, Q.; Hou, M. The Effects of Saline Water Drip Irrigation on Tomato Yield, Quality, and Blossom-End Rot Incidence–A 3a Case Study in the South of China. *PLoS ONE* **2015**, *10*, e0142204. [CrossRef] [PubMed]
60. Zushi, K.; Matsuzoe, N. Utilization of Correlation Network Analysis to Identify Differences in Sensory Attributes and Organoleptic Compositions of Tomato Cultivars Grown under Salt Stress. *Sci. Hortic.* **2011**, *129*, 18–26. [CrossRef]
61. Kang, S.M.; Shahzad, R.; Bilal, S.; Khan, A.L.; Park, Y.G.; Lee, K.E.; Asaf, S.; Khan, M.A.; Lee, I.J. Indole-3-Acetic-Acid and ACC Deaminase Producing Leclercia Adecarboxylata MO1 Improves *Solanum lycopersicum* L. Growth and Salinity Stress Tolerance by Endogenous Secondary Metabolites Regulation. *BMC Microbiol.* **2019**, *19*, 80. [CrossRef] [PubMed]
62. Akram, W.; Aslam, H.; Ahmad, S.R.; Anjum, T.; Yasin, N.A.; Khan, W.U.; Ahmad, A.; Guo, J.; Wu, T.; Luo, W.; et al. Bacillus Megaterium Strain A12 Ameliorates Salinity Stress in Tomato Plants through Multiple Mechanisms. *J. Plant Interact.* **2019**, *14*, 506–518. [CrossRef]
63. Orozco-Mosqueda, M.D.C.; Duan, J.; DiBernardo, M.; Zetter, E.; Campos-García, J.; Glick, B.R.; Santoyo, G. The Production of ACC Deaminase and Trehalose by the Plant Growth Promoting Bacterium *Pseudomonas* sp. UW4 Synergistically Protect Tomato Plants against Salt Stress. *Front. Microbiol.* **2019**, *10*, 1392. [CrossRef] [PubMed]

64. Vaishnav, A.; Singh, J.; Singh, P.; Rajput, R.S.; Singh, H.B.; Sarma, B.K. *Sphingobacterium* sp. BHU-AV3 Induces Salt Tolerance in Tomato by Enhancing Antioxidant Activities and Energy Metabolism. *Front. Microbiol.* **2020**, *11*, 443. [CrossRef]
65. Ha-Tran, D.M.; Nguyen, T.T.M.; Hung, S.-H.; Huang, E.; Huang, C.-C. Roles of Plant Growth-Promoting Rhizobacteria (PGPR) in Stimulating Salinity Stress Defense in Plants: A Review. *Int. J. Mol. Sci.* **2021**, *22*, 3154. [CrossRef]
66. Adams, P. Plant Nutrition Demystified. *Acta Hortic.* **1999**, *481*, 341–344. [CrossRef]
67. Kanai, S.; Moghaieb, R.E.; El-Shemy, H.A.; Panigrahi, R.; Mohapatra, P.K.; Ito, J.; Nguyen, N.T.; Saneoka, H.; Fujita, K. Potassium Deficiency Affects Water Status and Photosynthetic Rate of the Vegetative Sink in Green House Tomato Prior to Its Effects on Source Activity. *Plant Sci.* **2011**, *180*, 368–374. [CrossRef]
68. Pujos, A.; Morard, P. Effects of Potassium Deficiency on Tomato Growth and Mineral Nutrition at the Early Production Stage. *Plant Soil* **1997**, *189*, 189–196. [CrossRef]
69. Kataoka, K.; Sugimoto, K.; Ohashi, H.; Yamada, H. Effect of Organo-Mineral Fertilizer on Tomato Fruit Production and Incidence of Blossom-End Rot under Salinity. *Hortic. J.* **2017**, *86*, 357–364. [CrossRef]
70. Ronga, D.; Caradonia, F.; Setti, L.; Hagassou, D.; Azevedo, C.V.G.; Milc, J.; Pedrazzi, S.; Allesina, G.; Arru, L.; Francia, E. Effects of Innovative Biofertilizers on Yield of Processing Tomato Cultivated in Organic Cropping Systems in Northern Italy. *Acta Hortic.* **2019**, *1233*, 129–135. [CrossRef]
71. Reitz, N.F.; Mitcham, E.J. Validation and Demonstration of a Pericarp Disc System for Studying Blossom-End Rot of Tomatoes. *Plant Methods* **2021**, *17*, 28. [CrossRef] [PubMed]
72. Park, S.; Cheng, N.H.; Pittman, J.K.; Yoo, K.S.; Park, J.; Smith, R.H.; Hirschi, K.D. Increased Calcium Levels and Prolonged Shelf Life in Tomatoes Expressing Arabidopsis H^+/Ca^{2+} Transporters. *Plant Physiol.* **2005**, *139*, 1194–1206. [CrossRef] [PubMed]
73. Corriveau, J.; Gaudreau, L.; Caron, J.; Jenni, S.; Gosselin, A. Testing Irrigation, Day/Night Foliar Spraying, Foliar Calcium and Growth Inhibitor as Possible Cultural Practices to Reduce Tipburn in Lettuce. *Can. J. Plant Sci.* **2012**, *92*, 889–899. [CrossRef]
74. de Freitas, S.T.; Handa, A.K.; Wu, Q.; Park, S.; Mitcham, E.J. Role of Pectin Methylesterases in Cellular Calcium Distribution and Blossom-End Rot Development in Tomato Fruit. *Plant J.* **2012**, *71*, 824–835. [CrossRef]
75. Fenn, M.A.; Giovannoni, J.J. Phytohormones in Fruit Development and Maturation. *Plant J.* **2021**, *105*, 446–458. [CrossRef]
76. Jiroutova, P.; Oklestkova, J.; Strnad, M. Crosstalk between Brassinosteroids and Ethylene during Plant Growth and under Abiotic Stress Conditions. *Int. J. Mol. Sci.* **2018**, *19*, 3283. [CrossRef]
77. Bielach, A.; Hrtyan, M.; Tognetti, V.B. Plants under Stress: Involvement of Auxin and Cytokinin. *Int. J. Mol. Sci.* **2017**, *18*, 1427. [CrossRef]
78. Mostofa, M.G.; Li, W.; Nguyen, K.H.; Fujita, M.; Tran, L.-S.P. Strigolactones in Plant Adaptation to Abiotic Stresses: An Emerging Avenue of Plant Research. *Plant Cell Environ.* **2018**, *41*, 2227–2243. [CrossRef] [PubMed]
79. de Jong, M.; Mariani, C.; Vriezen, W.H. The Role of Auxin and Gibberellin in Tomato Fruit Set. *J. Exp. Bot.* **2009**, *60*, 1523–1532. [CrossRef] [PubMed]
80. Castro, P.R.C. Plant Growth Regulators in Tomato Crop Production. *Acta Hortic.* **1980**, *100*, 99–104. [CrossRef]
81. Bangerth, F. Calcium Related Physiological Disorders of Plants. *Annu. Rev. Phytopathol.* **1979**, *17*, 97–122. [CrossRef]
82. de Freitas, S.; Shackel, K.A.; Mitcham, E.J. Abscisic Acid Triggers Whole-Plant and Fruit-Specific Mechanisms to Increase Fruit Calcium Uptake and Prevent Blossom End Rot Development in Tomato Fruit. *J. Exp. Bot.* **2011**, *62*, 2645–2656. [CrossRef] [PubMed]
83. de Freitas, S.T.; Martinelli, F.; Feng, B.; Reitz, N.F.; Mitcham, E.J. Transcriptome Approach to Understand the Potential Mechanisms Inhibiting or Triggering Blossom-End Rot Development in Tomato Fruit in Response to Plant Growth Regulators. *J. Plant Growth Regul.* **2018**, *37*, 183–198. [CrossRef]
84. Kuronuma, T.; Watanabe, H. Identification of the Causative Genes of Calcium Deficiency Disorders in Horticulture Crops: A Systematic Review. *Agriculture* **2021**, *11*, 906. [CrossRef]
85. Wang, W.; Wang, J.; Wei, Q.; Li, B.; Zhong, X.; Hu, T.; Hu, H.; Bao, C. Transcriptome-Wide Identification and Characterization of Circular RNAs in Leaves of Chinese Cabbage (*Brassica rapa* L. ssp. *pekinensis*) in Response to Calcium Deficiency-Induced Tip-Burn. *Sci. Rep.* **2019**, *9*, 14544.
86. Gaion, L.A.; Muniz, J.C.; Barreto, R.F.; D'Amico-Damião, V.; de Mello Prado, R.; Carvalho, R.F. Amplification of Gibberellins Response in Tomato Modulates Calcium Metabolism and Blossom End Rot Occurrence. *Sci. Hortic.* **2019**, *246*, 498–505. [CrossRef]
87. White, P.J.; Broadley, M.R. Calcium in Plants. *Ann. Bot.* **2003**, *92*, 487–511. [CrossRef]
88. Barker, A.V.; Ready, K.M. Ethylene Evolution by Tomatoes Stressed by Ammonium Nutrition. *J. Am. Soc. Hortic. Sci.* **1994**, *119*, 706–710. [CrossRef]
89. Besada, C.; Gil, R.; Bonet, L.; Quiñones, A.; Intrigliolo, D.; Salvador, A. Chloride Stress Triggers Maturation and Negatively Affects the Postharvest Quality of Persimmon Fruit. Involvement of Calyx Ethylene Production. *Plant Physiol. Biochem.* **2016**, *100*, 105–112. [CrossRef] [PubMed]
90. Bajaj, S.; Targolli, J.; Liu, L.F.; Ho, T.H.D.; Wu, R. Transgenic Approaches to Increase Dehydration-Stress Tolerance in Plants. *Mol. Breed.* **1999**, *5*, 493–503. [CrossRef]
91. Birlanga, V.; Acosta-Motos, J.R.; Pérez-Pérez, J.M. Genotype-Dependent Tipburn Severity during Lettuce Hydroponic Culture Is Associated with Altered Nutrient Leaf Content. *Agronomy* **2021**, *11*, 616. [CrossRef]
92. Jenni, S.; Hayes, R.J. Genetic Variation, Genotype × Environment Interaction, and Selection for Tipburn Resistance in Lettuce in Multi-Environments. *Euphytica* **2010**, *171*, 427–439. [CrossRef]

93. Jenni, S.; Yan, W. Genotype by Environment Interactions of Heat Stress Disorder Resistance in Crisphead Lettuce. *Plant Breed.* **2009**, *128*, 374–380. [CrossRef]
94. Koyama, R.; Sanada, M.; Itoh, H.; Kanechi, M.; Inagaki, N.; Uno, Y. In Vitro Evaluation of Tipburn Resistance in Lettuce (*Lactuca sativa* L.). *Plant Cell Tissue Organ Cult.* **2012**, *108*, 221–227. [CrossRef]
95. Ryder, E.J.; Waycott, W. Crisphead Lettuce Resistant to Tipburn Cultivar Tiber and Eight Breeding Lines. *HortScience* **1998**, *33*, 903–904. [CrossRef]
96. Macias-González, M.; Truco, M.J.; Bertier, L.D.; Jenni, S.; Simko, I.; Hayes, R.J.; Michelmore, R.W. Genetic Architecture of Tipburn Resistance in Lettuce. *Theor. Appl. Genet.* **2019**, *132*, 2209–2222. [CrossRef]
97. Macias-González, M.; Truco, M.J.; Han, R.; Jenni, S.; Michelmore, R.W. High Resolution Genetic Dissection of the Major QTL for Tipburn Resistance in Lettuce, *Lactuca sativa*. *G3 Genes Genomes Genet.* **2021**, *11*, jkab097. [CrossRef]
98. Holmes, S.C.; Wells, D.E.; Pickens, J.M.; Kemble, J.M. Selection of Heat Tolerant Lettuce (*Lactuca sativa* L.) Cultivars Grown in Deep Water Culture and Their Marketability. *Horticulturae* **2019**, *5*, 50. [CrossRef]
99. Carassay, L.R.; Bustos, D.A.; Golberg, A.D.; Taleisnik, E. Tipburn in Salt-Affected Lettuce (*Lactuca sativa* L.) Plants Results from Local Oxidative Stress. *J. Plant Physiol.* **2012**, *169*, 285–293. [CrossRef] [PubMed]
100. Wang, Y.; Martins, L.B.; Sermons, S.; Balint-Kurti, P. Genetic and Physiological Characterization of a Calcium Deficiency Phenotype in Maize. *G3 Genes Genomes Genet.* **2020**, *10*, 1963–1970. [CrossRef] [PubMed]
101. Uozumi, A.; Ikeda, H.; Hiraga, M.; Kanno, H.; Nanzyo, M.; Nishiyama, M.; Kanahama, K.; Kanayama, Y. Tolerance to Salt Stress and Blossom-End Rot in an Introgression Line, IL8-3, of Tomato. *Sci. Hortic.* **2012**, *138*, 1–6. [CrossRef]
102. Watanabe, T.; Tomizaki, R.; Watanabe, R.; Maruyama, H.; Shinano, T.; Urayama, M.; Kanayama, Y. Ionomic Differences between Tomato Introgression Line IL8–3 and Its Parent Cultivar M82 with Different Trends to the Incidence of Blossom-End Rot. *Sci. Hortic.* **2021**, *287*, 110266. [CrossRef]
103. Changmai, T.; Gertphol, S.; Chulak, P. Smart Hydroponic Lettuce Farm Using Internet of Things. In Proceedings of the 2018 10th International Conference on Knowledge and Smart Technology: Cybernetics in the Next Decades, KST 2018, Chiangmai, Thailand, 21 January–3 February 2018; Institute of Electrical and Electronics Engineers Inc.: Piscataway, NJ, USA, 2018; pp. 231–236.
104. Duarte-Galvan, C.; Romero-Troncoso, R.d.J.; Torres-Pacheco, I.; Guevara-Gonzalez, R.G.; Fernandez-Jaramillo, A.A.; Contreras-Medina, L.M.; Carrillo-Serrano, R.V.; Millan-Almaraz, J.R. FPGA-Based Smart Sensor for Drought Stress Detection in Tomato Plants Using Novel Physiological Variables and Discrete Wavelet Transform. *Sensors* **2014**, *14*, 18650–18669. [CrossRef] [PubMed]
105. Bongiovanni, R.; Lowenberg-Deboer, J. Precision Agriculture and Sustainability. *Precis. Agric.* **2004**, *5*, 359–387. [CrossRef]
106. Sambo, P.; Nicoletto, C.; Giro, A.; Pii, Y.; Valentinuzzi, F.; Mimmo, T.; Lugli, P.; Orzes, G.; Mazzetto, F.; Astolfi, S.; et al. Hydroponic Solutions for Soilless Production Systems: Issues and Opportunities in a Smart Agriculture Perspective. *Front. Plant Sci.* **2019**, *10*, 923. [CrossRef]
107. Yin, H.; Cao, Y.; Marelli, B.; Zeng, X.; Mason, A.J.; Cao, C. Soil Sensors and Plant Wearables for Smart and Precision Agriculture. *Adv. Mater.* **2021**, *33*, 2007764. [CrossRef]
108. Srivani, P.; Yamuna Devi, C.; Manjula, S.H. A Controlled Environment Agriculture with Hydroponics: Variants, Parameters, Methodologies and Challenges for Smart Farming. In Proceedings of the 2019 Fifteenth International Conference on Information Processing (ICINPRO), Bengaluru, India, 20–22 December 2019; pp. 1–8.
109. Hasan, M.; Tanawala, B.; Patel, K.J. Deep Learning Precision Farming: Tomato Leaf Disease Detection by Transfer Learning. In Proceedings of the 2nd International Conference on Advanced Computing and Software Engineering, Sultanpur, India, 8–9 February 2019; pp. 171–175.
110. Rubanga, D.P.; Hatanaka, K.; Shimada, S. Development of a Simplified Smart Agriculture System for Small-Scale Greenhouse Farming. *Sens. Mater.* **2019**, *31*, 831–843. [CrossRef]
111. Liu, Y.; Zhou, Y.; Wang, T.; Pan, J.; Zhou, B.; Muhammad, T.; Zhou, C.; Li, Y. Micro-Nano Bubble Water Oxygation: Synergistically Improving Irrigation Water Use Efficiency, Crop Yield and Quality. *J. Clean. Prod.* **2019**, *222*, 835–843. [CrossRef]
112. Wilson, M.A.; Tran, N.H.; Milev, A.S.; Kannangara, G.S.K.; Volk, H.; Lu, G.Q.M. Nanomaterials in Soils. *Geoderma* **2008**, *146*, 291–302. [CrossRef]
113. Corradini, E.; de Moura, M.R.; Mattoso, L.H.C. A Preliminary Study of the Incorparation of NPK Fertilizer into Chitosan Nanoparticles. *Express Polym. Lett.* **2010**, *4*, 509–515. [CrossRef]
114. Manjunatha, S.B.; Biradar, D.P.; Aladakatti, Y.R. Nanotechnology and Its Applications in Agriculture: A Review. *J. Farm Sci.* **2016**, *29*, 1–13.
115. Lai, F.; Wissing, S.A.; Müller, R.H.; Fadda, A.M. *Artemisia arborescens* L Essential Oil-Loaded Solid Lipid Nanoparticles for Potential Agricultural Application: Preparation and Characterization. *AAPS PharmSciTech* **2006**, *7*, E10. [CrossRef]
116. Liu, R.; Lal, R. Potentials of Engineered Nanoparticles as Fertilizers for Increasing Agronomic Productions. *Sci. Total Environ.* **2015**, *514*, 131–139. [CrossRef]
117. Khodakovskaya, M.V.; Kim, B.S.; Kim, J.N.; Alimohammadi, M.; Dervishi, E.; Mustafa, T.; Cernigla, C.E. Carbon Nanotubes as Plant Growth Regulators: Effects on Tomato Growth, Reproductive System, and Soil Microbial Community. *Small* **2013**, *9*, 115–123. [CrossRef]
118. Lucas García, J.A.; Probanza, A.; Ramos, B.; Palomino, M.; Gutiérrez Mañero, F.J. Effect of Inoculation of *Bacillus licheniformis* on Tomato and Pepper. *Agron. EDP Sci.* **2004**, *24*, 169–176.

119. Sheikhalipour, P.; Bolandndnaza, S.A.; Panahandeh, J. Influence of KSB, PSB and NFB on Fruit Quality and Potassium Contents in Tomato. *Int. J. Adv. Biol. Biomed. Res.* **2016**, *4*, 170–178.
120. Lee, S.W.; Ahn, I.P.; Sim, S.Y.; Lee, S.Y.; Seo, M.W.; Kim, S.; Park, S.Y.; Lee, Y.H.; Kang, S. *Pseudomonas* sp. LSW25R, Antagonistic to Plant Pathogens, Promoted Plant Growth, and Reduced Blossom-End Rot of Tomato Fruits in a Hydroponic System. *Eur. J. Plant Pathol.* **2010**, *126*, 1–11. [CrossRef]
121. Lee, S.; Lee, J. Beneficial Bacteria and Fungi in Hydroponic Systems: Types and Characteristics of Hydroponic Food Production Methods. *Sci. Hortic.* **2015**, *195*, 206–215. [CrossRef]

agronomy

MDPI

Review

Oxidative Stress, Ageing and Methods of Seed Invigoration: An Overview and Perspectives

Ademola Emmanuel Adetunji [1,*], Tomi Lois Adetunji [2], Boby Varghese [1], Sershen [3,4] and Norman W. Pammenter [1]

1 School of Life Sciences, University of KwaZulu-Natal, Durban 4001, South Africa; varghese@ukzn.ac.za (B.V.); pammente@ukzn.ac.za (N.W.P.)
2 Unit for Environmental Sciences and Management (UESM), Faculty of Natural and Agricultural Sciences, North-West University, Private Bag X6001, Potchefstroom 2520, South Africa; lois.olatunji@gmail.com
3 Department for Biodiversity and Conservation Biology, University of the Western Cape, Private Bag X17, Bellville 7535, South Africa; sershenn@gmail.com
4 Institute of Natural Resources, P.O. Box 100396, Scottsville 3209, South Africa
* Correspondence: adetunjiademola@hotmail.com

Abstract: The maintenance of seed quality during the long-term conservation of plant genetic resources is crucial for averting the projected food crises that are linked to the changing climate and rising world population. However, ageing-induced loss of seed vigour and viability during storage remains an inevitable process that compromises productivity in several orthodox-seeded crop species. Seed ageing under prolonged storage, which can occur even under optimal conditions, induces several modifications capable of causing loss of intrinsic physiological quality traits, including germination capacity and vigour, and stand establishment. The problems posed by seed ageing have motivated the development of various techniques for mitigating their detrimental effects. These invigoration techniques generally fall within one of two categories: (1) priming or pre-hydrating seeds in a solution for improved post-harvest performance, or (2) post-storage reinvigoration which often involves soaking seeds recovered from storage in a solution. Seed priming methods are generally divided into classical (hydropriming, osmopriming, redox priming, biostimulant priming, etc.) and advanced (nanopriming, magnetopriming and priming using other physical agents) techniques. With the increasing popularity of seed invigoration techniques to achieve the much-desired enhanced productivity and resilience in the face of a changing climate, there is an urgent need to explore these techniques effectively (in addition to other important practices such as plant breeding, fertilizer application, and the control of pests and diseases). This review aims to provide an overview of ageing in orthodox seeds and invigoration techniques that can enhance desirable agronomic and physiological characters.

Keywords: gene bank; germination; orthodox seeds; priming; reactive oxygen species

Citation: Adetunji, A.E.; Adetunji, T.L.; Varghese, B.; Sershen; Pammenter, N.W. Oxidative Stress, Ageing and Methods of Seed Invigoration: An Overview and Perspectives. *Agronomy* **2021**, *11*, 2369. https://doi.org/10.3390/agronomy11122369

Academic Editors: Sara Álvarez and José Ramón Acosta-Motos

Received: 15 October 2021
Accepted: 3 November 2021
Published: 23 November 2021

Publisher's Note: MDPI stays neutral with regard to jurisdictional claims in published maps and institutional affiliations.

1. Introduction

Given that global food demand is rising, it is necessary to ensure the conservation of genetic resources to preserve ecosystem resilience and to protect plant biodiversity for future agricultural food production [1,2]. Over a billion people are estimated to be added to the already large world population by 2050 [3]. If no pragmatic response is implemented, the challenge of food security will worsen with the increasing impact of hunger and poverty, particularly in developing countries.

The worrisome, widespread drop in crop yield due to a combination of factors, including, but not limited to, soil degradation and drastic changes in the climate [4,5], and the negative crop production projections across the globe [6,7], all point to a need for another Green Revolution with much more yield and better conservation of resources than the first [8,9]. For instance, by the mid-twenty-first century, an increase of up to 60% in food production is estimated to be needed to feed the growing population [10,11]. This underscores the need to prioritise various approaches and research interventions towards increased crop production.

Attempts being made to address the identified needs include the development of approaches, such as conservation agriculture, sustainable intensification [12,13], and climate-smart agriculture [10,11], with the aim of raising productivity, decreasing emissions and reducing susceptibility to environmental stresses (i.e., improved resilience). The benefits of the application of cutting-edge techniques, based on investigative research efforts, in various areas of agricultural science, including agroecology, ecophysiology, soil science and plant physiology [14], have been recognised. Addressing the identified needs can be achieved using dynamic approaches involving the application of modern biotechnological and physiological research techniques, among others, geared towards addressing low- crop yield-related challenges, often attributable to the low quality of genetic resources, such as seeds—the principal yield determining factor [15], which forms the subject of this study.

Seed, as a genetic resource, may be regarded as the insurance system for world food schemes. The depletion of this resource exposes the schemes to higher risks, which could ultimately lead to catastrophic failure. Without a systematic approach for the conservation of seed genetic and physiological quality, achieving the much-desired increased productivity and greater resilience in the face of the rising world population and changing climate is a mirage. Moreover, seeds are considered the main basis for the sustenance of humans as plants form over 80% of the human diet; promoting high-quality seed delivery is thus essential for enhancing crop production and plant tolerance to environmental challenges [15]. Achieving food security largely depends on the seed security of seed-producing communities in all cropping seasons [15]. The application of advances in plant physiology, particularly the various techniques of pre-hydration treatment (which uses priming technology to invigorate debilitated germplasms), in addition to other important components, such as plant breeding for adaptation to climate change and higher yields, and cultural practices (e.g., irrigation, fertilizer application, and control of pests and diseases), is needed to improve seed performance, crop yields, maximum yield, and to enable planting on less favourable land, by making seeds better able to withstand sub-optimal conditions, thereby reducing crop losses. Accordingly, agriculture in this century and subsequently can be more productive and provide for improved conservation of plant genetic resources compared to previous periods. Heightened efforts in this regard, therefore, will ensure that the prospect of reaching millions of the poor with crop production research benefits is achieved [8].

2. Storage of Orthodox Seeds in Gene Banks

In terms of conserving plant genetic resources, seed capacity for prolonged storage is particularly essential for gene banks. As far back as 1908, Ewart had grouped seed longevity into short-, medium- and long-term, providing insights into the duration of seed storage before severe viability loss [16]. Later, several experiments testing seed longevity were conducted under artificial and natural sowing conditions. The Beal [17–19] and the Vienna [20,21] germination studies include the oldest (over 100 years) seed longevity studies performed under natural conditions [22]. Other pioneering seed longevity studies [23–25] have shown that moisture content, temperature, relative humidity, and oxygen are the critical factors influencing seed viability and vigour during storage; however, genetic factors and pre-storage conditions are also important [16].

At moisture levels as low as 5% (fresh mass basis) or less, and at sub-zero temperatures (usually −18 °C) in dry conditions, the mature seeds of some species classified as orthodox can be stored for long periods [26–29]. This is also the easiest method of conserving most spermatophyte genetic resources in conventional gene banks [30], but seeds do not retain their initial quality with extended storage, gradually deteriorating, and inevitably proceeding towards death [31,32]. For instance, seeds that were initially stored in a gene bank at 5 °C but were later moved to −18 °C and stored for between 15–19 years suffered a decline in germination capacity from 91% to 11% in *Brassica oleracea* and 97% to 2% in

Lactuca sativa [33]. The post-harvest loss of physiological quality of seeds, even when seeds are stored in gene banks, has thus remained a major issue demanding attention for long-term storage [34]. As seeds deteriorate, vigour is first lost, after which comes a loss of viability [35].

Moreover, some species classified as recalcitrant (not covered in this review) have desiccation-sensitive seeds and are not amenable to short or long-term storage under the conditions mentioned above [30]. With the development of cryostorage techniques, involving germplasm storage at ultra-low temperatures (−120 to −196 degrees, [36]), the life span of seeds (including both orthodox and recalcitrant species) can be further extended, but not indefinitely [33,34,37]. This implies that though the degree and rate of deterioration of seeds stored under the enhanced conditions of the conventional seed gene banks can be reduced to an appreciable level [32], seed deterioration cannot be completely halted. Walters et al. (2004) showed, by measuring changes in the viability of seeds of several plant species over 20 years of cryostorage, that cryogenic temperatures could not sufficiently stop seed deterioration. They further suggested that there could be as much as a 300% variation in longevity among species and within accessions stored in these conditions, as the degree of longevity in cryostorage depends on the inherent properties of seeds and seed handling, such as the pre-storage temperature and the year of harvest. This implies that cryostorage temperatures do not completely halt all biological activities; molecules are still sufficiently mobile at these low temperatures to allow ageing reactions to proceed [37].

3. Germination-Related Physiology

Under favourable conditions of moisture, warmth, and oxygen, quiescent but viable seed is vivified, forming an actively metabolising structure, in a process described as germination [38]. The progress of germination can be roughly assessed by measuring respiration or water uptake [39], while the completion of germination can be taken to be when the system no longer depends on its stored food [38], or is visibly marked by the protrusion of the radicle [39]. In instances where the radicle may grow before penetrating the surrounding tissues, germination can be taken to have been completed from the time a sustained increase in seed fresh weight is recognised [39]. So, the initiation of germinative activities gradually and eventually leads to the formation of normal, growing seedlings [38].

In cases where a viable seed fails to germinate under favourable germination conditions, dormancy is said to have set in as such seeds require additional conditions, such as a specific light, or temperature regime, or exposure to chemical or physical treatments [39]. Dormant seeds that have been hydrated undergo almost all the metabolic processes that take place during the germination of nondormant seeds, yet the protrusion of the radicle does not occur [40]. Their germination takes place later when the additional conditions required for release from dormancy are met. Three identified stages of seed germination include water imbibition (first stage), nutrient conversion (second stage), and cell elongation and cell multiplication (third stage) [41]. The events following germination, such as the mobilisation of food reserves from the endosperm, supply the much-needed energy for seedling growth until the seedlings become photoautotrophic [42]. The seed germination pattern usually follows a sigmoid curve whereby a few seeds germinate earlier than the others in a population, followed by a rapid rise in percentage germination, and then relatively late germination of a few seeds is recorded. Seed-germination curves are generally right-skewed as the occurrence of more germinations is recorded in the first half of the germination period than the second. Whilst the shape of the curves are generally similar, notable differences in germination patterns are observed among populations [39].

4. Oxidative Stress in Plants

Oxidative stress is widely described as a physiological state (response) in cells, tissues and organs, as a result of increased pro-oxidative activities (through the generation of reactive oxygen species (ROS)) compared to antioxidative (enzymic and non-enzymic) activities [43]. This is a consequence of aerobic metabolism during which aerobic organisms produce incompletely paired, oxygen-containing radicals formed by the unavoidable leakage of electrons onto molecular oxygen during electron transport in the mitochondria, chloroplast, and cell membranes [44]. The generation of ROS may be triggered by severe abiotic and biotic stress conditions [45]. The physiological responses are often characterised by a gradual accretion of various oxidised biomolecules, such as nucleic acids, proteins, lipids, polysaccharides, and metabolites, causing deleterious changes in normal biochemical, mechanical, and physical functions of cell components [43].

Oxidative stress functions, therefore, as an injurious factor. The main mechanisms involve altering the balance between the levels of generated and quenched ROS owing to the upset of regular cellular metabolism, and ROS biosynthesis as a component of developmental processes such as the signalling responses needed for adaptation and defence or programmed cell death [43,46]. Demidchik and Maathuis [47] stated that plants could employ ROS accumulation for encoding and recognising various stress factors, including xenobiotic stressors such as nanoparticles and herbicides that were not recognised before. Stress factors often engender secondary metabolic effects to be overcome by plant tissues for survival and restoration of growth and development [48]. For instance, salinity [49], drought [44], desiccation [50], light [51], temperature [52], and pathogens [46] can induce oxidative stress by increasing the production of free radicals and reactive oxygen species (ROS). Uncontrolled production of ROS can upset the balance of ROS generated during aerobic events and the antioxidative defence system [53], leading to oxidative stress [54].

4.1. Biochemical Effects of Ageing and Oxidative Stress in Seeds

Oxidative stress has been implicated in the loss of vigour in plant tissues [55]. Loss of vigour in plant tissues is a fundamental physiological phenomenon observed when plant tissues are exposed to environmental stress of any type (abiotic and biotic) under suboptimal external (both agricultural and natural) conditions [56]. It is a pressing global challenge for modern agriculture. Both abiotic and biotic factors can cause oxidative stress [57], which is widely described as the physiological state (response) brought about by increased pro-oxidative activities (through a gradual generation and accumulation of reactive oxygen species (ROS)) over antioxidative (enzymic and non-enzymic) activities [43,58]. In *B. oleracea*, for instance, seed deterioration has been related to changes in the levels of electrolyte leakage [59], proline, proteins, soluble sugars, and phenolic compounds [60]. However, Golovina et al. [61,62] reported no change in protein secondary structure in some seeds of orthodox species stored for 20 to 30 years, including *Allium cepa*, *Raphanus sativus*, *Cucumis melo*, *Capsicum annuum*, and *Brassica napus*, despite the loss of membrane integrity. In *L. sativa*, seed deterioration has been attributed to changes in the levels of lipid hydroperoxides [63] and volatile products such as aldehydes and alcohols [64]. Of these environmental conditions, abiotic stress is recognised to constitute a major drawback to crop farming [65,66], accounting for an approximately 51–82% loss of potential crop yield worldwide [56,67]. In many cases, abiotic stressors engender secondary metabolic effects that need to be overcome by plant tissues for survival and restoration of growth and development [48]. For instance, salinity [49], drought [44], light [51] and temperature [52] induce oxidative stress by increasing the production of free radicals and ROS, thereby offsetting the balance of ROS generated during metabolic events and the defence system [53]. The impacts of oxidative stress are usually expressed in relation to the overall growth of the plant, including vigour, yield, biomass accumulation or primary assimilation events [56], as well as in terms of the quality of seed [68].

Seeds, due to their rich genetic diversity, are considered the most efficient natural means of protecting the variability of genetic material, in comparison to somatic tissues. The challenge of loss of vigour creates a severe threat [69] which risks the conservation of millions of genetic materials kept in several world gene banks [70], making the understanding of loss of vigour, and consequential seed ageing in storage, vital for plant physiologists. Seed ageing has been intimately linked to oxidative stress involving ROS which are highly reactive, toxic, and capable of causing degradative reactions of several biomolecules over an extended storage period [69,71]. Biological molecules, including carbohydrates, lipids, proteins, and polynucleic acids, such as DNA and RNA, are believed to be the main ROS targets during oxidative stress [43,72]. The physiological lesions that result include loss of membrane integrity (through lipid peroxidation), reduced respiration, enzyme inactivation and degradation, and genetic degradation [31,73,74], leading to severely damaging effects on seed vigour, viability and germinability, especially.

4.2. Oxidation of Major Biological Molecules

4.2.1. Lipids

Among other affected biomolecules, ROS-mediated oxidation of polyunsaturated fatty acids (lipid peroxidation) is the most harmful as it can cause chain reactions involving the formation and spread of ROS [75]. Lipid peroxidation is suggested to be a significant bioindicator of oxidative stress [76]. The damaging effect is irreversible, causing severe degradation of the membrane, inactivation of enzymes, total loss of membrane-bound protein activities, and cell death [43]. Lipid peroxidation has been implicated in the loss of viability during storage of seeds of many crop species [77,78] and has been shown to lead to swelling of mitochondria, increased membrane viscosity and heightened bilayer permeability (measured as increased solute leakage) [31,79]. Products of lipid oxidation can also cause DNA damage and interrupt the normal functioning of several cellular systems [79,80].

With respect to mechanisms, lipid peroxidation can occur via non-enzymic and enzymic processes [75,81]. The non-enzymic, ROS-mediated, process of lipid peroxidation entails an activation (initiation) stage involving ROS generation, a distribution (propagation) stage involving ROS chain reactions, and a termination stage in which non-radical products are formed [75,82]. The peroxidation initiation stage is activated by the removal of a hydrogen atom from a methylene ($-CH_2-$) group, leaving behind $-^{\bullet}CH-$ (lipid radical), by sufficiently reactive species such as alkoxyl ($RO^{\bullet}$), hydroxyl radicals ($HO^{\bullet}$), peroxyl radicals ($ROO^{\bullet}$), hydroperoxyl ($HO_2^{\bullet}$) and peroxynitrite, but not superoxide ($O_2^{\bullet -}$) or hydrogen peroxide (H_2O_2). Saturated and monounsaturated fatty acids, for example, oleic acid, with one double bond and 18 carbon atoms, can be subjected to oxidation reactions but not the chain reaction of lipid peroxidation, as they are less vulnerable [75,83]. However, the polyunsaturated fatty acids of cellular membrane phospholipids contain double bonds, which make them susceptible to peroxidation by the facilitation of hydrogen atom removal [43,82,84]. The lipid radicals formed then trigger O_2-mediated chain reactions involving the formation of lipid peroxyl radicals ($LOO^{\bullet}$), which in turn abstract hydrogen atoms from nearby fatty acid molecules forming a stable intermediate lipid hydroperoxide (LOOH) and another lipid radical in the propagation phase [43,83]. The process is limited by the termination reaction phase producing non-radical products. In addition, non-radical peroxidation of lipids can occur by polyunsaturated fatty acids reacting with singlet oxygen ($^1O_2$) forming LOOH without production of intermediate radicals [85–88]. Though reasonably stable, lipid peroxides may be decomposed by metal complexes in a reaction catalysed by transition metals producing radicals that can reinitiate peroxidation via redox cycling of the metal ions, forming products, such as 4-hydroxy-2-nonenal (4HNE), 4-hydroxyhexenal (4-HHE), and malonaldehyde (MDA), which are useful and extensively studied biomarkers of lipid peroxidation [75,82,83,89]. These aldehydes, in turn, bind with DNA or protein, causing more severe damage [90]. Loss of membrane integrity, breakdown of organelles, oxidation and impairment of DNA, RNA, and proteins result where there is severe lipid peroxidation reaction [76,88]. Of these aldehydes, 4-HNE is considered the key product of the peroxidation of omega-6 fatty acids, such as linoleic acid (C18:2, n-6) and arachidonic acid (C20:4, n-6). The production of 4-HHE, the aldehyde thought to induce the permeability of mitochondrial inner membrane [91] and upset metabolic events [92], has been reported from the peroxidation of omega-3 fatty acids, such as α-linolenic acid (C18:3, n-3) and docosahexaenoic acid (C22:6, n-3) [83].

In enzymic peroxidation, dioxygenases including lipoxygenases (LOX enzymes) are considered the key oxidising enzymes of polyunsaturated fatty acids, with linoleic acids (C18:2 and C18:3) as the major substrates [81,93]. In plants, LOX enzymes can add an oxygen molecule at carbon 9 or 13 of C18-fatty acids [81], forming 9- and 13-hydroperoxyl derivatives of linoleic acid, respectively [94]. The involvement of LOX enzymes in the ageing-induced lipid peroxidation of seeds has been investigated in several species where it was demonstrated that absence or lowering of LOX enzyme activity decreased the levels of MDA (*Zea mays* [95]), MDA and LOOH (*Oryza sativa* seeds [96]) promoted storability and germination (*Oryza sativa* seeds [97]), and improved vigour and viability (*Nicotiana tabacum* [98]).

4.2.2. Proteins

Since reactive oxidants can be indiscriminately generated in cells, especially at a heightened rate during abiotic or biotic stress, proteins are also a major target biomolecule [99,100], as they are abundant and readily reactive with several oxidants [101]. Proteins constitute about 68% of oxidised biomolecules [99,102]; thus, protein oxidation is a useful bioindicator of oxidative stress [100]. Defined as a covalent alteration of proteins by reactive oxidants or oxidative stress spinoffs [100], the ROS-mediated oxidation of proteins has been described extensively [103–106]. Protein oxidation often occurs even under normal physiological circumstances indicating that it is not always an injurious plant process [75,107]. Avery (2011) suggested that some proteins are more vulnerable to oxidation than others due to factors such as more easily oxidised amino acid residue content, metal-binding sites, the localisation of protein within cells, molecular conformation, and degradation rate. It is becoming increasingly clear that newly synthesised proteins are highly susceptible to post-synthesis oxidative degradation, suggesting that attaining and conforming to a stable multimeric protein complex may be protective against oxidative injury [108,109]. The oxidation of protein can facilitate the build-up of toxic non-native proteins capable of inducing programmed cell death in severe cases [43,75]. The production of unstable intermediates and the formation of stable products are useful for the estimation of protein damage [101]. ROS-induced protein injury can vary since protein properties are not all equivalent. The extremely reactive ROS, $HO^{\bullet}$, usually generated from H_2O_2 via the Fenton reaction, often leads to non-specific oxidation, unlike the specific type caused by other ROS [75]. Other ROS causing oxidation of proteins include the radicals of alkoxyl ($RO^{\bullet}$), hydroperoxyl ($HO_2{}^{\bullet}$), peroxyl ($RO_2{}^{\bullet}$), superoxide ($O_2{}^{\bullet-}$) and non-radical species, such as hypochlorous acid (HOCl), hydrogen peroxide (H_2O_2), ozone (O_3), peroxynitrite ($ONOO^-$), and singlet oxygen ($^1O_2$) [75,110]. While the oxidation of certain amino acids (e.g., those containing sulphur) is reversible, most ROS-mediated modifications are characterised by irreversible loss or inactivation of the parent amino acid residue, and catalytic, metabolic, regulatory, structural, or other activities and functions leading to protein damage or elimination [43,101,111]. Irreversible amino acid modifications, such as to arginine and lysine, tryptophan and tyrosine, the production of dityrosine, and protein-to-protein cross-linking, are in most cases accountable for the permanent shutdown of function of the affected proteins which are later degraded [111]. The reversible types of amino acid modification, such as S-nitrosylation and glutathionylation, may play a redox regulatory role protecting cysteine from irreversible oxidation, as well as modulating protein function [100,111]. The main oxidative modifications of proteins are outlined in Table 1.

Table 1. Commonly reported ROS-induced modifications of polyunsaturated fatty acids (PUFA), proteins, carbohydrates, and DNA.

Examples of commonly reported ROS-induced modifications of PUFA [100]	
PUFA	Oxidised product
Linoleic acid (18:2)	4-HNE
Linolenic acid (18:3)	Cyclic oxylipin, hydroxyoctadecatrieonic acid, MDA

Table 1. *Cont.*

Examples of commonly reported ROS-induced modifications of proteins [43,100,112,113]	
Amino acid	Oxidised product
Cysteine	Cysteic acid (cysteine sulfonic acid)
Methionine	Methionine sulfone
Arginine, Lysine, Proline, Threonine	Carbonyls (ketones, aldehydes): aminoadipic semialdehyde, pyrrolidone, acrolein, 4-HNE, MDA, glu γ-semialdehyde, 2-amino-3-ketobutyric acid
Glutamyl (glutathione, glutamine, glutamate)	Pyruvic acid, oxalic acid
Histidine	2-Oxohistidine, 4-HNE, aspartate, asparagine
Phenylalanine	Hydroxyphenylalanines
Tryptophan	Kynurenine
Tyrosine	3-Nitrotyrosine
Examples of commonly reported ROS-induced modifications of carbohydrates [100,114]	
Sugar	Oxidised product
Aldohexose, polyol	Aldopentose, formic acid
Examples of commonly reported ROS-induced modifications of DNA [100]	
DNA	Oxidised product
Purines (e.g., guanine)	8-Hydroxyguanine, FapyGua

The most common mechanisms of ROS-mediated protein damage involve the direct metal-catalysed oxidation (primary carbonylation) of S-containing amino acid residues such as:

(1) cysteine (Cys) to produce cystine (disulfide), which is further oxidised through cysteine sulfenic acid to form cysteine sulfinic acid; these initial stages are reversible until the highest oxidation and damaging level where cysteic acid is irreversibly formed [43,100,111];

(2) methionine (Met) to produce methionine sulfoxide. This stage is also reversible, but the final stage of Met oxidation to sulfone seems to be damaging and irreversible [100]; and

(3) most of the other amino acids, especially arginine (Arg), lysine (Lys), proline (Pro), and threonine (Thr) form stable aldehydes or ketones (carbonyls) in an irreversible reaction [75,100,112] that is not particular to any oxidants [113,115]. Thus, the extent of reactive oxidant-induced modification of proteins is generically measured as protein carbonyl [113,116,117].

Carbonyl formation (protein carbonylation, (PC)) demands higher energy inputs than the oxidation of other AA residues and leads to deleterious alterations of protein structure and function [43]. Secondary carbonylation reactions may occur by the reaction of proteins with aggressive lipid peroxidation products, such as 4HNE and MDA [43,75]. In addition, carbonyl formation can result from protein glycation or glycoxidation [118,119], which may be a confounding factor in using carbonylation as an exclusive oxidation biomarker [113], or by direct protein backbone oxidation forming protein fragments with an N-terminal α-ketoacyl amino acid residue [111]. All these processes severely alter or inhibit the physiological and enzymatic activities of protein [43]. Heightened PC has been reported for several plant oxidative stresses [120] induced by salinity [121,122], dehydration [116,123], heavy metals [124,125], pathogen attack [126,127], and ROS-induced seed ageing [128,129].

4.2.3. Carbohydrates

Studies on the oxidative modification of carbohydrates have received less attention even though carbohydrates are considered more abundant than the other plant biomolecules [43]. As with other biomolecules, the oxidative modification of carbohydrates may be injurious to living systems [43]. Free polyols, such as mannitol, pinitol and sorbitol [130], and sugars, are oxidised by $HO^{\bullet}$, mainly forming formic acid [100]. Miller [131] stated that arabinogalactan, cellulose, pectin, and similar polysaccharides in the cell wall could be broken down by $HO^\bullet$. The auxin-mediated extension of cells induces the generation of ROS, which is used by cell wall-bound peroxidases to produce $HO^\bullet$ near scission sites [100,132]. Moreover, cell wall Cu^{2+} reduced to Cu^+ by $O_2^{\bullet-}$ and ascorbate can produce $HO^\bullet$ by reacting with apoplastic H_2O_2 [133]. The $HO^\bullet$ formed causes non-enzymatic separation of pectins and xyloglucans, leading to loosening of the cell wall [43,133]. Similar Fenton reactions of H_2O_2 with Cu or Fe might substantially increase under stress conditions, leading to deleterious effects [43,134]. On the other hand, simple sugars, disaccharides [43,135] and some osmoprotectants (e.g., mannitol, sorbitol, proline, and myo-inositol) may be capable of scavenging ROS, such as $HO^\bullet$ [130]. Increased levels of carbohydrate, such as mannitol, sucrose, and glucose have been correlated with oxidative stress resistance in several species of plant [136,137]; however, there is a dearth of information on the direct connection between the physiology of plants and the ROS-induced oxidation of carbohydrates [43].

4.2.4. Polynucleotides

The oxidative modification of DNA is often implicated in the ageing of seeds [69,138] and, in some cases, perennial plants [43]. Essentially, ROS attack on DNA causes chemical modification of bases, fragmentation of deoxysugar, and breaking of strands [139]. Again, $HO^\bullet$, being the most reactive, are particularly harmful to polynucleic acids (DNA and RNA) [43]. $HO^\bullet$ attaches to double bonds of nucleotide bases and abstracts H^+ from 2′-deoxyribose (resulting in sugar damage) [139] and $-CH_3-$ of thymine [43]. $HO^\bullet$ can also oxidise purines, forming products such as 7-hydro-8-oxoguanine (8-oxoG). The formation of 2,6-diamino-4-hydroxy-5-formamidopyrimidine (FapyGua) has also been reported as a product of polynucleic acid oxidation [43,139]. Guanine is often attacked by $^1O_2$, but not $O_2^{\bullet-}$ and H_2O_2, to form 8-Hydroxyguanine [100]. ROS-modification of DNA can be both direct and indirect. Often, MDA (a breakdown product of PUFA) conjugation with guanine leads to the formation of an additional ring [100,140]. DNA impairment has both cytotoxic and genotoxic effects [141]. Besides mutations, DNA oxidation can cause alterations of cytosine methylation required for the regulation of gene expression [100]. Repair mechanisms of the oxidative damage of plant DNA include directly reversing the impairment caused as well as replacing the base or even the entire nucleotide [100,142,143]. A defence system, both in cytosol and organelles, may also be implemented as a form of protection [43]. Under oxidative stress, however, nuclear ROS-scavengers (glutathione and peroxiredoxin) inadequately protect the DNA [100,142]. Enzymes such as catalase and ascorbate peroxidase in the cytosol are required to protect nuclear DNA in such conditions [43,144].

4.3. Cellular Generation of Reactive Oxygen Species (ROS)

Reactive oxygen species are produced at several locations in the cells, such as chloroplasts, mitochondria, the plasma membrane, peroxisomes, apoplast, the endoplasmic reticulum, and the cell wall. Conventionally, it is believed that ROS are unavoidably produced during metabolic processes of aerobic systems [145]. Several possible sources of ROS have been identified in plants, including reactions involving normal plant metabolisms such as photosynthesis [46] and mitochondrial respiration [72]. There are other ROS sources as well, which are produced from abiotic stress-induced pathways. For example, during photorespiration, the oxidation of glycolate by glycolate oxidase in peroxisomes accounts for the majority of ROS, such as hydrogen peroxide [46]. Recently, more plant ROS sources have been recognised, such as plasma membrane-bound peroxidases, amine oxidases, and nicotinamide adenine dinucleotide phosphate (NADPH) oxidases, involved in events such as apoptosis and defence against pathogens [146]. While a low level of cellular ROS is formed under normal growth conditions, ROS formation is heightened under stress conditions [46].

Various enzymes (e.g., oxygenases) and non-enzymic processes "fix" oxygen atoms into various biological molecules [147]. Partly reduced forms of molecular oxygen (O_2), resulting from O_2 excitation to produce singlet oxygen ($^1O_2$), or from the transfer of one electron to O_2 forming the superoxide radical ($O_2^{\bullet -}$), two electrons to O_2 forming hydrogen peroxide (H_2O_2), or three electrons to O_2 forming the hydroxyl radical ($HO^{\bullet}$) [46], are more readily reactive than atmospheric oxygen. Hence, they are termed reactive oxygen species [58]. These ROS can cause unrestrained oxidation of various biomolecules leading to oxidative cellular damage [46]. Metabolically active organelles, such as the mitochondria, peroxisomes and chloroplasts, processing extremely oxidising reactions, or that have high electron flow rates, are the primary ROS sources within cells [148]. Ubiquinone-cytochrome complexes I and III of the electron transport chain (ETC) are the main sites of $O_2^{\bullet -}$ production in mitochondria, while photosystems I and II are the main sites of $^1O_2$ and $O_2^{\bullet -}$ production in chloroplasts [148].

4.3.1. The Dual Capacity of ROS

Though ROS are harmful when in excess, they are beneficial in cellular processes, such as signalling, cellular differentiation and proliferation [72], ion transport, and gene expression [149], when produced in moderation. While plants can employ the ROS steady-state concentration to monitor stress levels within cells, this must be tightly controlled to avoid over-accretion of ROS that can cause cell death [146]. The ROS-induced death of cells can occur as a result of oxidative modifications of biomolecules, such as enzymes, DNA, RNA, proteins, and membrane lipids (the classical concept). On the other hand, heightened ROS levels can trigger programmed cell death, which has been shown by anti-apoptotic genes suppression of paraquat-induced oxidative stress cell death in *Nicotiana tabacum* [46]. Further, some cell death, earlier believed to be directly caused by oxidative stress, is now regarded as programmed cell death, consistent with the view that ROS can have beneficial effects on plants, promoting physiological function, cellular proliferation, and viability [72]. In essence, plants require a regulatory system to ensure low ROS concentration, and another to allow for the quenching of surplus ROS production [46]. Balancing the different steady-state ROS level and generated ROS types, as driven by the interaction of different ROS-generating and ROS-quenching systems, is also important. The balance may be altered significantly depending on the physiological state of the plant and the combination of various biochemical, developmental, and environmental stimuli [46]. Apart from aggravating cellular impairment, ROS can stimulate the expression of defence genes. ROS, such as $O_2^{\bullet -}$ or H_2O_2, can separately or jointly induce various genes, thereby allowing for more ROS signalling flexibility. Furthermore, reports on plant responses to abiotic stress show that ROS may be involved in regular signalling for adaptation to stress [146].

4.3.2. ROS Scavenging in Plant Cells

The main plant defence system against ROS involves the activities of antioxidants –compounds that can protect cells from oxidative injury even when present in low quantity [149]. These antioxidants can be either enzymatic or non-enzymatic. Major enzymic antioxidants include

(1) Superoxide dismutases (SODs): These are ubiquitous metalloenzymes involved in essential defence against superoxide [149] via a redox cycle where the active site metal is deoxidised by one $O_2^{\bullet -}$ radical and re-oxidised by another [150]. The three (3) forms of identified SOD, defined by the active site metals, are iron-SOD, copper and zinc-SOD, and manganese-SOD [151]. SOD catalyses the dismutation of $O_2^{\bullet -}$ to O_2 and H_2O_2 [147], which can then be broken down by other essential enzymes—the catalases.

(2) Catalases (CATs): These are peroxisome-localised, heme-group-containing enzymes [152], though their presence has also been reported in mitochondria [153]. They are involved in the breakdown of H_2O_2 to H_2O and O_2 [147]. They are recognised as essential defence enzymes against ROS-induced oxidative stress [154,155]. In addition, they are involved in plant defence and metabolism as well as the perception of cellular signals [156].

(3) Glutathione reductases (GRs): These flavoproteins occur mostly in the chloroplasts but have also been reported in the cytosol, mitochondria, and peroxisomes [157]. They are extremely specific and are involved in the reduction of oxidised glutathione (GSSG) back to the reduced form (GSH) using NADPH as the reductant [147], thereby sustaining a high GSH to GSSG ratio [158]. They sustain the reduced state of GSH through the ascorbate–glutathione cycle and are involved in maintaining the –SH group and act as a substrate for glutathione-S-transferases. In conjunction with superoxide dismutase and ascorbate–glutathione pathway enzymes, GRs constitute an important ROS scavenger [158]. They have been demonstrated to enhance oxidative stress tolerance in transgenic *Nicotiana tabacum* [157].

(4) Ascorbate peroxidases (APXs): These heme-containing enzymes are also involved in the decomposition of H_2O_2 using ascorbate as a reductant [159]. Different isoforms have been reported in the cytosol, chloroplast, mitochondria, thylakoid, stroma, and peroxisome [152,159,160]. Increased APX activity has been reported under abiotic stress such as light [161], drought and heat [162], and heavy metal contamination [163].

(5) Glutathione peroxidases (GPXs): These are non-heme-containing antioxidant enzymes [159] using glutathione as a reductant [164]. They are ubiquitous and predicted to be localised in cytosol, chloroplast, endoplasmic reticulum, mitochondria and plastids [165]. They have been demonstrated to play a role in lipid hydroperoxide detoxification, plant defence, and response to biotic [166] and abiotic stresses [164].

A balance between the activities of antioxidant enzymes such as APX, CAT, and SOD is necessary to determine the steady-state ROS (e.g., $O_2^{\bullet-}$ and H_2O_2) level [46]. In addition to metal ion sequestration, this balanced activity is considered crucial to forestalling the production of the extremely toxic $HO^\bullet$ through the metal-dependent Fenton or Haber–Weiss reactions [46]. APX and CAT are thought to be of different groups of H_2O_2 scavengers due to their different affinity for H_2O_2 (μM and mM range, respectively). Ascorbate peroxidase can reduce H_2O_2 to very low concentrations and is conceivably involved in ROS modulation for signalling, while the main role of CAT is to scavenge excess ROS under stress [46]. Since CAT is not reductant dependent to play its role, it might not be sensitive to cell redox status, contrary to the other systems [46]. Interestingly, some intricate interactions between the mechanisms generating ROS and those scavenging ROS have been reported in transgenics having repressed ROS-quenching systems. Plants having repressed APX formation have their CAT, GR, and SOD induced to compensate for the absence of APX, while plants having inhibited CAT compensate for it by inducing other antioxidant enzymes, such as GPX and APX, suggesting some level of redundancy [167].

The non-enzymic antioxidants also play vital roles in the antioxidant defence system, which forms a strong basis for their use as indicators of stress [148]. Major non-enzymic antioxidants include:

(1) Ascorbic acid (AA): AA is known to be abundant and one of the most potent antioxidants involved in ROS (e.g., $O_2^{\bullet-}$ [168]) detoxification and prevention [148,149]. This water-soluble antioxidant is found in all cellular compartments and at higher concentrations in photosynthetic cells [46]. AA is mostly present in its reduced form [169]. It is crucial for the maintenance of membrane structure and capable of completely preventing lipid peroxidation initiation, scavenging ROS, such as singlet oxygen, hydroperoxyl radicals, superoxide, and peroxynitrite, and protects other substrates from oxidative impairment [148,149]. In addition, AA has been documented to be involved in ROS scavenging by controlling redox balance in cells [170]. AA has been reported to enhance abiotic stress tolerance [169,171]. AA is involved in the modulation of the synthesis of tocopherol [172] and the regulation of plant defence responses over and above developmental processes [173].

(2) Glutathione (GSH): In addition to AA, GSH is another non-enzymic antioxidant involved in the detoxification of ROS [168]. Both GSH and AA are involved in the ascorbate–glutathione cycle, where ascorbate peroxidase plays a role in the direct removal of H_2O_2 [174], singlet oxygen [148] and hydroxyl radical [175]. AA is most

abundant in its reduced and active form and found in various cellular compartments, including the cytosol, mitochondria, endoplasmic reticulum, vacuole, peroxisomes, apoplast, and chloroplasts [176]. GSH provides a substrate for several reactions forming oxidised glutathione (GSSG). Balanced GSH to GSSG levels are key to maintaining a redox state in cells [177]. A decline in GSH levels during stress often leads to an imbalanced redox state, thereby causing system deterioration [178]. Heightened biosynthesis of GSH in chloroplasts, instead of protecting cells, may cause oxidative impairment, perhaps by adjusting the general redox state of chloroplasts [179]. It has been reported that the ratio of reduced to oxidised antioxidants can signal the modulation of ROS-scavenging mechanisms [46,51]. GSH plays a major role in protection from oxidative attack on biological membranes [149] and participates in various physiological events, including sulphate transport regulation, xenobiotics detoxification, and signal transduction [148]. Heightened GSH level has been linked with plants' ability to withstand oxidative stress [180].

(3) Tocopherol (vitamin E): This lipophilic phenolic compound exists in eight similarly potent forms as alpha(α)-, beta(β)-, gamma(γ)- and delta(δ)- tocotrienols and tocopherols [181]. It forms part of the biological membrane, playing both non-antioxidant and radical-chain-breaker functions [181]. It is regarded as a potential ROS and lipid radical scavenger [182]. Reduction in tocopherol levels following seed ageing suggests that it is involved in protection against oxidative stress-induced impairments [183], thus making it a useful indicator of seed deterioration [149]. Its synthetic analogue, trolox, has also been reported to be similarly capable of preventing oxidative impairment [184]. Trolox has some advantages in being moderately soluble in water [185]. Unlike α-tocopherol, trolox may be integrated directly into both lipid and water parts of cells [184], thus making it suitable for conducting studies involving both living systems and model systems [185]. The antioxidant power of trolox has been reported to be much more than that of α-tocopherol [184]. Other synthesised analogues include Vitamin E acetate, α-tocopherylphosphate, and α-tocopherylsuccinate [181].

(4) β-carotene: Besides tocopherols, carotenoids play an important role in the photoprotection of phototrophs by eliminating surplus energy as heat, directly scavenging reactive oxidants [148], including $^1O_2$ free radicals, and protecting cells from oxidative impairment by suppressing lipid peroxidation [149]. Their antioxidant property is attributed to their extended conjugated double bond system [186]. Low β-carotene levels have been shown to protect membrane lipids from peroxidative reactions [149].

(5) Gallic acid (GA): In plants, GA is a relatively ubiquitous [187] endogenous polyphenolic compound with several biological activities, including reacting with active oxidants, preventing their formation and accumulation [188]. GA occurs in the free or conjugate (as esterified hydrolysable tannins [56]) form in several plants [189]. Though polyphenols, such as quercetin [190], as well as GA [191], may act as prooxidants depending on concentration and condition [192], GA is primarily used as an antioxidant [193] due to its capacity to scavenge ROS, such as H_2O_2 [194].

5. Seed Invigoration Treatments

5.1. A Brief History of Seed Pre-Hydration Treatment

Seeds are continually faced with multiple challenges relating to production, post-harvest storage, and subsequent quality. Moreover, in view of the effects of global warming as a symptom of climate change, different stress factors may cause poor seed performance in terms of reduced germination, uneven seedling emergence, poor seedling establishment, or destructive alteration of root cell architecture, leading to a substantial yield loss [195]. Hence, dating back to the ancient Greeks, concerted efforts towards the improvement of seed performance have led to the development of different pre-sowing treatment techniques that can augment germination and synchronise seedling emergence under different suboptimal growth conditions [196]. Seed pre-hydration was discovered by Theophrastus, Democritus (5th century B.C.), and Mago (4th-3rd century B.C) [197]. It was suggested that seed pre-hydration treatments in water or milk enhanced the germination of cucumber seeds (Theophrastus, D.H.P. Book VII, 1: 6). Democritus suggested steeping all seeds in some "roof tiles" plant extract before sowing (Plinius, N.H. Book XVIII, XLV: 159). Some other mentions include pre-hydrating almond seeds in a solution of honey or manure according to Carthaginian Mago (N.H. Book XVII, XI: 63), pre-hydrating pulses in "nitre" (Theophrastus, D.H.P. Book II, IV: 2), seed ripening of mistletoe in bird droppings (Plinius, N.H. Book XVI, XCII: 247), and pre-soaking *B. oleracea* seeds in houseleek extract to provide resistance to various insects (Plinius, N.H. Book XIX, LVIII: 180). The need to dry seeds artificially "to make them fertile" was also mentioned by Plinius (Plinius, N.H. Book XIX, XXXVI: 120) [197].

In the 16th century, Olivier de Serres described the steeping of grains (*Hordeum*, *Secale* and *Triticum* spp.) in manure solution for 24 h followed by drying back as a pre-sowing technique for enhanced seedling performance [196,198]. In the 19th century (1855), Charles Darwin [199] experimented with a seawater pre-hydration treatment and reported enhanced germination in treated cress and *L. sativa* seeds. May et al. [200] demonstrated that drying seeds for some time after hydration bestowed beneficial effects leading to increased germination rates under normal and adverse conditions. In 1963, Ells James presented the modern seed priming concept, pointing out the vital parameters of seed pre-hydration treatment and reporting that an increased rate of seedling emergence was observed in tomato seed exposed to the nutrient solution [196,198]. Heydecker [201] recognised the term seed "priming" as used by Malnassy [202], describing it as a seed pre-sowing treatment that can improve performance under suboptimal conditions [203]. Furthermore, Heydecker [201] described seed priming as a pre-hydration treatment in an osmotic solution that permits imbibition in the first germination phase before conversion of nutrient and radical protrusion. Such seeds are sometimes dried back ('hardening' [204]) to their initial moisture level and sown or stored [205]. In addition, the use of specific terms, such as halopriming (imbibing in salt solutions), and osmotic priming (imbibing in other osmotic solutions), were proposed [201] to specify the priming agent. The technique thus far is recognized and widely used to improve seed performance in the field of agriculture [206]. During pre-hydration treatment, the absorption of water is controlled to allow for the activation of pre-germinative metabolism without permitting radicle emergence by limiting the seed moisture content [203,207]. The resultant seedlings assume a physiological (primed) state which enables faster growth and/or better activation of plant defence responses [66,208].

5.2. Seed Pre-Hydration and Pre-Germinative Metabolism

In the 'primed state', the hydration-induced specific metabolic changes are responsible for the ensuing beneficial effects of seed pre-hydration treatments [196]. Upon seed imbibition, major cell functions and processes are activated, such as de novo proteins and nucleic acid biosynthesis, ATP formation, phospholipid and sterol accumulation, DNA repair and antioxidant system activation—the 'pre-germinative metabolism' [196]. Severe oxidative impairment of biomolecules such as lipids, nucleic acids, and proteins may occur in the early germination stages, during maturation on the mother plant, as well as in post-harvest storage, and under various stress conditions [54,78]. For seed vigour to be preserved and germination to be successful, embryonic DNA repair mechanisms must be well preserved. A good repair of impaired DNA allows for the resumption of cell cycle progression and DNA replication, while a defective repair system causes oxidative cell death [54,209]. DNA impairments in seed embryo are repaired during early imbibition and are essential for performance in terms of germination and storability [210]. Thus, DNA repair is a vital part of 'pre-germinative metabolism' triggered during imbibition and accompanied by unrestrained ROS activities [196] capable of causing mutation in the meristematic tissues of the embryo [211]. All major DNA repair pathways, such as base- and nucleotide-excision repair, are triggered at the early imbibition phase for the maintenance of genome integrity [209]. Efficient ligase-dependent re-joining of strand breaks is key to most DNA repairs, and DNA ligase VI, found only in plants, has been described as a major determining factor of seed quality and storability in *Arabidopsis thaliana* [210].

With regards to the regulatory roles of reactive oxidants in the germination of seeds, Møller et al. [100] suggested that comparatively long-lived oxidants, such as H_2O_2 take the signal to a distant target, whereas short-lived oxidants, such as $HO^{\bullet}$, likely act near their production site; the product of oxidation (acting as a secondary messenger) then takes the signal to the target transcription factors. Besides signalling mediated by ROS, severe lesions to biomolecules can result from ROS activities. Though DNA impairments can be 'addressed' by certain repair functions, RNA is extremely sensitive to ROS-induced oxidative impairment owing to the lack of a specified mechanism of repair [212], while protein damage can be reversible (as in the oxidation of cysteine and methionine) and/or irreversible (as in carbonylation) [75,100,111].

Nevertheless, the enhanced activity of antioxidant (defence) enzymes such as APX, CAT, SOD, and GR allows for the control of ROS levels during imbibition [213,214]. The ROS scavenging antioxidant potential of the seed is critical for the enhancement of germination and stress tolerance [78]. In addition, gene expression profiling encoding antioxidant enzymes is a useful index of seed antioxidant response during germination. These safeguarding functions are triggered during pre-hydration treatments, thereby allowing seeds to undergo major metabolic and physiological pre-germinative phase changes up to the first cellular division, leading to improved germination and increased seedling vigour upon sowing [196].

5.3. The Seed Priming Technology Overview

The priming concept usually refers to several approaches towards seed invigoration, all involving controlled hydration of seeds [215]. The seed priming technique is used to improve the overall post-harvest performance of seed [216,217], including longevity (storability) [218,219], and ability to withstand unfriendly environmental conditions [66,220]. Priming enhances seed germination in three phases (Figure 1) [40]: imbibition, germination, and growth [198]. During the first phase (imbibition), characterised by rapid water uptake owing to low seed water potential, respiratory activity and protein synthesis, through existing DNA and mRNA, are promoted. Phase II (germination) is a lag phase involving the initiation of various physiological functions relating to germination, including protein and mitochondria synthesis, degradation of stored food and reorganisation of cellular membranes, to support radicle protrusion and growth of the seedling, which commences in Phase III (growth phase) [198,221]. The key determinant of seed priming is the controlled uptake of water up to Phase II, prior to radicle emergence [198,221], which allows for vital physiological events, such as damaged DNA and mitochondria repair [40]. Priming duration can vary from less than 24 h [222] to days [223] or weeks [224], depending on cultivars, species, and seed lot [225]. Phase II is more sensitive to environmental factors than Phase III. Hence, primed seeds that have undergone Phase II may be able to germinate better than unprimed seeds under suboptimal conditions [198].

In many cases, primed seeds are dried back to a particular moisture level and stored [226] or sown by the conventional method [227,228]. Seed drying back is thought to confer a 'hardening' effect [229]. In the hardening technique, multiple (two to three soakings with drying back) cycles, are suggested to yield a better result, although one cycle is enough for most species [226,230,231]. Seed hardening induced by pre-sowing treatments is attributed to some cytoplasmic, physico-chemical changes, such as decreased lipophilic and increased hydrophilic colloids, greater protoplasmic elasticity and viscosity, increased hydration of colloids, increased bond water level, and increased protein coagulation temperature [231]. However, there have been reports of delayed germination and/or emergence in primed seeds that are dried back, relative to primed but not dried back seeds, owing to the extra time needed for rehydration, though other beneficial effects of priming are conserved [203,232]. Additionally, deterioration of seeds in storage has been reported when primed seeds were dried back in different species, such as *Lycopersicon esculentum* [233], *Cichorium endivia* [234], and *L. sativa* [235]. Tarquis and Bradford [236] stated that though pre-hydration treatments caused an increased germination rate, drying back predisposed *L. sativa* seeds to loss of storability. This effect varies depending on initial seed quality [203]. Thus, it has been suggested that the storage of primed seeds cannot extend beyond a few weeks as mechanisms for repair of impaired DNA become reduced [237]. In a study on *Mimosa bimucronata*, Brancalion et al. [238] added that priming benefits were partly lost in dried back seeds, recording lower performance in terms of percentage germination, seedling vigour, uniformity and germination speed index, and higher electrical conductivity relative to primed but not dried back seeds.

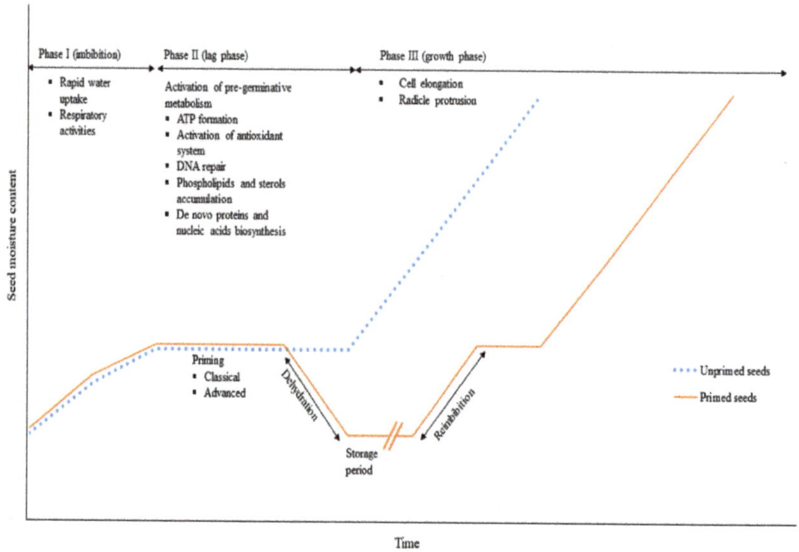

Figure 1. Imbibition curves showing the three phases of germination in unprimed and primed seeds.

5.4. *Seed Priming Methods*

Seed priming methods are generally divided into classical (hydropriming, osmopriming, redox priming, hormonal priming, cellular chemical priming, nutrient priming, and plant biostimulant biopriming [66,198,207]), and advanced (nanopriming [239], magnetopriming, irradiation with microwaves or ionising radiations and some other physical priming agents [240]) techniques, some of which are described below.

5.4.1. Classical Seed Priming Techniques

Hydropriming

Hydropriming is an age-old seed invigoration method popular with farmers as it is simple and economical. Hydropriming is of two types: drum-priming and on-farm priming [207]. Drum-priming involves seed hydration by water vapour generated from a gentle rotation of a drum at a particular temperature [204]. In on-farm priming, seeds are pre-soaked in water for a period before sowing [207,241]. The hydropriming technique is particularly useful under stressful conditions, such as high heat and salinity and water deficit stress, as seed hydration and water uptake efficiency in these conditions are enhanced [198]. However, maintaining optimum humidity and temperature is critical to preventing radicle protrusion, as hydropriming can allow for uncontrolled water uptake [242]. In contrast to unprimed (direct) sowing, the benefits of hydropriming have been demonstrated in several studies, including a 3–4 times increase in biomass allocation and seedling length of *Cicer arietinum* under drought stress conditions [243], rapid emergence and increased seedling vigour in rice seeds subjected to water-stress [227], and increased germination of three-years-stored seeds of napa cabbage (*Brassica rapa*) which correlated with decreased electrical leakage, as well as enhanced antioxidant enzymes (superoxide dismutase and peroxidase) activities, and soluble sugar level [244].

Osmopriming

In this pre-sowing treatment method, seeds are subjected to controlled hydration in an osmotic solution of low water potential generated from the addition of osmotica such as polyethylene glycol, sorbitol, glycerol, and mannitol to priming water [66,207]. The low water potential of the osmotic solution is a crucial factor enabling seeds to be partially hydrated for pre-germinative metabolism but inhibited protrusion of the radicle [245]. In addition, the use of various salt solutions (halopriming) has been widely reported, and their beneficial effects elucidated [246]. For instance, Singh et al. [247] osmoprimed *Vigna unguiculata* seeds with KNO_3 solution and reported improved germination, plant height, and biomass accumulation compared with unprimed and hydroprimed seeds. Fatokun et al. [248] reported enhanced seedling emergence, photosynthetic and growth parameters of *Pisum sativum* and *Cucurbita pepo* seeds aged to 50% viability after priming with a mixture of $CaCl_2$ and $MgCl_2$ solutions relative to the unprimed seeds. Priming of *B. oleracea* seeds using varying levels (1%, 2% and 3%) of inorganic salts, such as KCl, KH_2PO_4, KNO_3, $MgCl_2$, $MgSO_4$, and NaCl, significantly increased germination, seedling vigour, biomass accumulation and reduced mean germination time [249]. Priming of artificially deteriorated *Brassica napus* seeds with $CaCl_2$ promoted seedling vigour [250]. Carrozzi et al. [251] reported that priming with $MgSO_4$ increased germination of *L. sativa* seeds stored for a year. Osmopriming is a low-cost priming option and allows for better water conservation [252].

Redox Priming

This seed invigoration method refers to priming with antioxidative compounds [66]. Plant cell redox state is key to the regulation of growth, development, and stress tolerance [253,254]. Plant redox status is disturbed in response to external stimuli, and the severity of disturbance is determined by the kind of stimulus, the amount, and the duration of, tissue exposure [66]. Maintaining an appropriate redox environment [255] is thought to help in minimising the severity of stress-induced damage [66]. During oxidative stress, antioxidants are well-known redox buffers capable of reacting with ROS and functioning as a metabolic interface that moderate the proper induction of acclimation responses or programmed cell death [256]. Among the compounds of major importance in the antioxidant pathway of plants, glutathione plays a significant role in the cellular redox signalling networks influencing growth, development, and defence [66,178]. Glutathione and tocopherol used as seed pre-hydration treatments resulted in increased seedling length in *Helianthus annuus* [257], gallic acid improved seedling vigour in *B. oleracea* [258], and trolox enhanced photosynthetic rate in *L. sativa* [259]. In addition, pre-hydration treatment of seeds with other antioxidant solutions has been reported to improve seed performance in several species. For instance, pre-hydration with ascorbic acid solution improved agronomic and biochemical vigour of *Pisum sativum* seeds [260] and improved germinability and tolerance to deterioration of *Elymus sibiricus* artificially aged for 48 h [261]. As mentioned by Afzal et al. [262], seed pre-hydration treatment with AA and tocopherol enhanced vigour and storability of *H. annuus* [263], maize, mustard [264] and *Oryza sativa* [265].

Plant Biostimulant Priming

Biostimulants are substances sourced from biological materials, e.g., microbial bioeffectors and extracts from plants and animals. They are diverse, ranging from single isolated compounds to complex matrices with various groups of biologically active constituents. The application of biostimulants to offset abiotic stress effects is well-established and represents one of the most promising techniques for attenuating stress impact in plants. It has attracted much interest both in research and commerce [266]. In addition to improving plant tolerance against several abiotic stressors, this eco-friendly innovation enhances nutrient use efficiency, plant growth, and crop productivity [267]. The stimulatory effects of plant biostimulants, such as smoke-water on *Sceletium tortuosum* seeds [268], karrikinolide on *Lactuca sativa* [269] and *Aristolochia debilis* [270] seeds, commercial brown seaweed extract (Kelpak®) on *Abelmoschus esculentus* [271] and *Ceratotheca triloba* [272] seeds, and yeast extract on *Oryza sativa* seeds [273], have been reported.

5.4.2. Advanced Seed Priming Techniques

Nanopriming

The use of nanomaterials in agriculture is somewhat recent relative to their application in biomedical and industrial sectors [239], and it is considered a promising approach that can transform food production and agriculture [274,275] to meet the demand for food security in view of the envisaged rise in world population [276,277]. Nanotechnology employs not more than 100 nm size of biocompatible nanoparticles [198], often synthesised with plant extracts of desirable phytochemical properties as the nanopriming agents (phytosynthesised nanoparticles) [239]. For example, Mahakham et al. [239] primed *Oryza sativa* seeds stored for three years using phytosynthesised silver particles obtained from silver nitrate ($AgNO_3$) solution mixed with *Citrus hystrix* leaves extract (as reducing and stabilising agents). They reported enhanced performance in terms of germination and seedling vigour. Further, they proposed the mechanisms of action of nanopriming-induced invigoration of seed to include nanopore formation for the enhancement of water uptake, optimising ROS/antioxidant systems in seeds, production of $HO^{\bullet}$ for loosening of the cell wall and weakening of endosperm to enhance seed germination as well as nanocatalyst-enhanced hydrolysis of starch. In another study, Sundaria et al. [277] demonstrated increased germination and shoot length in IITR26 and WL711 wheat (*Triticum aestivum*) genotypes, respectively, using iron oxide synthesised nanoparticles as a priming agent. Further, they demonstrated and proposed nanopriming for wheat grain biofortification with iron, which is a potential strategy for overcoming iron deficiency in humans.

Seed Priming with Physical Agents

Thus far, many studies have shown that plant metabolic and developmental processes are sensitive to magnetic fields [278–280]. Magnetic fields are now being used for the invigoration of seeds and enhancement of agricultural productivity [207,281]. Several beneficial effects of magnetopriming (priming with the magnetic field) have been documented in various studies for different plant species. For instance, Baby et al. [282] reported improved germination, vigour, seedling biomass, the performance index of chlorophyll a fluorescence, and reduced level of $O_2^{\bullet -}$ in leaves of *Glycine max* seeds primed with a static magnetic field. Besides increased germination and germination speed, field emergence, vigour and seedling biomass, other beneficial effects, such as improved membrane integrity and reduced electrolyte leakage, were reported in *Helianthus annuus* seeds subjected to magnetopriming [283]. Further, they ascribed high germination rates and vigour to magnetopriming-induced rise in α-amylase, protease, and dehydrogenase activities.

Gamma radiation [240,284], UV radiation [285,286], X-rays [287,288], and microwaves [289,290] are some other commonly used physical priming agents [195,198].

6. Conclusions

There is little doubt that ageing-induced loss of crop seed vigour and viability is a serious threat to food security, particularly in countries where farmers are dependent on seed storage. Seed deterioration during long-term storage also poses a significant threat to germplasm conservation. Seed ageing, therefore, represents a challenge for the agri-food sector and seed industry, threatening the world's ability to meet global food demand. With increasing populations, especially in the developing nations, crop production losses owing to poor seed vigour have already resulted in market instability and enormous pressure on governments. This is expected to worsen when combined with the effects of climate change and human mass migrations. As a result, the United Nations (UN), in the 2030 Agenda for Sustainable Development, placed great emphasis on food security, improved nutrition, and sustainable agriculture [291]. Many resources have also been invested in research on seed storage, priming, and invigoration, to increase crop production.

In this regard, slowing down the deterioration of seeds and enhancing seed viability and vigour have become crucial for seed preservation, given the inevitability of seed viability loss even under enhanced storage conditions in gene banks. Seed treatments before storage for enhancing ageing resistance are useful and urgently needed, especially where long-term storage facilities are not available, or seeds are stored using poor and/or ageing infrastructure. Focused research involving the use of state-of-the-art techniques on seed invigoration will be useful in elucidating the mechanisms of ageing-induced loss of seed vigour and promising invigorative methods.

Author Contributions: Writing—original draft preparation, A.E.A.; writing—review and editing, T.L.A.; supervision, B.V., S. and N.W.P. All authors have read and agreed to the published version of the manuscript.

Funding: This work was supported by the National Research Foundation (Grant Holder Bursary), South Africa (grant number CPRR13092145823). The APC was partially funded by the University of KwaZulu-Natal, South Africa.

Conflicts of Interest: The authors declare no conflict of interest.

References

1. Hoban, S.M.; Hauffe, H.C.; Pérez-Espona, S.; Arntzen, J.W.; Bertorelle, G.; Bryja, J.; Frith, K.; Gaggiotti, O.E.; Galbusera, P.; Godoy, J.A.; et al. Bringing genetic diversity to the forefront of conservation policy and management. *Conserv. Genet. Resour.* **2013**, *5*, 593–598. [CrossRef]
2. Jacobsen, S.E.; Sørensen, M.; Pedersen, S.M.; Weiner, J. Feeding the world: Genetically modified crops versus agricultural biodiversity. *Agron. Sustain. Dev.* **2013**, *33*, 651–662. [CrossRef]
3. United Nations World Population Prospects 2019. Available online: https://population.un.org/wpp/Download/Standard/Population/ (accessed on 14 October 2021).
4. Challinor, A.J.; Watson, J.; Lobell, D.B.; Howden, S.M.; Smith, D.R.; Chhetri, N. A meta-analysis of crop yield under climate change and adaptation. *Nat. Clim. Chang.* **2014**, *4*, 287–291. [CrossRef]
5. Khan, M.A.; Tahir, A.; Khurshid, N.; Husnain, M.I.u.; Ahmed, M.; Boughanmi, H. Economic effects of climate change-induced loss of agricultural production by 2050: A case study of Pakistan. *Sustainability* **2020**, *12*, 1216. [CrossRef]
6. Zinyengere, N.; Crespo, O.; Hachigonta, S. Crop response to climate change in southern Africa: A comprehensive review. *Glob. Planet. Chang.* **2013**, *111*, 118–126. [CrossRef]
7. Fahad, S.; Bajwa, A.A.; Nazir, U.; Anjum, S.A.; Farooq, A.; Zohaib, A.; Sadia, S.; Nasim, W.; Adkins, S.; Saud, S.; et al. Crop production under drought and heat stress: Plant responses and management options. *Front. Plant Sci.* **2017**, *8*, 1147. [CrossRef]
8. Conway, G.; Toenniessen, G. Feeding the world in the twenty-first century. *Nature* **1999**, *402*, C55–C58. [CrossRef]
9. Mann, C.C. Crop scientists seek a new revolution. *Science* **1999**, *283*, 310–314. [CrossRef]
10. Lipper, L.; Thornton, P.; Campbell, B.M.; Baedeker, T.; Braimoh, A.; Bwalya, M.; Caron, P.; Cattaneo, A.; Garrity, D.; Henry, K.; et al. Climate-smart agriculture for food security. *Nat. Clim. Chang.* **2014**, *4*, 1068–1072. [CrossRef]
11. Thornton, P.K.; Whitbread, A.; Baedeker, T.; Cairns, J.; Claessens, L.; Baethgen, W.; Bunn, C.; Friedmann, M.; Giller, K.E.; Herrero, M.; et al. A framework for priority-setting in climate smart agriculture research. *Agric. Syst.* **2018**, *167*, 161–175. [CrossRef]
12. Garnett, T.; Appleby, M.C.; Balmford, A.; Bateman, I.J.; Benton, T.G.; Bloomer, P.; Burlingame, B.; Dawkins, M.; Dolan, L.; Fraser, D.; et al. Sustainable intensification in agriculture: Premises and policies. *Science* **2013**, *341*, 33–34. [CrossRef]
13. Pretty, J.; Bharucha, Z.P. Sustainable intensification in agricultural systems. *Ann. Bot.* **2014**, *114*, 1571–1596. [CrossRef]
14. Cassman, K.G. Ecological intensification of cereal production systems: Yield potential, soil quality, and precision agriculture. *Proc. Natl. Acad. Sci. USA* **1999**, *96*, 5952–5959. [CrossRef]
15. FAO Seeds. Available online: http://www.fao.org/seeds/en/ (accessed on 31 August 2020).
16. Solberg, S.Ø.; Yndgaard, F.; Andreasen, C.; von Bothmer, R.; Loskutov, I.G.; Asdal, Å. Long-term storage and longevity of orthodox seeds: A systematic review. *Front. Plant Sci.* **2020**, *11*, 1007. [CrossRef]
17. Beal, W.J. The vitality of seeds buried in the soil. In *Proceedings of the Society for the Promotion of Agricultural Science*; 1911; pp. 21–23.
18. Beal, W.J. The viability of seeds. *Bot. Gaz.* **1905**, *40*, 140–143. [CrossRef]
19. Kivilaan, A.; Bandurski, R.S. The one hundred-year period for Dr. Beal's seed viability experiment. *Am. J. Bot.* **1981**, *68*, 1290–1292. [CrossRef]

20. Steiner, A.M.; Ruckenbauer, P. Germination of 110-year-old cereal and weed seeds, the Vienna Sample of 1877. Verification of effective ultra-dry storage at ambient temperature. *Seed Sci. Res.* **1995**, *5*, 195–199. [CrossRef]
21. Ruckenbauer, P. Keimfahiger Winterweizen aus dem Jahre 1877.—Beobachtungen und versuche (germinating winter wheat of the year 1877.—Observations and experiments). *Die Bodenkultur.* **1971**, *22*, 372–386.
22. Telewski, F.W.; Zeevaart, J.A.D. The 120-yr period for Dr. Beal's seed viability experiment. *Am. J. Bot.* **2002**, *89*, 1285–1288. [CrossRef] [PubMed]
23. Roberts, E.H.; Ellis, R.H. Water and seed survival. *Ann. Bot.* **1989**, *63*, 39. [CrossRef]
24. Vertucci, C.W.; Roos, E.E. Theoretical basis of protocols for seed storage. *Plant Physiol.* **1990**, *94*, 1019–1023. [CrossRef]
25. Ellis, R.H.; Hong, T.D.; Roberts, E.H. Seed moisture content, storage, viability and vigour. *Seed Sci. Res.* **1991**, *1*, 275–279. [CrossRef]
26. Ellis, R.H.; Roberts, E.H. Improved equations for the prediction of seed longevity. *Ann. Bot.* **1980**, *45*, 13–30. [CrossRef]
27. Ibrahim, A.E.; Roberts, E.H. Viability of lettuce seeds: I. Survival in hermetic storage. *J. Exp. Bot.* **1983**, *34*, 620–630. [CrossRef]
28. Still, D.W. The development of seed quality in brassicas. *Horttechnology* **1999**, *9*, 335–340. [CrossRef]
29. Walters, C.; Towill, L. Seeds and pollen. In *Agriculture Handbook. The Commercial Storage of Fruits, Vegetables, and Florist and Nursery Stocks*; Gross, K.C., Wang, C.Y., Saltveit, M., Eds.; United States Department of Agriculture: Fort Collins, CO, USA, 2004; pp. 735–743; ISBN 3015046128.
30. Berjak, P.; Pammenter, N.W. Biotechnological aspects of non-orthodox seeds: An African perspective. *S. Afr. J. Bot.* **2004**, *70*, 102–108. [CrossRef]
31. Basra, S.M.A.; Ahmad, N.; Khan, M.M.; Iqbal, N.; Cheema, M.A. Assessment of cottonseed deterioration during accelerated ageing. *Seed Sci. Technol.* **2003**, *31*, 531–540. [CrossRef]
32. Poonguzhali, S. Improving vigour and viability of blackgram cv.co 6 [*Vigna mungo* (L) Hepper] through seed priming with inorganics. *Legum. Res. Int. J.* **2016**, *39*, 820–829. [CrossRef]
33. Walters, C.; Wheeler, L.M.; Grotenhuis, J.M. Longevity of seeds stored in a genebank: Species characteristics. *Seed Sci. Res.* **2005**, *15*, 1–20. [CrossRef]
34. Chmielarz, P. Cryopreservation of dormant European ash (*Fraxinus excelsior*) orthodox seeds. *Tree Physiol.* **2009**, *29*, 1279–1285. [CrossRef] [PubMed]
35. Shaban, M. Review on physiological aspects of seed deterioration. *Int. J. Agric. Crop Sci.* **2013**, *6*, 627–631.
36. Chmielarz, P. Cryopreservation of orthodox seeds of *Alnus glutinosa*. *CryoLetters* **2010**, *31*, 139–146. [PubMed]
37. Walters, C.; Wheeler, L.; Stanwood, P.C. Longevity of cryogenically stored seeds. *Cryobiology* **2004**, *48*, 229–244. [CrossRef]
38. Brown, R. Physiology of seed germination. In *Differenzierung und Entwicklung/Differentiation and Development. Handbuch der Pflanzenphysiologie/Encyclopedia of Plant Physiology*; Lang, A., Ed.; Springer: Berlin/Heidelberg, Germany, 1965; Volume 15, pp. 2541–2555.
39. Bewley, J.D.; Black, M. *Seeds: Physiology of Development and Germination*, 2nd ed.; Springer US: Boston, MA, USA, 1994; ISBN 978-0-306-44748-8.
40. Bewley, J.D. Seed germination and dormancy. *Plant Cell* **1997**, *9*, 1055–1066. [CrossRef] [PubMed]
41. Welbaum, G.E.; Bradford, K.J.; Yim, K.-O.; Booth, D.T.; Oluoch, M.O. Biophysical, physiological and biochemical processes regulating seed germination. *Seed Sci. Res.* **1998**, *8*, 161–172. [CrossRef]
42. Pritchard, S.L.; Charlton, W.L.; Baker, A.; Graham, I.A. Germination and storage reserve mobilization are regulated independently in *Arabidopsis*. *Plant J.* **2002**, *31*, 639–647. [CrossRef]
43. Demidchik, V. Reactive oxygen species and their role in plant oxidative stress. In *Plant Stress Physiology*; Shabala, S., Ed.; CABI: Wallingford, UK, 2017; pp. 64–96. ISBN 9781780647296.
44. Sharma, P.; Dubey, R.S. Drought induces oxidative stress and enhances the activities of antioxidant enzymes in growing rice seedlings. *Plant Growth Regul.* **2005**, *46*, 209–221. [CrossRef]
45. Demidchik, V. Mechanisms of oxidative stress in plants: From classical chemistry to cell biology. *Environ. Exp. Bot.* **2015**, *109*, 212–228. [CrossRef]
46. Mittler, R. Oxidative stress, antioxidants and stress tolerance. *Trends Plant Sci.* **2002**, *7*, 405–410. [CrossRef]
47. Demidchik, V. Reactive oxygen species, oxidative stress and plant ion channels. In *Ion Channels and Plant Stress Responses. Signaling and Communication in Plants*; Signaling and Communication in Plants; Demidchik, V., Maathuis, F., Eds.; Springer: Berlin/Heidelberg, Germany, 2010; pp. 207–232.
48. Zhu, J.; Gong, Z.; Zhang, C.; Song, C.P.; Damsz, B.; Inan, G.; Koiwa, H.; Zhu, J.K.; Hasegawa, P.M.; Bressan, R.A. OSM1/SYP61: A syntaxin protein in *Arabidopsis* controls abscisic acid-mediated and non-abscisic acid-mediated responses to abiotic stress. *Plant Cell* **2002**, *14*, 3009–3028. [CrossRef]
49. Hasegawa, P.M.; Bressan, R.A.; Zhu, J.-K.; Bohnert, H.J. Plant cellular and molecular responses to high salinity. *Annu. Rev. Plant Physiol. Plant Mol. Biol.* **2000**, *51*, 463–499. [CrossRef]
50. Varghese, B.; Sershen; Berjak, P.; Varghese, D.; Pammenter, N.W. Differential drying rates of recalcitrant *Trichilia dregeana* embryonic axes: A study of survival and oxidative stress metabolism. *Physiol. Plant.* **2011**, *142*, 326–338. [CrossRef] [PubMed]
51. Karpinski, S.; Escobar, C.; Karpinska, B.; Creissen, G.; Mullineaux, P.M. Photosynthetic electron transport regulates the expression of cytosolic ascorbate peroxidase genes in Arabidopsis during excess light stress. *Plant Cell* **1997**, *9*, 627–640. [CrossRef] [PubMed]
52. Ali, M.B.; Hahn, E.J.; Paek, K.Y. Effects of temperature on oxidative stress defense systems, lipid peroxidation and lipoxygenase activity in Phalaenopsis. *Plant Physiol. Biochem.* **2005**, *43*, 213–223. [CrossRef]

53. Møller, I.M. Plant mitochondria and oxidative stress: Electron transport, NADPH turnover, and metabolism of reactive oxygen species. *Annu. Rev. Plant Physiol. Plant Mol. Biol.* **2001**, *52*, 561–591. [CrossRef]
54. Kranner, I.; Minibayeva, F.V.; Beckett, R.P.; Seal, C.E. What is stress? Concepts, definitions and applications in seed science. *New Phytol.* **2010**, *188*, 655–673. [CrossRef]
55. Hendry, G.A.F. Oxygen, free radical processes and seed longevity. *Seed Sci. Res.* **1993**, *3*, 141–153. [CrossRef]
56. Taiz, L.; Zeiger, E. *Plant Physiology*, 5th ed.; Sinauer Assosiates Inc.: Sunderland, MA, USA, 2010; ISBN 9780878938667.
57. Saha, H.; Mitra, M.; Deepa Sankar, P. Oxidative stress and approaches to enhance abiotic stress tolerance in plants. *Res. J. Pharm. Biol. Chem. Sci.* **2014**, *5*, 724–734.
58. Bartosz, G. Oxidative stress in plants. *Acta Physiol. Plant.* **1997**, *19*, 47–64. [CrossRef]
59. Mirdad, Z.; Powell, A.A.; Matthews, S. Prediction of germination in artificially aged seeds of *Brassica* spp. using the bulk conductivity test. *Seed Sci. Technol.* **2006**, *34*, 273–286. [CrossRef]
60. Boniecka, J.; Kotowicz, K.; Skrzypek, E.; Dziurka, K.; Rewers, M.; Jedrzejczyk, I.; Wilmowicz, E.; Berdychowska, J.; Dąbrowska, G.B. Potential biochemical, genetic and molecular markers of deterioration advancement in seeds of oilseed rape (*Brassica napus* L.). *Ind. Crops Prod.* **2019**, *130*, 478–490. [CrossRef]
61. Golovina, E.A.; Wolkers, W.F.; Hoekstra, F.A. Behaviour of membranes and proteins during natural seed ageing. In *Basic and Applied Aspects of Seed Biology: Proceedings of the Fifth International Workshop on Seeds, Reading, 1995*; Ellis, R.H., Black, M., Murdoch, A.J., Hong, T.D., Eds.; Springer: Dordrecht, The Netherlands, 1997; pp. 787–796. ISBN 978-94-011-5716-2.
62. Golovina, E.A.; Wolkers, W.F.; Hoekstra, F.A. Long-term stability of protein secondary structure in dry seeds. *Comp. Biochem. Physiol. Part A Physiol.* **1997**, *117*, 343–348. [CrossRef]
63. Smith, M.T. Membrane Changes and Lipid Peroxidation during Ageing in Seeds of *Lactuca sativa* L. Ph.D. Thesis, University of Natal, Durban, South Africa, 1986.
64. Mira, S.; González-Benito, M.E.; Hill, L.M.; Walters, C. Characterization of volatile production during storage of lettuce (*Lactuca sativa*) seed. *J. Exp. Bot.* **2010**, *61*, 3915–3924. [CrossRef]
65. Mittler, R. Abiotic stress, the field environment and stress combination. *Trends Plant Sci.* **2006**, *11*, 15–19. [CrossRef]
66. Jisha, K.C.; Vijayakumari, K.; Puthur, J.T. Seed priming for abiotic stress tolerance: An overview. *Acta Physiol. Plant.* **2013**, *35*, 1381–1396. [CrossRef]
67. Ghosh, D.; Xu, J. Abiotic stress responses in plant roots: A proteomics perspective. *Front. Plant Sci.* **2014**, *5*, 6. [CrossRef] [PubMed]
68. Gouveia, G.C.C.; da Silva Binotti, F.F.; Costa, E. Priming effect on the physiological potential of maize seeds under abiotic stress1. *Pesqui. Agropecuária Trop.* **2017**, *47*, 328–335. [CrossRef]
69. Kurek, K.; Plitta-Michalak, B.; Ratajczak, E. Reactive oxygen species as potential drivers of the seed aging process. *Plants* **2019**, *8*, 174. [CrossRef]
70. Zhao, L.; Wang, S.; Fu, Y.B.; Wang, H. *Arabidopsis* seed stored mRNAs are degraded constantly over aging time, as revealed by new quantification methods. *Front. Plant Sci.* **2020**, *10*, 1764. [CrossRef]
71. Saed-Moucheshi, A.; Shekoofa, A.; Pessarakli, M. Reactive oxygen species (ROS) generation and detoxifying in plants. *J. Plant Nutr.* **2014**, *37*, 1573–1585. [CrossRef]
72. Mittler, R. ROS are good. *Trends Plant Sci.* **2017**, *22*, 11–19. [CrossRef]
73. Kibinza, S.; Vinel, D.; Côme, D.; Bailly, C.; Corbineau, F. Sunflower seed deterioration as related to moisture content during ageing, energy metabolism and active oxygen species scavenging. *Physiol. Plant.* **2006**, *128*, 496–506. [CrossRef]
74. Mira, S.; Estrelles, E.; González-Benito, M.E.; Corbineau, F. Biochemical changes induced in seeds of Brassicaceae wild species during ageing. *Acta Physiol. Plant.* **2011**, *33*, 1803–1809. [CrossRef]
75. Anjum, N.A.; Sofo, A.; Scopa, A.; Roychoudhury, A.; Gill, S.S.; Iqbal, M.; Lukatkin, A.S.; Pereira, E.; Duarte, A.C.; Ahmad, I. Lipids and proteins—Major targets of oxidative modifications in abiotic stressed plants. *Environ. Sci. Pollut. Res.* **2015**, *22*, 4099–4121. [CrossRef]
76. Farmer, E.E.; Mueller, M.J. ROS-mediated lipid peroxidation and RES-activated signaling. *Annu. Rev. Plant Biol.* **2013**, *64*, 429–450. [CrossRef]
77. Feng, J.; Shen, Y.; Shi, F.; Li, C. Changes in seed germination ability, lipid peroxidation and antioxidant enzyme activities of *Ginkgo biloba* seed during desiccation. *Forests* **2017**, *8*, 286. [CrossRef]
78. Sahu, B.; Sahu, A.K.; Thomas, V.; Naithani, S.C. Reactive oxygen species, lipid peroxidation, protein oxidation and antioxidative enzymes in dehydrating Karanj (*Pongamia pinnata*) seeds during storage. *S. Afr. J. Bot.* **2017**, *112*, 383–390. [CrossRef]
79. Al-maskri, A.; Kharr, M.M.; Ai-mantheriand, O.; Al-habs, K. Effect of accelerated aging on lipid peroxidation, leakage and seedling vigor (RGR) in cucumber (*Cucumis sativus* L.) seeds. *Park. J Agric. Sci.* **2002**, *39*, 330–337.
80. Alexeyev, M.F. Is there more to aging than mitochondrial DNA and reactive oxygen species? *FEBS J.* **2009**, *276*, 5768–5787. [CrossRef]
81. Oenel, A.; Fekete, A.; Krischke, M.; Faul, S.C.; Gresser, G.; Havaux, M.; Mueller, M.J.; Berger, S. Enzymatic and non-enzymatic mechanisms contribute to lipid oxidation during seed aging. *Plant Cell Physiol.* **2017**, *58*, 925–933. [CrossRef] [PubMed]
82. Gutteridge, J.M.C. Lipid peroxidation and antioxidants as biomarkers of tissue damage. *Clin. Chem.* **1995**, *41*, 1819–1828. [CrossRef]

83. Catalá, A. An overview of lipid peroxidation with emphasis in outer segments of photoreceptors and the chemiluminescence assay. *Int. J. Biochem. Cell Biol.* **2006**, *38*, 1482–1495. [CrossRef]
84. Porter, N.A. Chemistry of lipid peroxidation. In *Methods in Enzymology*; Packer, L., Ed.; Academic Press: Cambridge, MA, USA, 1984; Volume 105, pp. 273–282. ISBN 012182005X.
85. Halliwell, B.; Chirico, S. Lipid peroxidation: Its mechanism, measurement, and significance. *Am. J. Clin. Nutr.* **1993**, *57*, 715S–725S. [CrossRef]
86. Krieger-Liszkay, A.; Fufezan, C.; Trebst, A. Singlet oxygen production in photosystem II and related protection mechanism. *Photosynth. Res.* **2008**, *98*, 551–564. [CrossRef]
87. Przybyla, D.; Göbel, C.; Imboden, A.; Hamberg, M.; Feussner, I.; Apel, K. Enzymatic, but not non-enzymatic, 1O2-mediated peroxidation of polyunsaturated fatty acids forms part of the EXECUTER1-dependent stress response program in the flu mutant of *Arabidopsis thaliana*. *Plant J.* **2008**, *54*, 236–248. [CrossRef] [PubMed]
88. Nowicka, B.; Gruszka, J.; Kruk, J. Function of plastochromanol and other biological prenyllipids in the inhibition of lipid peroxidation—A comparative study in model systems. *Biochim. Biophys. Acta-Biomembr.* **2013**, *1828*, 233–240. [CrossRef] [PubMed]
89. Esterbauer, H.; Schaur, R.J.; Zollner, H. Chemistry and biochemistry of 4-hydroxynonenal, malonaldehyde and related aldehydes. *Free Radic. Biol. Med.* **1991**, *11*, 81–128. [CrossRef]
90. Bentinger, M.; Brismar, K.; Dallner, G. The antioxidant role of coenzyme Q. *Mitochondrion* **2007**, *7*, S41–S50. [CrossRef]
91. Kristal, B.S.; Park, B.K.; Yu, B.P. 4-Hydroxyhexenal is a potent inducer of the mitochondrial permeability transition. *J. Biol. Chem.* **1996**, *271*, 6033–6038. [CrossRef]
92. Yin, L.; Mano, J.; Wang, S.; Tsuji, W.; Tanaka, K. The involvement of lipid peroxide-derived aldehydes in aluminum toxicity of tobacco roots. *Plant Physiol.* **2010**, *152*, 1406–1417. [CrossRef]
93. Feussner, I.; Wasternack, C. The lipoxygenase pathway. *Annu. Rev. Plant Biol.* **2002**, *53*, 275–297. [CrossRef]
94. Andreou, A.; Feussner, I. Lipoxygenases—Structure and reaction mechanism. *Phytochemistry* **2009**, *70*, 1504–1510. [CrossRef] [PubMed]
95. Li, J.; Zhang, Y.; Yu, Z.; Wang, Y.; Yang, Y.; Liu, Z.; Jiang, J.; Song, M.; Wu, Y. Superior storage stability in low lipoxygenase maize varieties. *J. Stored Prod. Res.* **2007**, *43*, 530–534. [CrossRef]
96. Gayen, D.; Ali, N.; Ganguly, M.; Paul, S.; Datta, K.; Datta, S.K. RNAi mediated silencing of lipoxygenase gene to maintain rice grain quality and viability during storage. *Plant Cell. Tissue Organ Cult.* **2014**, *118*, 229–243. [CrossRef]
97. Song, M.; Wu, Y.; Zhang, Y.; Liu, B.M.; Jiang, J.Y.; Xu, X.; Yu, Z.L. Mutation of rice (*Oryza sativa* L.) LOX-1/2 near-isogenic lines with ion beam implantation and study of their storability. *Nucl. Instrum. Methods Phys. Res. B* **2007**, *265*, 495–500. [CrossRef]
98. Li, Z.; Gao, Y.; Lin, C.; Pan, R.; Ma, W.; Zheng, Y.; Guan, Y.; Hu, J. Suppression of LOX activity enhanced seed vigour and longevity of tobacco (*Nicotiana tabacum* L.) seeds during storage. *Conserv. Physiol.* **2018**, *6*, coy047. [CrossRef]
99. Davies, M.J. Singlet oxygen-mediated damage to proteins and its consequences. *Biochem. Biophys. Res. Commun.* **2003**, *305*, 761–770. [CrossRef]
100. Møller, I.M.; Jensen, P.E.; Hansson, A. Oxidative modifications to cellular components in plants. *Annu. Rev. Plant Biol.* **2007**, *58*, 459–481. [CrossRef]
101. Hawkins, C.L.; Morgan, P.E.; Davies, M.J. Quantification of protein modification by oxidants. *Free Radic. Biol. Med.* **2009**, *46*, 965–988. [CrossRef]
102. Rinalducci, S.; Murgiano, L.; Zolla, L. Redox proteomics: Basic principles and future perspectives for the detection of protein oxidation in plants. *J. Exp. Bot.* **2008**, *59*, 3781–3801. [CrossRef] [PubMed]
103. Starke-Reed, P.E.; Oliver, C.N. Protein oxidation and proteolysis during aging and oxidative stress. *Arch. Biochem. Biophys.* **1989**, *275*, 559–567. [CrossRef]
104. Stadtman, E. Protein oxidation and aging. *Science* **1992**, *257*, 1220–1224. [CrossRef]
105. Oracz, K.; Bouteau, H.E.M.; Farrant, J.M.; Cooper, K.; Belghazi, M.; Job, C.; Job, D.; Corbineau, F.; Bailly, C. ROS production and protein oxidation as a novel mechanism for seed dormancy alleviation. *Plant J.* **2007**, *50*, 452–465. [CrossRef] [PubMed]
106. Kumar, A.; Prasad, A.; Sedlářová, M.; Pospíšil, P. Organic radical imaging in plants: Focus on protein radicals. *Free Radic. Biol. Med.* **2019**, *130*, 568–575. [CrossRef]
107. Johansson, E.; Olsson, O.; Nyström, T. Progression and specificity of protein oxidation in the life cycle of *Arabidopsis thaliana*. *J. Biol. Chem.* **2004**, *279*, 22204–22208. [CrossRef] [PubMed]
108. Avery, S.V. Molecular targets of oxidative stress. *Biochem. J.* **2011**, *434*, 201–210. [CrossRef] [PubMed]
109. Medicherla, B.; Goldberg, A.L. Heat shock and oxygen radicals stimulate ubiquitin-dependent degradation mainly of newly synthesized proteins. *J. Cell Biol.* **2008**, *182*, 663–673. [CrossRef]
110. Dalle-Donne, I.; Rossi, R.; Giustarini, D.; Milzani, A.; Colombo, R. Protein carbonyl groups as biomarkers of oxidative stress. *Clin. Chim. Acta* **2003**, *329*, 23–38. [CrossRef]
111. Ghezzi, P.; Bonetto, V. Redox proteomics: Identification of oxidatively modified proteins. *Proteomics* **2003**, *3*, 1145–1153. [CrossRef]
112. Shacter, E. Quantification and significance of protein oxidation in biological samples. *Drug Metab. Rev.* **2000**, *32*, 307–326. [CrossRef] [PubMed]
113. Hawkins, C.L.; Davies, M.J. Detection, identification, and quantification of oxidative protein modifications. *J. Biol. Chem.* **2019**, *294*, 19683–19708. [CrossRef] [PubMed]

114. Isbell, H.S.; Frush, H.L.; Martin, E.T. Reactions of carbohydrates with hydroperoxides. *Carbohydr. Res.* **1973**, *26*, 287–295. [CrossRef]
115. Headlam, H.A.; Davies, M.J. Markers of protein oxidation: Different oxidants give rise to variable yields of bound and released carbonyl products. *Free Radic. Biol. Med.* **2004**, *36*, 1175–1184. [CrossRef] [PubMed]
116. Moran, J.F.; Becana, M.; Iturbe-Ormaetxe, I.; Frechilla, S.; Klucas, R.V.; Aparicio-Tejo, P. Drought induces oxidative stress in pea plants. *Planta* **1994**, *194*, 346–352. [CrossRef]
117. Morscher, F.; Kranner, I.; Arc, E.; Bailly, C.; Roach, T. Glutathione redox state, tocochromanols, fatty acids, antioxidant enzymes and protein carbonylation in sunflower seed embryos associated with after-ripening and ageing. *Ann. Bot.* **2015**, *116*, 669–678. [CrossRef] [PubMed]
118. Milkovska-Stamenova, S.; Schmidt, R.; Frolov, A.; Birkemeyer, C. GC-MS Method for the quantitation of carbohydrate intermediates in glycation systems. *J. Agric. Food Chem.* **2015**, *63*, 5911–5919. [CrossRef]
119. Shumilina, J.; Kusnetsova, A.; Tsarev, A.; Janse van Rensburg, H.C.; Medvedev, S.; Demidchik, V.; Van den Ende, W.; Frolov, A. Glycation of plant proteins: Regulatory roles and interplay with sugar signalling? *Int. J. Mol. Sci.* **2019**, *20*, 2366. [CrossRef]
120. Lounifi, I.; Arc, E.; Molassiotis, A.; Job, D.; Rajjou, L.; Tanou, G. Interplay between protein carbonylation and nitrosylation in plants. *Proteomics* **2013**, *13*, 568–578. [CrossRef]
121. Roychoudhury, A.; Basu, S.; Sengupta, D.N. Amelioration of salinity stress by exogenously applied spermidine or spermine in three varieties of indica rice differing in their level of salt tolerance. *J. Plant Physiol.* **2011**, *168*, 317–328. [CrossRef]
122. Tanou, G.; Filippou, P.; Belghazi, M.; Job, D.; Diamantidis, G.; Fotopoulos, V.; Molassiotis, A. Oxidative and nitrosative-based signaling and associated post-translational modifications orchestrate the acclimation of citrus plants to salinity stress. *Plant J.* **2012**, *72*, 585–599. [CrossRef]
123. Pyngrope, S.; Bhoomika, K.; Dubey, R.S. Oxidative stress, protein carbonylation, proteolysis and antioxidative defense system as a model for depicting water deficit tolerance in Indica rice seedlings. *Plant Growth Regul.* **2013**, *69*, 149–165. [CrossRef]
124. Rellán-Álvarez, R.; Ortega-Villasante, C.; Álvarez-Fernández, A.; del Campo, F.F.; Hernández, L.E. Stress responses of *Zea mays* to cadmium and mercury. *Plant Soil* **2006**, *279*, 41–50. [CrossRef]
125. Song, H.; Xu, X.; Wang, H.; Tao, Y. Protein carbonylation in barley seedling roots caused by aluminum and proton toxicity is suppressed by salicylic acid. *Russ. J. Plant Physiol.* **2011**, *58*, 653–659. [CrossRef]
126. Xu, X.; Qin, G.; Tian, S. Effect of microbial biocontrol agents on alleviating oxidative damage of peach fruit subjected to fungal pathogen. *Int. J. Food Microbiol.* **2008**, *126*, 153–158. [CrossRef] [PubMed]
127. Sundaram, S.; Rathinasabapathi, B. Transgenic expression of fern *Pteris vittata* glutaredoxin PvGrx5 in *Arabidopsis thaliana* increases plant tolerance to high temperature stress and reduces oxidative damage to proteins. *Planta* **2010**, *231*, 361–369. [CrossRef] [PubMed]
128. Rajjou, L.; Lovigny, Y.; Groot, S.P.C.; Belghazi, M.; Job, C.; Job, D. Proteome-wide characterization of seed aging in *Arabidopsis*: A comparison between artificial and natural aging protocols. *Plant Physiol.* **2008**, *148*, 620–641. [CrossRef]
129. Yin, G.; Xin, X.; Fu, S.; An, M.; Wu, S.; Chen, X.; Zhang, J.; He, J.; Whelan, J.; Lu, X. Proteomic and carbonylation profile analysis at the critical node of seed ageing in *Oryza sativa*. *Sci. Rep.* **2017**, *7*, 40611. [CrossRef]
130. Smirnoff, N.; Cumbes, Q.J. Hydroxyl radical scavenging activity of compatible solutes. *Phytochemistry* **1989**, *28*, 1057–1060. [CrossRef]
131. Miller, A.R. Oxidation of cell wall polysaccharides by hydrogen peroxide: A potential mechanism for cell wall breakdown in plants. *Biochem. Biophys. Res. Commun.* **1986**, *141*, 238–244. [CrossRef]
132. Schopfer, P.; Liszkay, A.; Bechtold, M.; Frahry, G.; Wagner, A. Evidence that hydroxyl radicals mediate auxin-induced extension growth. *Planta* **2002**, *214*, 821–828. [CrossRef]
133. Fry, S.C.; Miller, J.G.; Dumville, J.C. A proposed role for copper ions in cell wall loosening. In *Progress in Plant Nutrition: Plenary Lectures of the XIV International Plant Nutrition Colloquium*; Springer: Dordrecht, The Netherlands, 2002; Volume 247, pp. 57–67.
134. Connolly, E.L.; Guerinot, M. Lou Iron stress in plants. *Genome Biol.* **2002**, *3*, 1–4. [CrossRef]
135. Couée, I.; Sulmon, C.; Gouesbet, G.; El Amrani, A. Involvement of soluble sugars in reactive oxygen species balance and responses to oxidative stress in plants. *J. Exp. Bot.* **2006**, *57*, 449–459. [CrossRef] [PubMed]
136. Jouve, L.; Hoffmann, L.; Hausman, J.F. Polyamine, carbohydrate, and proline content changes during salt stress exposure of aspen (*Populus tremula* L.): Involvement of oxidation and osmoregulation metabolism. *Plant Biol.* **2004**, *6*, 74–80. [CrossRef]
137. Patel, T.K.; Williamson, J.D. Mannitol in plants, fungi, and plant-fungal interactions. *Trends Plant Sci.* **2016**, *21*, 486–497. [CrossRef] [PubMed]
138. El-Maarouf-Bouteau, H.; Mazuy, C.; Corbineau, F.; Bailly, C. DNA alteration and programmed cell death during ageing of sunflower seed. *J. Exp. Bot.* **2011**, *62*, 5003–5011. [CrossRef]
139. Roldán-Arjona, T.; Ariza, R.R. Repair and tolerance of oxidative DNA damage in plants. *Mutat. Res.* **2009**, *681*, 169–179. [CrossRef] [PubMed]
140. Jeong, Y.-C. Pyrimido[1,2-a]-purin-10(3H)-one, M1G, is less prone to artifact than base oxidation. *Nucleic Acids Res.* **2005**, *33*, 6426–6434. [CrossRef]
141. Britt, A.B. DNA damage and repair in plants. *Annu. Rev. Plant Physiol. Plant Mol. Biol.* **1996**, *47*, 75–100. [CrossRef]
142. Larsen, N.B.; Rasmussen, M.; Rasmussen, L.J. Nuclear and mitochondrial DNA repair: Similar pathways? *Mitochondrion* **2005**, *5*, 89–108. [CrossRef]

143. Yoshiyama, K.; Sakaguchi, K.; Kimura, S. DNA damage response in plants: Conserved and variable response compared to animals. *Biology* **2013**, *2*, 1338–1356. [CrossRef]
144. Vanderauwera, S.; Suzuki, N.; Miller, G.; van de Cotte, B.; Morsa, S.; Ravanat, J.-L.; Hegie, A.; Triantaphylides, C.; Shulaev, V.; Van Montagu, M.C.E.; et al. Extranuclear protection of chromosomal DNA from oxidative stress. *Proc. Natl. Acad. Sci. USA* **2011**, *108*, 1711–1716. [CrossRef] [PubMed]
145. Mullarky, E.; Cantley, L.C. Diverting glycolysis to combat oxidative stress. In *Innovative Medicine*; Nakao, K., Minato, N., Uemoto, S., Eds.; Springer: Tokyo, Japan, 2015; pp. 3–23.
146. Dat, J.; Vandenabeele, S.; Vranová, E.; Van Montagu, M.; Inzé, D.; Van Breusegem, F. Dual action of the active oxygen species during plant stress responses. *Cell. Mol. Life Sci.* **2000**, *57*, 779–795. [CrossRef] [PubMed]
147. Perl-Treves, R.; Perl, A. Oxidative stress: An introduction. In *Oxidative Stress in Plants*; Inzé, D., Van Montagu, M., Eds.; Taylor & Francis: London, UK, 2002; pp. 1–32.
148. Gill, S.S.; Tuteja, N. Reactive oxygen species and antioxidant machinery in abiotic stress tolerance in crop plants. *Plant Physiol. Biochem.* **2010**, *48*, 909–930. [CrossRef]
149. Govindaraj, M.; Masilamani, P.; Albert, V.A.; Bhaskaran, M. Role of antioxidant in seed quality—A review. *Agric. Rev.* **2017**, *38*, 180–190. [CrossRef]
150. Fridovich, I. Superoxide radical and superoxide dismutases. In *Autoxidation in Food and Biological Systems*; Springer: Boston, MA, USA, 1981; Volume 110, pp. 250–272.
151. del Río, L.A.; Corpas, F.J.; López-Huertas, E.; Palma, J.M. Plant superoxide dismutases: Function under abiotic stress conditions. In *Antioxidants and Antioxidant Enzymes in Higher Plants*; Springer International Publishing: Cham, Switzerland, 2018; pp. 1–26. ISBN 9783319750880.
152. Anjum, N.A.; Sharma, P.; Gill, S.S.; Hasanuzzaman, M.; Khan, E.A.; Kachhap, K.; Mohamed, A.A.; Thangavel, P.; Devi, G.D.; Vasudhevan, P.; et al. Catalase and ascorbate peroxidase—Representative H_2O_2-detoxifying heme enzymes in plants. *Environ. Sci. Pollut. Res.* **2016**, *23*, 19002–19029. [CrossRef]
153. Shugaev, A.G.; Lashtabega, D.A.; Shugaeva, N.A.; Vyskrebentseva, E.I. Activities of antioxidant enzymes in mitochondria of growing and dormant sugar beet roots. *Russ. J. Plant Physiol.* **2011**, *58*, 387–393. [CrossRef]
154. Bailly, C.; Benamar, A.; Corbineau, F.; Côme, D. Changes in malondialdehyde content and in superoxide dismutase, catalase and glutathione reductase activities in sunflower seeds as related to deterioration during accelerated aging. *Physiol. Plant.* **1996**, *97*, 104–110. [CrossRef]
155. Kibinza, S.; Bazin, J.; Bailly, C.; Farrant, J.M.; Corbineau, F.; El-Maarouf-Bouteau, H. Catalase is a key enzyme in seed recovery from ageing during priming. *Plant Sci.* **2011**, *181*, 309–315. [CrossRef]
156. Liu, Y.; Yao, Y.; Hu, X.; Xing, S.; Xu, L. Cloning and allelic variation of two novel catalase genes (SoCAT-1 and SsCAT-1) in *Saccharum officinarum* L. and *Saccharum spontaneum* L. *Biotechnol. Biotechnol. Equip.* **2015**, *29*, 431–440. [CrossRef]
157. Yoshimura, K.; Miyao, K.; Gaber, A.; Takeda, T.; Kanaboshi, H.; Miyasaka, H.; Shigeoka, S. Enhancement of stress tolerance in transgenic tobacco plants overexpressing *Chlamydomonas* glutathione peroxidase in chloroplasts or cytosol. *Plant J.* **2004**, *37*, 21–33. [CrossRef]
158. Yousuf, P.Y.; Hakeem, K.U.R.; Chandna, R.; Ahmad, P. Role of glutathione reductase in plant abiotic stress. In *Abiotic Stress Responses in Plants*; Ahmad, P., Prasad, M.N.V., Eds.; Springer: New York, NY, USA, 2012; pp. 149–158. ISBN 978-1-4614-0633-4.
159. Ozyigit, I.I.; Filiz, E.; Vatansever, R.; Kurtoglu, K.Y.; Koc, I.; Öztürk, M.X.; Anjum, N.A. Identification and comparative analysis of H_2O_2-scavenging enzymes (ascorbate peroxidase and glutathione peroxidase) in selected plants employing bioinformatics approaches. *Front. Plant Sci.* **2016**, *7*, 301. [CrossRef] [PubMed]
160. Chew, O.; Whelan, J.; Millar, A.H. Molecular definition of the ascorbate-glutathione cycle in *Arabidopsis* mitochondria reveals dual targeting of antioxidant defenses in plants. *J. Biol. Chem.* **2003**, *278*, 46869–46877. [CrossRef] [PubMed]
161. Yang, Y.; Han, C.; Liu, Q.; Lin, B.; Wang, J. Effect of drought and low light on growth and enzymatic antioxidant system of *Picea asperata* seedlings. *Acta Physiol. Plant.* **2008**, *30*, 433–440. [CrossRef]
162. Koussevitzky, S.; Suzuki, N.; Huntington, S.; Armijo, L.; Sha, W.; Cortes, D.; Shulaev, V.; Mittler, R. Ascorbate peroxidase 1 plays a key role in the response of *Arabidopsis thaliana* to stress combination. *J. Biol. Chem.* **2008**, *283*, 34197–34203. [CrossRef]
163. Anjum, N.A.; Gill, S.S.; Gill, R.; Hasanuzzaman, M.; Duarte, A.C.; Pereira, E.; Ahmad, I.; Tuteja, R.; Tuteja, N. Metal/metalloid stress tolerance in plants: Role of ascorbate, its redox couple, and associated enzymes. *Protoplasma* **2014**, *251*, 1265–1283. [CrossRef]
164. Bela, K.; Horváth, E.; Gallé, Á.; Szabados, L.; Tari, I.; Csiszár, J. Plant glutathione peroxidases: Emerging role of the antioxidant enzymes in plant development and stress responses. *J. Plant Physiol.* **2015**, *176*, 192–201. [CrossRef]
165. Rouhier, N.; Jacquot, J.-P. The plant multigenic family of thiol peroxidases. *Free Radic. Biol. Med.* **2005**, *38*, 1413–1421. [CrossRef]
166. Navrot, N.; Collin, V.; Gualberto, J.; Gelhaye, E.; Hirasawa, M.; Rey, P.; Knaff, D.B.; Issakidis, E.; Jacquot, J.P.; Rouhier, N. Plant glutathione peroxidases are functional peroxiredoxins distributed in several subcellular compartments and regulated during biotic and abiotic stresses. *Plant Physiol.* **2006**, *142*, 1364–1379. [CrossRef]
167. Rizhsky, L.; Hallak-Herr, E.; Van Breusegem, F.; Rachmilevitch, S.; Barr, J.E.; Rodermel, S.; Inzé, D.; Mittler, R. Double antisense plants lacking ascorbate peroxidase and catalase are less sensitive to oxidative stress than single antisense plants lacking ascorbate peroxidase or catalase. *Plant J.* **2002**, *32*, 329–342. [CrossRef]
168. Foyer, C.; Lelandais, M.; Galap, C.; Kunert, K.J. Effects of elevated cytosolic glutathione reductase activity on the cellular glutathione pool and photosynthesis in leaves under normal and stress conditions. *Plant Physiol.* **1991**, *97*, 863–872. [CrossRef]

169. Khan, T.; Mazid, M.; Mohammad, F. A review of ascorbic acid potentialities against oxidative stress induced in plants. *J. Agrobiol.* **2011**, *28*, 97–111. [CrossRef]
170. Tommasi, F.; Paciolla, C.; de Pinto, M.C.; De Gara, L. A comparative study of glutathione and ascorbate metabolism during germination of *Pinus pinea* L. seeds. *J. Exp. Bot.* **2001**, *52*, 1647–1654. [CrossRef] [PubMed]
171. Alamri, S.A.; Siddiqui, M.H.; Al-Khaishany, M.Y.Y.; Nasir Khan, M.; Ali, H.M.; Alaraidh, I.A.; Alsahli, A.A.; Al-Rabiah, H.; Mateen, M. Ascorbic acid improves the tolerance of wheat plants to lead toxicity. *J. Plant Interact.* **2018**, *13*, 409–419. [CrossRef]
172. Ahmad, I.; Basra, S.M.A.; Wahid, A. Exogenous application of ascorbic acid, salicylic acid and hydrogen peroxide improves the productivity of hybrid maize at low temperature stress. *Int. J. Agric. Biol.* **2014**, *16*, 825–830.
173. Conklin, P.L.; Barth, C. Ascorbic acid, a familiar small molecule intertwined in the response of plants to ozone, pathogens, and the onset of senescence. *Plant Cell Environ.* **2004**, *27*, 959–970. [CrossRef]
174. Noctor, G.; Foyer, C.H. Ascorbate and glutathione: Keeping active oxygen under control. *Annu. Rev. Plant Physiol. Plant Mol. Biol.* **1998**, *49*, 249–279. [CrossRef]
175. Larson, R.A. The antioxidants of higher plants. *Phytochemistry* **1988**, *27*, 969–978. [CrossRef]
176. Jiménez, A.; Hernández, J.A.; Pastori, G.; del Río, L.A.; Sevilla, F. Role of the ascorbate-glutathione cycle of mitochondria and peroxisomes in the senescence of pea leaves. *Plant Physiol.* **1998**, *118*, 1327–1335. [CrossRef]
177. Foyer, C.H.; Noctor, G. Redox homeostasis and antioxidant signaling: A metabolic interface between stress perception and physiological responses. *Plant Cell* **2005**, *17*, 1866–1875. [CrossRef]
178. Tausz, M.; Šircelj, H.; Grill, D. The glutathione system as a stress marker in plant ecophysiology: Is a stress-response concept valid? *J. Exp. Bot.* **2004**, *55*, 1955–1962. [CrossRef]
179. Creissen, G.; Firmin, J.; Fryer, M.; Kular, B.; Leyland, N.; Reynolds, H.; Pastori, G.; Wellburn, F.; Baker, N.; Wellburn, A.; et al. Elevated glutathione biosynthetic capacity in the chloroplasts of transgenic tobacco plants paradoxically causes increased oxidative stress. *Plant Cell* **1999**, *11*, 1277–1291. [CrossRef] [PubMed]
180. Pietrini, F.; Iannelli, M.A.; Pasqualini, S.; Massacci, A. Interaction of cadmium with glutathione and photosynthesis in developing leaves and chloroplasts of *Phragmites australis* (Cav.) Trin. ex Steudel. *Plant Physiol.* **2003**, *133*, 829–837. [CrossRef]
181. Zingg, J.-M.; Azzi, A. Non-antioxidant activities of vitamin E. *Curr. Med. Chem.* **2004**, *11*, 1113–1133. [CrossRef]
182. Holländer-Czytko, H.; Grabowski, J.; Sandorf, I.; Weckermann, K.; Weiler, E.W. Tocopherol content and activities of tyrosine aminotransferase and cystine lyase in *Arabidopsis* under stress conditions. *J. Plant Physiol.* **2005**, *162*, 767–770. [CrossRef]
183. Senaratna, T.; Gusse, J.F.; McKersie, B.D. Age-induced changes in cellular membranes of imbibed soybean seed axes. *Physiol. Plant.* **1988**, *73*, 85–91. [CrossRef]
184. Hamad, I.; Arda, N.; Pekmez, M.; Karaer, S.; Temizkan, G. Intracellular scavenging activity of Trolox (6-hydroxy-2,5,7,8-tetramethylchromane-2-carboxylic acid) in the fission yeast, *Schizosaccharomyces pombe*. *J. Nat. Sci. Biol. Med.* **2010**, *1*, 16–21. [CrossRef] [PubMed]
185. Lúcio, M.; Nunes, C.; Gaspar, D.; Ferreira, H.; Lima, J.L.F.C.; Reis, S. Antioxidant activity of vitamin E and Trolox: Understanding of the factors that govern lipid peroxidation studies in vitro. *Food Biophys.* **2009**, *4*, 312–320. [CrossRef]
186. Collins, A.R. Carotenoids and genomic stability. *Mutat. Res.-Fundam. Mol. Mech. Mutagen.* **2001**, *475*, 21–28. [CrossRef]
187. Haslam, E.; Cai, Y. Plant polyphenols (vegetable tannins): Gallic acid metabolism. *Nat. Prod. Rep.* **1994**, *11*, 41–66. [CrossRef]
188. Handique, J.G.; Baruah, J.B. Polyphenolic compounds: An overview. *React. Funct. Polym.* **2002**, *52*, 163–188. [CrossRef]
189. Ow, Y.-Y.; Stupans, I. Gallic acid and gallic acid derivatives: Effects on drug metabolizing enzymes. *Curr. Drug Metab.* **2005**, *4*, 241–248. [CrossRef]
190. Metodiewa, D.; Jaiswal, A.K.; Cenas, N.; Dickancaité, E.; Segura-Aguilar, J. Quercetin may act as a cytotoxic prooxidant after its metabolic activation to semiquinone and quinoidal product. *Free Radic. Biol. Med.* **1999**, *26*, 107–116. [CrossRef]
191. Sakagami, H.; Satoh, K. Prooxidant action of two antioxidants: Ascorbic acid and gallic acid. *Anticancer Res.* **1997**, *17*, 221–224. [PubMed]
192. Verma, S.; Singh, A.; Mishra, A. Gallic acid: Molecular rival of cancer. *Environ. Toxicol. Pharmacol.* **2013**, *35*, 473–485. [CrossRef]
193. Nakatani, N. Natural Antioxidants from Spices. In *Phenolic Compounds in Food and Their Effects on Health II: Antioxidants and Cancer Prevention*; Huang, M.T., Ho, C.T., Lee, C., Eds.; American Chemical Society: Washington, DC, USA, 1992; pp. 72–86.
194. Yen, G.C.; Der Duh, P.; Tsai, H.L. Antioxidant and pro-oxidant properties of ascorbic acid and gallic acid. *Food Chem.* **2002**, *79*, 307–313. [CrossRef]
195. Rakshit, A.; Singh, H.B. *Advances in Seed Priming*; Rakshit, A., Singh, H.B., Eds.; Springer: Singapore, 2018; ISBN 978-981-13-0031-8.
196. Paparella, S.; Araújo, S.S.; Rossi, G.; Wijayasinghe, M.; Carbonera, D.; Balestrazzi, A. Seed priming: State of the art and new perspectives. *Plant Cell Rep.* **2015**, *34*, 1281–1293. [CrossRef]
197. Evenari, M. Seed physiology: Its history from antiquity to the beginning of the 20th century. *Bot. Rev.* **1984**, *50*, 119–142. [CrossRef]
198. Waqas, M.; Korres, N.E.; Khan, M.D.; Nizami, A.; Deeba, F.; Ali, I.; Hussain, H. Advances in the concept and methods of seed priming. In *Priming and Pretreatment of Seeds and Seedlings*; Hasanuzzaman, M., Fotopoulos, V., Eds.; Springer: Singapore, 2019; pp. 11–41. ISBN 978-981-13-8624-4.
199. Darwin, C.R. Effect of salt-water on the germination of seeds. *Gardeners' Chron. Agric. Gazet.* **1855**, *47*, 773.
200. May, L.H.; Milthorpe, E.J.; Milthorpe, F.L. Pre-sowing hardening of plants to drought. In *Field Crop Abstracts*; CABI: Wallingford, UK, 1962; pp. 93–98.

201. Heydecker, W. Germination of an idea: The priming of seeds. Ph.D. Thesis, School of Agriculture Research, University of Nottingham, Nottingham, UK, 1973; pp. 50–67.
202. Malnassy, P.G. Physiological and biochemical studies on a treatment hastening the germination of seeds at low temperatures. Ph.D. Thesis, The State University of New Jersey, Rutgers, NJ, USA, 1971.
203. Sivasubramaniam, K.; Geetha, R.; Sujatha, K.; Raja, K.; Sripunitha, A.; Selvarani, R. Seed Priming: Triumphs and tribulations. *Physiology* **2011**, *98*, 197–209.
204. Rowse, H.R. Methods of priming seed. U.S. Patent 5,119,589, 1992.
205. Bradford, K.J. Manipulation of seed water relations via osmotic priming to improve germination under stress conditions. *HortScience* **1986**, *21*, 1105.
206. Parera, C.A.; Cantliffe, D.J. Presowing Seed Priming. In *Horticultural Reviews*; John Wiley & Sons, Inc.: Oxford, UK, 1994; Volume 16, pp. 109–141. ISBN 9780470650561.
207. Singh, V.K.; Singh, R.; Tripathi, S.; Devi, R.S.; Srivastava, P.; Singh, P.; Kumar, A.; Bhadouria, R. Seed priming: State of the art and new perspectives in the era of climate change. In *Climate Change and Soil Interactions*; Prasad, M.N.V., Pietrzykowski, M., Eds.; Elsevier: Amsterdam, The Netherlands, 2020; pp. 143–170. ISBN 9780128180327.
208. Beckers, G.J.; Conrath, U. Priming for stress resistance: From the lab to the field. *Curr. Opin. Plant Biol.* **2007**, *10*, 425–431. [CrossRef]
209. Waterworth, W.M.; Bray, C.M.; West, C.E. The importance of safeguarding genome integrity in germination and seed longevity. *J. Exp. Bot.* **2015**, *66*, 3549–3558. [CrossRef]
210. Waterworth, W.M.; Masnavi, G.; Bhardwaj, R.M.; Jiang, Q.; Bray, C.M.; West, C.E. A plant DNA ligase is an important determinant of seed longevity. *Plant J.* **2010**, *63*, 848–860. [CrossRef]
211. Vonarx, E.J.; Mitchell, H.L.; Karthikeyan, R.; Chatterjee, I.; Kunz, B.A. DNA repair in higher plants. *Mutat. Res.* **1998**, *400*, 187–200. [CrossRef]
212. El-Maarouf-Bouteau, H.; Meimoun, P.; Job, C.; Job, D.; Bailly, C. Role of protein and mRNA oxidation in seed dormancy and germination. *Front. Plant Sci.* **2013**, *4*, 77. [CrossRef]
213. Bailly, C.; Benamar, A.; Corbineau, F.; Côme, D. Antioxidant systems in sunflower (*Helianthus annuus* L.) seeds as affected by priming. *Seed Sci. Res.* **2000**, *10*, 35–42. [CrossRef]
214. Hsu, C.C.; Chen, C.L.; Chen, J.J.; Sung, J.M. Accelerated aging-enhanced lipid peroxidation in bitter gourd seeds and effects of priming and hot water soaking treatments. *Sci. Hortic.* **2003**, *98*, 201–212. [CrossRef]
215. Farooq, M.; Basra, S.M.A.; Afzal, I.; Khaliq, A. Optimization of hydropriming techniques for rice seed invigoration. *Seed Sci. Technol.* **2006**, *34*, 507–512. [CrossRef]
216. Ghassemi-Golezani, K.; Esmaeilpour, B. The Effect of salt priming on the performance of differentially matured cucumber (*Cucumis sativus*) seeds. *Not. Bot. Hort. Agrobot.* **2008**, *36*, 67–70. [CrossRef]
217. Mirmazloum, I.; Kiss, A.; Erdélyi, É.; Ladányi, M.; Németh, É.Z.; Radácsi, P. The effect of osmopriming on seed germination and early seedling characteristics of *Carum carvi* L. *Agriculture* **2020**, *10*, 94. [CrossRef]
218. Rajjou, L.; Debeaujon, I. Seed longevity: Survival and maintenance of high germination ability of dry seeds. *Comptes Rendus-Biol.* **2008**, *331*, 796–805. [CrossRef]
219. Chandra, J.; Sershen; Varghese, B.; Keshavkant, S. The potential of ROS inhibitors and hydrated storage in improving the storability of recalcitrant *Madhuca latifolia* seeds. *Seed Sci. Technol.* **2019**, *47*, 33–45. [CrossRef]
220. Ashraf, M.A.; Akbar, A.; Askari, S.H.; Iqbal, M.; Rasheed, R.; Hussain, I. Recent advances in abiotic stress tolerance of plants through chemical priming: An overview. In *Advances in Seed Priming*; Rakshit, A., Singh, H.B., Eds.; Springer: Singapore, 2018; pp. 51–79. ISBN 9789811300325.
221. Varier, A.; Vari, A.K.; Dadlani, M. The subcellular basis of seed priming. *Curr. Sci.* **2010**, *99*, 450–456.
222. Cantliffe, D.J. Priming of lettuce seed for early and uniform emergence under conditions of environmental stress. *Acta Hortic.* **1981**, 29–38. [CrossRef]
223. Bradford, K.J.; Steiner, J.J.; Trawatha, S.E. Seed priming influence on germination and emergence of pepper seed lots. *Crop Sci.* **1990**, *30*, 718–721. [CrossRef]
224. Khan, A.A.; Peck, N.H.; Samimy, C. Seed osmoconditioning: Physiological and biochemical changes. *Isr. J. Bot.* **1980**, *29*, 133–144. [CrossRef]
225. Taylor, A.G.; Klein, D.E.; Whitlow, T.H. SMP: Solid matrix priming of seeds. *Sci. Hortic.* **1988**, *37*, 1–11. [CrossRef]
226. Mondal, S.; Vijai, P.; Bose, B. Role of seed hardening in rice variety swarna (MTU 7029). *Res. J. Seed Sci.* **2011**, *4*, 157–165. [CrossRef]
227. Matsushima, K.-I.; Sakagami, J.-I. Effects of seed hydropriming on germination and seedling vigour during emergence of rice under different soil moisture conditions. *Am. J. Plant Sci.* **2013**, *4*, 1584–1593. [CrossRef]
228. Forti, C.; Ottobrino, V.; Bassolino, L.; Toppino, L.; Rotino, G.L.; Pagano, A.; Macovei, A.; Balestrazzi, A. Molecular dynamics of pre-germinative metabolism in primed eggplant (*Solanum melongena* L.) seeds. *Hortic. Res.* **2020**, *7*, 87. [CrossRef]
229. Basra, S.M.A.; Farooq, M.; Tabassam, R.; Ahmad, N. Physiological and biochemical aspects of pre-sowing seed treatments in fine rice (*Oryza sativa* L.). *Seed Sci. Technol.* **2005**, *33*, 623–628. [CrossRef]
230. Farooq, M.; Basra, S.M.A.; Abid Karim, H.; Afzal, I. Optimization of seed hardening techniques for rice seed invigoration. *Emir. J. Food Agric.* **2004**, *16*, 48–58. [CrossRef]
231. Solaimalai, A.; Subburamu, K. Seed hardening for field crops—A review. *Agric. Rev.* **2004**, *25*, 129–140.

232. Akbar, M. Studies on Seed Priming and Fungicide in Pearl Millet under Dry Land Conditions. Ph.D. Thesis, NWFP Agricultural University, Peshawar, Pakistan, 2008.
233. Alvarado, A.D.; Bradford, K.J. Priming and storage of tomato (*Lycopersicon lycopersicum*) seeds. I. Effects of storage temperature on germination rate and viability. *Seed Sci. Technol.* **1988**, *16*, 601–612.
234. Bekendam, J.; van Pijlen, J.G.; Kraak, H.L. The effect of priming on the rate and uniformity of germination of endive seed. *Acta Hortic.* **1987**, 209–218. [CrossRef]
235. Weges, R. Physiological Analysis of Methods to Relieve Dormancy of Lettuce Seeds. Ph.D. Thesis, Wageningen Agricultural University, Wageningen, The Netherlands, 1987.
236. Tarquis, A.M.; Bradford, K.J. Prehydration and priming treatments that advance germination also increase the rate of deterioration of lettuce seeds. *J. Exp. Bot.* **1992**, *43*, 307–317. [CrossRef]
237. van Pijlen, J.G.; Groot, S.P.C.; Kraak, H.L.; Bergervoet, J.H.W.; Bino, R.J. Effects of pre-storage hydration treatments on germination performance, moisture content, DNA synthesis and controlled deterioration tolerance of tomato (*Lycopersicon esculentum* Mill.) seeds. *Seed Sci. Res.* **1996**, *6*, 57–63. [CrossRef]
238. Brancalion, P.H.S.; Novembre, A.D.L.C.; Rodrigues, R.R.; Tay, D. Priming of *Mimosa bimucronata* seeds—A tropical tree species from Brazil. *Acta Hortic.* **2008**, *782*, 163–168. [CrossRef]
239. Mahakham, W.; Sarmah, A.K.; Maensiri, S.; Theerakulpisut, P. Nanopriming technology for enhancing germination and starch metabolism of aged rice seeds using phytosynthesized silver nanoparticles. *Sci. Rep.* **2017**, *7*, 8263. [CrossRef] [PubMed]
240. de Sousa Araújo, S.; Paparella, S.; Dondi, D.; Bentivoglio, A.; Carbonera, D.; Balestrazzi, A. Physical methods for seed invigoration: Advantages and challenges in seed technology. *Front. Plant Sci.* **2016**, *7*, 646. [CrossRef]
241. Harris, D.; Raghuwanshi, B.S.; Gangwar, J.S.; Singh, S.C.; Joshi, K.D.; Rashid, A.; Hollington, P.A. Participatory evaluation by farmers of on-farm seed priming in wheat in India, Nepal and Pakistan. *Exp. Agric.* **2001**, *37*, 403–415. [CrossRef]
242. Taylor, A.G.; Allen, P.S.; Bennett, M.A.; Bradford, K.J.; Burris, J.S.; Misra, M.K. Seed enhancements. *Seed Sci. Res.* **1998**, *8*, 245–256. [CrossRef]
243. Kaur, S.; Gupta, A.K.; Kaur, N. Effect of osmo- and hydropriming of chickpea seeds on seedling growth and carbohydrate metabolism under water deficit stress. *Plant Growth Regul.* **2002**, *37*, 17–22. [CrossRef]
244. Yan, M. Hydropriming promotes germination of aged napa cabbage seeds. *Seed Sci. Technol.* **2015**, *43*, 303–307. [CrossRef]
245. Bennett, M.A.; Fritz, V.A.; Callan, N.W. Impact of seed treatments on crop stand establishment. *Horttechnology* **2018**, *2*, 345–349. [CrossRef]
246. Adetunji, A.E.; Sershen; Varghese, B.; Pammenter, N.W. Effects of inorganic salt solutions on vigour, viability, oxidative metabolism and germination enzymes in aged cabbage and lettuce seeds. *Plants* **2020**, *9*, 1164. [CrossRef]
247. Singh, A.; Dahiru, R.; Musa, M.; Sani Haliru, B. Effect of osmopriming duration on germination, emergence, and early growth of cowpea (*Vigna unguiculata* (L.) Walp.) in the sudan savanna of Nigeria. *Int. J. Agron.* **2014**, *2014*, 1–4. [CrossRef]
248. Fatokun, K.; Beckett, R.P.; Varghese, B.; Cloete, J.; Pammenter, N.W. Influence of cathodic water invigoration on the emergence and subsequent growth of controlled deteriorated pea and pumpkin seeds. *Plants* **2020**, *9*, 955. [CrossRef]
249. Batool, A.; Ziaf, K.; Amjad, M. Effect of halo-priming on germination and vigor index of cabbage (*Brassica oleracea* var. capitata). *J. Environ. Agric. Sci.* **2015**, *2*, 7.
250. Abdolahi, M.; Andelibi, B.; Zangani, E.; Shekari, F.; Jamaati-E-Somarin, S. Effect of accelerated aging and priming on seed germination of rapeseed (*Brassica napus* L.) cultivars. *Int. Res. J. Appl. Basic Sci.* **2012**, *3*, 499–508.
251. Carrozzi, L.E.; Creus, C.M.; Barassi, C.A.; Monterubbianesi, G.; Di Benedetto, A. Reparation of aged lettuce (*Lactuca sativa*) seeds by osmotic priming and Azospirillum brasilense inoculation. *Botany* **2012**, *90*, 1093–1102. [CrossRef]
252. Moradi, A.; Younesi, O. Effects of osmo- and hydro-priming on seed parameters of grain sorghum (*Sorghum bicolor* L.). *Aust. J. Basic Appl. Sci.* **2009**, *3*, 1696–1700.
253. Gupta, D.K.; Palma, J.M.; Corpas, F.J. *Redox State as a Central Regulator of Plant-Cell Stress Responses*; Gupta, D.K., Palma, J.M., Corpas, F.J., Eds.; Springer International Publishing: Cham, Switzerland, 2016; ISBN 978-3-319-44080-4.
254. Kumar, S.R.; Mohanapriya, G.; Sathishkumar, R. Abiotic stress-induced redox changes and programmed cell death in plants—A path to survival or death? In *Redox State as a Central Regulator of Plant-Cell Stress Responses*; Springer: Cham, Switzerland, 2016; pp. 233–252. ISBN 9783319440811.
255. Schafer, F.Q.; Buettner, G.R. Redox state and redox environment in biology. In *Signal Transduction by Reactive Oxygen and Nitrogen Species: Pathways and Chemical Principles*; Kluwer Academic Publishers: Dordrecht, The Netherlands, 2006; pp. 1–14.
256. Miller, G.; Suzuki, N.; Ciftci-Yilmaz, S.; Mittler, R. Reactive oxygen species homeostasis and signalling during drought and salinity stresses. *Plant Cell Environ.* **2010**, *33*, 453–467. [CrossRef]
257. Draganić, I.; Lekić, S. Seed priming with antioxidants improves sunflower seed germination and seedling growth under unfavorable germination conditions. *Turkish J. Agric. For.* **2012**, *36*, 421–428. [CrossRef]
258. Adetunji, A.E.; Sershen; Varghese, B.; Pammenter, N. Effects of exogenous application of five antioxidants on vigour, viability, oxidative metabolism and germination enzymes in aged cabbage and lettuce seeds. *S. Afr. J. Bot.* **2021**, *137*, 85–97. [CrossRef]
259. Adetunji, A.E.; Sershen; Varghese, B.; Pammenter, N.W. Exogenous antioxidants enhance seedling growth and yield of artificially aged cabbage and lettuce seeds. *Horticulturae* **2021**, *7*, 274. [CrossRef]
260. Burguieres, E.; McCue, P.; Kwon, Y.I.; Shetty, K. Effect of vitamin C and folic acid on seed vigour response and phenolic-linked antioxidant activity. *Bioresour. Technol.* **2007**, *98*, 1393–1404. [CrossRef]

261. Yan, H.-F.; Mao, P.-S.; Sun, Y.; Li, M.-L. Impacts of ascorbic acid on germination, antioxidant enzymes and ultrastructure of embryo cells of aged *Elymus sibiricus* seeds with different moisture contents. *Int. J. Agric. Biol.* **2016**, *18*, 176–183. [CrossRef]
262. Afzal, I.; Basra, S.M.A.; Hameed, A.; Farooq, M. Physiological enhancements for alleviation of salt stress in wheat. *Pakistan J. Bot.* **2006**, *38*, 1649–1659.
263. Bhattacharjee, A.; Gupta, K. Effect of dikegulac-sodium, a growth retardant, on the viability of sunflower seeds. *Seed Sci. Technol.* **1985**, *13*, 165–174.
264. Dey, G.; Mukherjee, R.K. Invigoration of dry seeds with physiologically active chemicals in organic solvents. *Seed Sci. Technol.* **1988**, *16*, 145–153.
265. Bhattacharjee, A.; Bhattacharyya, R.N. Prolongation of seed viability of *Oryza sativa* L. *Seed Sci. Technol.* **1989**, *17*, 309–316.
266. Ugena, L.; Hýlová, A.; Podlešáková, K.; Humplík, J.F.; Doležal, K.; De Diego, N.; Spíchal, L. Characterization of biostimulant mode of action using novel multi-trait high-throughput screening of arabidopsis germination and rosette growth. *Front. Plant Sci.* **2018**, *9*, 1327. [CrossRef]
267. Rouphael, Y.; Colla, G. Editorial: Biostimulants in agriculture. *Front. Plant Sci.* **2020**, *11*, 40. [CrossRef]
268. Sreekissoon, A.; Finnie, J.F.; Van Staden, J. Effects of smoke water on germination, seedling vigour and growth of Sceletium tortuosum. *S. Afr. J. Bot.* **2021**, *139*, 427–431. [CrossRef]
269. Gupta, G.P. Role of global climate change in crop yield reductions. In *Journal of the Air Pollution Control Association*; Saxena, P., Srivastava, A., Eds.; Springer: Singapore, 2020; Volume 13, pp. 87–113. ISBN 9789811534805.
270. Zhou, J.; Teixeira da Silva, J.; Ma, G. Effects of smoke water and karrikin on seed germination of 13 species growing in China. *Open Life Sci.* **2014**, *9*, 1108–1116. [CrossRef]
271. Makhaye, G.; Aremu, A.O.; Gerrano, A.S.; Tesfay, S.; Du Plooy, C.P.; Amoo, S.O. Biopriming with seaweed extract and microbial-based commercial biostimulants influences seed germination of five *Abelmoschus esculentus* genotypes. *Plants* **2021**, *10*, 1327. [CrossRef] [PubMed]
272. Masondo, N.A.; Kulkarni, M.G.; Finnie, J.F.; Van Staden, J. Influence of biostimulants-seed-priming on *Ceratotheca triloba* germination and seedling growth under low temperatures, low osmotic potential and salinity stress. *Ecotoxicol. Environ. Saf.* **2018**, *147*, 43–48. [CrossRef]
273. Johnson, R.; Puthur, J.T. Biostimulant priming in *Oryza sativa*: A novel approach to reprogram the functional biology under nutrient-deficient soil. *Cereal Res. Commun.* **2021**, 1–8. [CrossRef]
274. Parisi, C.; Vigani, M.; Rodríguez-Cerezo, E. Agricultural nanotechnologies: What are the current possibilities? *Nano Today* **2015**, *10*, 124–127. [CrossRef]
275. Servin, A.; Elmer, W.; Mukherjee, A.; De la Torre-Roche, R.; Hamdi, H.; White, J.C.; Bindraban, P.; Dimkpa, C. A review of the use of engineered nanomaterials to suppress plant disease and enhance crop yield. *J. Nanoparticle Res.* **2015**, *17*, 92. [CrossRef]
276. Fraceto, L.F.; Grillo, R.; de Medeiros, G.A.; Scognamiglio, V.; Rea, G.; Bartolucci, C. Nanotechnology in agriculture: Which innovation potential does it have? *Front. Environ. Sci.* **2016**, *4*, 20. [CrossRef]
277. Sundaria, N.; Singh, M.; Upreti, P.; Chauhan, R.P.; Jaiswal, J.P.; Kumar, A. Seed priming with iron oxide nanoparticles triggers iron acquisition and biofortification in wheat (*Triticum aestivum* L.) grains. *J. Plant Growth Regul.* **2019**, *38*, 122–131. [CrossRef]
278. Hirota, N.; Nakagawa, J.; Kitazawa, K. Effects of a magnetic field on the germination of plants. *J. Appl. Phys.* **1999**, *85*, 5717–5719. [CrossRef]
279. Aladjadjiyan, A. Influence of stationary magnetic field on lentil seeds. *Int. Agrophys.* **2010**, *24*, 321–324.
280. Teixeira da Silva, J.A.; Dobránszki, J. Magnetic fields: How is plant growth and development impacted? *Protoplasma* **2016**, *253*, 231–248. [CrossRef] [PubMed]
281. Bilalis, D.J.; Katsenios, N.; Efthimiadou, A.; Karkanis, A.; Efthimiadis, P. Investigation of pulsed electromagnetic field as a novel organic pre-sowing method on germination and initial growth stages of cotton. *Electromagn. Biol. Med.* **2012**, *31*, 143–150. [CrossRef]
282. Baby, S.M.; Narayanaswamy, G.K.; Anand, A. Superoxide radical production and performance index of photosystem II in leaves from magnetoprimed soybean seeds. *Plant Signal. Behav.* **2011**, *6*, 1635–1637. [CrossRef] [PubMed]
283. Vashisth, A.; Nagarajan, S. Effect on germination and early growth characteristics in sunflower (*Helianthus annuus*) seeds exposed to static magnetic field. *J. Plant Physiol.* **2010**, *167*, 149–156. [CrossRef]
284. Marcu, D.; Cristea, V.; Daraban, L. Dose-dependent effects of gamma radiation on lettuce (*Lactuca sativa* var. capitata) seedlings. *Int. J. Radiat. Biol.* **2013**, *89*, 219–223. [CrossRef] [PubMed]
285. Ouhibi, C.; Attia, H.; Rebah, F.; Msilini, N.; Chebbi, M.; Aarrouf, J.; Urban, L.; Lachaal, M. Salt stress mitigation by seed priming with UV-C in lettuce plants: Growth, antioxidant activity and phenolic compounds. *Plant Physiol. Biochem.* **2014**, *83*, 126–133. [CrossRef] [PubMed]
286. Thomas, D.T.; Puthur, J.T. Amplification of abiotic stress tolerance potential in rice seedlings with a low dose of UV-B seed priming. *Funct. Plant Biol.* **2019**, *46*, 455–466. [CrossRef]
287. Al-Enezi, N.A.; Al-Bahrany, A.M.; Al-Khayri, J.M. Effect of X-irradiation on date palm seed germination and seedling growth. *Emir. J. Food Agric.* **2012**, *24*, 415–424.
288. De Micco, V.; Paradiso, R.; Aronne, G.; De Pascale, S.; Quarto, M.; Arena, C. Leaf anatomy and photochemical behaviour of *Solanum lycopersicum* L. Plants from seeds irradiated with low-LET ionising radiation. *Sci. World J.* **2014**, *2014*. [CrossRef]

289. Randhir, R.; Shetty, K. Microwave-induced stimulation of L-DOPA, phenolics and antioxidant activity in fava bean (*Vicia faba*) for Parkinson's diet. *Process Biochem.* **2004**, *39*, 1775–1784. [CrossRef]
290. Han, F. The effect of microwave treatment on germination, vigour and health of China aster (*Callistephus chinensis* Nees.) seeds. *J. Agric. Sci.* **2010**, *2*, 201–210. [CrossRef]
291. UN General Assembly. *Transforming Our World: The 2030 Agenda for Sustainable Development*; UN: New York, NY, USA, 2015; Report No. A/RES/70/1.

agronomy

Essay

Morphological and Physiological Responses of *Melia azedarach* Seedlings of Different Provenances to Drought Stress

Chao Han, Junna Chen, Zemao Liu, Hong Chen, Fangyuan Yu * and Wanwen Yu *

Collaborative Innovation Centre of Sustainable Forestry in Southern China, College of Forest Science, Nanjing Forestry University (NJFU), 159 Longpan Road, Nanjing 210037, China; hanc0909@njfu.edu.cn (C.H.); chenjn@njfu.edu.cn (J.C.); 727046890@njfu.edu.cn (Z.L.); hongchen@njfu.edu.cn (H.C.)
* Correspondence: fyyu@njfu.edu.cn (F.Y.); cc0212@njfu.edu.cn (W.Y.)

Abstract: *Melia azedarach* Linn. is a deciduous tree of the Melia genus in the Meliaceae family that is native to China. To study the mechanism of drought resistance in *Melia azedarach* and evaluate the drought resistance capacity of each provenance, we selected eight provenances (Shandong Kenli, Jiangsu Pizhou, Hubei Shayang, Jiangsu Xuanwu, Jiangxi Xihu, Jiangsu Jurong, Guangdong Luogang, and Henan Shihe) as the research subjects and set four levels of drought stress treatment (CK: 75% of field capacity, mild drought: 60% of field capacity, moderate drought: 45% of field capacity, and severe drought: 30% of field capacity). The results showed that the growth in the seedling height and the ground diameter, the leaf relative water content, transpiration rate (Tr), net photosynthetic rate (Pn), stomatal conductance (Gs), and the content of chlorophyll (Chl) decreased with the increasing stress levels, while the root–shoot ratio, water saturation deficit, and the contents of malondialdehyde (MDA) increased. The SOD in most provenances initially increased and then decreased, reaching a peak during moderate drought. At the late stage of treatment, the magnitude of the changes in the photosynthetic indicators was more pronounced than in the physiological indicators. Principal component analysis showed that the contribution of all four principal components under the three drought stresses was above 85%, which represented the majority of the original data. Combined with the affiliation function method and weights, the comprehensive evaluation value (D value) of the drought resistance was calculated for the eight provenances. Then, we obtained the order of drought resistance of the test materials under the three drought stresses, respectively. The combined results revealed that the drought resistance of Henan Shihe and Jiangxi Xihu was stronger, while the drought resistance of Guangdong Luogang and Hubei Shayang was weaker. Based on the above findings, we can select provenances with strong and weak drought resistance for transcriptome sequencing to screen drought-resistant genes for an in-depth study at the molecular level.

Keywords: *Melia azedarach*; provenance; water deficit; physiology response; comprehensive evaluation

Citation: Han, C.; Chen, J.; Liu, Z.; Chen, H.; Yu, F.; Yu, W. Morphological and Physiological Responses of *Melia azedarach* Seedlings of Different Provenances to Drought Stress. *Agronomy* **2022**, *12*, 1461. https://doi.org/10.3390/agronomy12061461

Academic Editors: Sara Álvarez and José Ramón Acosta-Motos

Received: 30 May 2022
Accepted: 13 June 2022
Published: 17 June 2022

1. Introduction

Water is not only the source of all life, but also the material for plant growth, since it affects the growth of plants and is an important component of the plant itself [1]. Drought stress in plants results from an insufficient amount of water available for the maintenance of normal physiological processes, such as photosynthesis, respiration, and cell, tissue, organ, and plant homeostasis [2]. According to most scientific sources, drylands account for 41% of the world's land area [3]. China is one of the countries with the highest frequency of drought; arid and semiarid areas account for 47% of China's territorial area, which threatens the sustainable development of agroforestry [4,5].

Melia azedarach [6] is a deciduous tree of the Melia genus, which also commonly known as the purple flower tree, forest tree, and golden Lingzi. It is a fast-growing and high-quality timber tree; it is also a good nectar plant and a vital plant pesticide. The timber, which resembles mahogany, is used to manufacture agricultural implements, furniture, plywood, etc. *Melia azedarach* is also of value for the health care and pharmaceutical industries, an effective composition due to its analgesic, anticancer, antiviral, antimalarial, antibacterial, antifeedent, and antifertility activity [7]. Furthermore, it is an important afforestation tree species, as are the surrounding greening tree species.

Melia azedarach is widely distributed. It is native to tropical Asia and has been introduced to the Philippines, the United States of America, Brazil, Argentina, many African countries, and many Arab countries [8]. In China, it is concentrated in the south and southwest, with a relatively concentrated distribution in the east and central regions, and a marginal distribution area in the north, southwest, and southern Shanxi and Gansu [9]. For this reason, *Melia azedarach*, as a tree native to China, has diverse provenances.

Currently, studies on *Melia azedarach* include the phenology, reproduction, cultivation, seeding, resistance, pharmacological action, comprehensive utilization, and the chemical constituents in the bark, leaf, and fruit of *Melia azedarach* [10–12], among which the resistance studies include low temperature stress, salt stress, heavy metal stress, water stress, etc. ROS production is a key characteristic of plant responses to various abiotic stresses, which leads to oxidative stress damage in plants. ROS can directly attack the membrane lipids, inactivate enzymes, and damage the nucleic acids leading, in some cases, to cell death [13,14]. As protection against ROS, plants have evolved an efficient defense system, with antioxidant enzymes and nonenzymatic compounds that can neutralize free radicals and reduce the potential damage of ROS. It was found that prolonged drought in *Melia azedarach* was very detrimental to its growth and the management of plantation forests [15]. It has been found that water stress induced stomatal closure, reduced net CO_2 assimilation rate, and the intercellular CO_2 in *Melia azedarach*, as well as the photosynthetic efficiency of PSII but not the pigment levels [16]. Water stress also upregulated the antioxidant enzymes and stimulated the production of antioxidant metabolites, preventing lipid peroxidation [16]. The study shows that *Melia azedarach* has the potential to acclimate to water shortage conditions. Although *Melia azedarach* is less distributed in the arid and semiarid regions of China, as a tree species that has been described as a "global problem solver", we need to consider how it responds when grown in an arid or semiarid region, and how the response varies from one provenance to another.

Thus, we selected eight provenances and studied the drought resistance mechanism of *Melia azedarach* from the aspects of growth, physiology, and biochemistry. The drought resistance capacity of these eight provenances was evaluated and compared according to principal component analysis, the affiliation function, and the weights. In the following paragraphs, KL, PZ, SY, XW, XH, JR, LG, and SH represent Shandong Kenli, Jiangsu Pizhou, Hubei Shayang, Jiangsu Xuanwu, Jiangxi Xihu, Jiangsu Jurong, Guangdong Luogang, and Henan Shihe, respectively.

2. Materials and Methods

2.1. Plant Material and Experimental Design

Seeds were collected from the different provenances (Table 1) in 2017 and sown in March 2018 at Hefeng Nursery in the Lishui district, Nanjing, Jiangsu province, China. On 18 March 2019, seedlings from each locality with relatively average growth were transplanted to Nanjing Forest University's Xiashu forest farm in Zhenjing, Jiangsu province (32°7′29″ N, 119°13′9″ E, elevation 20 m). This location has a north subtropical monsoon climate, with an annual average temperature of 15.2 °C, an average frost-free period of 233 days, an annual average precipitation of 1055.6 mm, and an annual average relative humidity of 79%, all of which contribute to a favorable natural environment for tree growth. In total, 260 seedlings were transplanted, each of which were grown in plastic culture pots (top diameter of 19 cm, bottom diameter of 14 cm, and a height of 17 cm) filled with a substrate weighing 3 kg ($V_{yellow\ mud}$:$V_{bog\ soil}$:$V_{perlite}$ = 5:2:1) and a piece of gauze placed on the bottom of the pot. To ensure the growth and consistency of the seedlings, proper root pruning and cutting treatments were carried out. We plucked off the undesirable shoots until mid-April. After slow seedling management, the seedlings with relatively consistent growth were treated with 4 levels of drought stress, and each treatment contained 4 pots.

Table 1. Location and climatic conditions of the *Melia azedarach.* L seed collection sites in 2017.

Original Area	Longitude	Latitude	Altitude/m	Annual Average Temperature/°C	Annual Precipitation/mm	Frost-Free Season/d
KL	118.58	37.42	3~5	12.8	555.9	206
PZ	117.99	34.35	18~32	14.2	868.6	210
SY	112.18	30.37	30~55	15.9	1250	255
XW	118.83	32.09	12~20	15.4	1106.5	237
XH	115.87	28.67	55~112	17.6	1458	277
JR	119.13	32.07	5~30	15.2	1055.6	233
LG	113.25	23.11	15~22	21	1720	190
SH	114.01	31.82	60~150	15.3	1100	225

Note: KL, PZ, SY, XW, XH, JR, LG, and SH represent Shandong Kenli, Jiangsu Pizhou, Hubei Shayang, Jiangsu Xuanwu, Jiangxi Xihu, Jiangsu Jurong, Guangdong Luogang, and Henan Shihe, respectively.

This experiment used the artificial water control method for potted plants, and four drought stress gradients were set in this experiment: control treatment (soil water content was 75% of field capacity), mild drought stress (soil water content was 60% of field capacity), moderate drought stress (soil water content was 45% of field capacity), and severe drought stress (soil water content was 30% of field capacity). The soil field capacity of the potting substrate was 32.5%, based on which the water weight under four drought stress treatments was calculated accordingly. Each pot was weighed individually beginning 22 July 2019. Until 2 August, the water content of each pot remained essentially constant within the intended parameters. After that, every other day at 3 p.m., an electronic scale was used to weigh each pot to replace the lost water and keep the water content within the predetermined range.

Drought stress treatment terminated on 26 August, and samples were taken on 2 August, 12 August, and 22 August, respectively. On branches in the middle of *Melia azedarach*, we chose leaves with essentially the same growth condition. Mixed samples were taken from four plants in each treatment, which were then separated into Ziploc bags by locality and treatment. All of the samples were put in an ice box and returned to the lab. Part of the fresh leaves was used for the indexes that needed to be determined immediately, while the rest was kept at −80 °C in an ultra-low temperature refrigerator. Three replicates were performed for each treatment from each provenance to determine the indicators.

2.2. Determination of Indexes

2.2.1. Determination of Field Water Capacity

The ring knife method was used to evaluate the field capacity (Sun et al., 2011) [17].

2.2.2. Measurement of Seedling Height and Ground Diameter

Four plants were chosen from each treatment of each provenance. On 2 August, the seedling height was measured as H0 with a tape measure, and the ground diameter was measured as G0 with a Vernier caliper. On 22 August, it was recorded as H1 and G1, for a seedling height growth of $H = H1 - H0$ and a ground diameter growth of $G = G1 - G0$.

2.2.3. Determination of Biomass

Three pots were chosen at random for destructive sampling of each provenance according to treatment at the end of the drought stress. We cleaned and dried the roots, weighed the roots, leaves, branches, and stems using electronic scales, and put them into envelopes with labels. They were heated for 30 min at 105 °C, then dried to a constant weight in an oven at 80 °C, and weighed individually.

2.2.4. Leaf Relative Water Content and Leaf Water Saturation Deficit

Leaf relative water content and leaf water saturation deficit were measured by the drying weighing method [18]. We took the fresh leaf, weighed its fresh weight, put the leaf in distilled water, and soaked it until saturation. Then, we put the leaves into the oven at 105 °C for 30 min; thereafter, we dried them at 80 °C to a constant weight and weighed their dry weight.

2.2.5. MDA Content

The thiobarbituric acid method [19] was used, with some modifications, with the weight of the fresh leaf sample, 0.2 g in this experiment, and centrifugation speed, 4000 r/min in this experiment.

2.2.6. Superoxide Dismutase Activity

The NBT photochemical reduction method was used [19].

First, 50% of the NBT photochemical reduction was expressed as an activity unit (U) of SOD enzyme; then, the SOD activity was calculated as follows:

$$\text{SOD total activity}\left(\text{U}\cdot\text{g}^{-1}\right) = \frac{(A_{\text{Ck}} - A_{\text{E}})\cdot V_{\text{T}}}{\frac{1}{2}\cdot A_{\text{Ck}}\cdot m\cdot Vs}$$

The total SOD activity was expressed in units of enzyme per gram of fresh weight of sample ($\text{U}\cdot\text{g}^{-1}$). A_{CK} is the absorbance of illumination pair care, A_{E} is the absorbance of the sample tube, V_{T} is the total volume of the sample solution (mL), V_{S} is the amount of the sample at the time of determination (mL), m is the fresh mass of the sample (g), 1/2 is the conversion factor for one unit of enzyme defined (50% inhibition of NBT photoreduction).

2.2.7. Chlorophyll Content

The determination of the chlorophyll content was slightly modified from X-K Wang's ethanol extraction method (We eliminated the grinding step and soaked the leaves in 95% ethanol protected from light until they turned white) [19]. The chlorophyll content was calculated according to the following formula:

$$\text{Ca} = 13.95\ \text{A}_{665} - 6.88\ \text{A}_{649}$$

$$\text{Cb} = 24.96\ \text{A}_{649} - 7.32\ \text{A}_{665}$$

$$\text{Chloroplast pigment content (mg/g)} = \frac{C\cdot V\cdot N}{m\cdot 1000}$$

C is the pigment content (mg/L); V is the volume of extracted liquid (mL); N is the dilution ratio; m is the sample mass (g); and 1000 means 1 L equals 1000 mL.

2.2.8. Photosynthetic Indexes

A CIRAS-3 photosynthesizer manufactured by pp Systems LTD in the UK was used. The light intensity and CO_2 concentration were set to 500 μmol m^{-2} s^{-1} and 400 $\pm$ 20 μmol mol^{-1}, respectively. Mature leaves in the middle of the plant with similar sizes were selected to determine the intercellular carbon dioxide concentration (*Ci*), net photosynthetic rate (*Pn*), transpiration rate (*Tr*), and stomatal conductance (*Gs*) at three periods (before, during, and after treatment), respectively. For each treatment, three plant pots were selected at random.

2.3. Data Statistics and Analysis

Data analysis was performed by SPSS 25. Data analysis between the different provenances was tested by one-way ANOVA, and the significance of differences was obtained by Duncan's test. The presence of outliers was judged by box plots, and the Shapiro-Wilk normality test was used. when the assumption of variance chi-squared was not satisfied, welch ANOVA was used. The graphs in the text were created using origin pro.

2.3.1. Principal Component Analysis

The data for performing principal component statistical analysis were the ratio of the measured values under the three drought stresses to the measured values of the control (drought tolerance coefficient).

The Kaiser–Meyer–Olkin (KMO) and Bartlett tests were first performed on the data, and the principal components were extracted according to the cumulative contribution rate greater than 85% or the eigenvalue greater than 1 for the principal component analysis to determine the composite indexes. The value of each composite indicator was calculated according to the following formula:

$$Y_{ik} = \sum_{j=1}^{n} X_{ij} \times f_{jk}$$

In the formula: Y_{ik} is the value of the kth composite index of the ith provenance, X_{ij} is the drought tolerance coefficient of the jth index of the ith provenance, and f_{jk} is the score coefficient of the jth index of the kth composite index, which is the corresponding eigenvector, calculated by the following formula:

$$f_j = \frac{b_{jk}}{\sqrt{\lambda k}}$$

In the formula: b_{jk} is the loading coefficient of the jth indicator in the kth principal component, and λk is the kth principal component of the eigenvalues.

2.3.2. Calculation of Affiliation Function Values

The value of the affiliation function for each composite indicator per provenance was obtained by the following formula:

$$U(Y_{ik}) = (Y_{ik} - X_{kmin})/(X_{kmax} - X_{kmin})$$

In the formula: Y_{ik} is the value of the kth composite index of the ith provenance, X_{kmin} is the minimum value of the kth composite index, and X_{kmax} is the maximum value of the kth composite index.

2.3.3. Comprehensive Evaluation of Drought Resistance

First, we calculated the weight of each composite index affiliation function value according to the following formula:

$$w_k = \frac{p_k}{\sum_{k=1}^{n} P_k} \quad k = 1, 2, \ldots \ldots n$$

In the formula: W_k denotes the importance of the kth composite indicator among all composite indicators; P_k is the contribution rate of the kth composite indicator of each provenance.

The degree of drought resistance by provenance is indicated by D:

$$\mathrm{D} = \sum_{k=1}^{n} [U(Y_{ik}) \times w_k] \quad k = 1, 2, \ldots \ldots n$$

3. Results

3.1. Seedling Height Increment and Ground Diameter Increase

With the deepening of the drought stress and the extension of the stress time, the seedling height growth of the eight provenances declined to varing degrees (Table 2). The variation in the seedling height growth for all test materials ranged from 13.43 to 24.63 cm, 11.68 to 17.35 cm, 2.93 to 14.70 cm, and 1.63 to 9.40 cm under the control, mild drought, moderate drought, and severe drought, respectively. The difference in seedling height growth of the XH was not significant under all four treatments, but the other seven provenances were significantly different between the control and under severe drought ($p < 0.05$). In terms of the decline relative to the control group, that of the XW decreased less (32.17%, 51.88%) and that of the JR decreased more (83.22%, 91.74%) under moderate and severe drought.

Table 2. The seedling height growth of *Melia azedarach* among different provenances under drought stress.

Provenance	Seedling Height Growth/cm				Relative Decline I/%	Relative Decline II/%	Relative Decline III/%
	CK	I	II	III			
KL	22.80 ± 5.25 a	11.75 ± 4.25 ab	8.98 ± 2.53 b	3.65 ± 1.93 b	48.47%	60.64%	83.99%
PZ	19.00 ± 0.53 a	14.88 ± 1.13 a	2.93 ± 2.33 b	3.48 ± 1.14 b	21.71%	84.61%	81.71%
SY	24.63 ± 5.44 a	13.93 ± 3.94 ab	10.40 ± 2.52 b	4.18 ± 0.65 b	43.45%	57.77%	83.05%
XW	19.53 ± 0.95 a	14.50 ± 2.10 ab	13.25 ± 3.57 ab	9.40 ± 2.39 b	25.77%	32.17%	51.88%
XH	24.55 ± 0.05 a	17.35 ± 3.45 a	14.70 ± 0.80 a	9.00 ± 8.10 a	29.33%	40.12%	63.34%
JR	19.67 ± 6.63 a	13.08 ± 4.51 ab	3.30 ± 1.19 b	1.63 ± 0.31 b	33.52%	83.22%	91.74%
LG	15.93 ± 1.26 a	14.65 ± 4.01 a	6.65 ± 2.02 b	4.55 ± 2.20 b	8.01%	71.43%	58.24%
SH	13.43 ± 3.42 a	11.68 ± 1.66 a	7.23 ± 3.40 ab	1.78 ± 0.63 b	13.00%	46.00%	87.00%

Note: Different lowercase letters in the same line indicate a significant difference between different treatments of the same variety ($p < 0.05$). I: mild drought; II: moderate drought; III: severe drought.

The ground diameter increase in the eight provenances basically showed a decreasing trend, except for the KL, where the ground diameter growth under moderate drought stress increased slightly compared to the mild drought. The LG had the largest increase in ground diameter in the control group (2.74 cm), nearly double that of the other provenances. Meanwhile, as the stress increased, the decrease in the LG's ground diameter was the highest in all stress treatments relative to the control group (68%, 78%, and 88%) (Table 3).

Table 3. Ground diameter increase in the *Melia azedarach* seedlings among different provenances under drought stress.

Provenance	Diameter Growth/cm				Relative Decline I/%	Relative Decline II/%	Relative Decline III/%
	CK	I	II	III			
KL	1.02 ± 0.30 a	0.80 ± 0.10 a	0.97 ± 0.40 a	0.31 ± 0.10 a	22%	5%	70%
PZ	1.10 ± 0.08 a	0.40 ± 0.23 b	0.34 ± 0.05 b	0.19 ± 0.07 b	64%	69%	83%
SY	0.89 ± 0.13 a	0.76 ± 0.25 a	0.53 ± 0.30 a	0.28 ± 0.06 a	15%	40%	69%
XW	1.39 ± 0.32 a	0.89 ± 0.08 b	0.44 ± 0.05 c	0.30 ± 0.10 c	36%	69%	79%
XH	1.05 ± 0.07 a	0.64 ± 0.21 a	0.50 ± 0.16 a	0.52 ± 0.10 a	40%	53%	51%
JR	1.10 ± 0.21 a	0.52 ± 0.05 b	0.47 ± 0.13 b	0.23 ± 0.04 b	53%	57%	79%
LG	2.74 ± 1.18 a	0.89 ± 0.26 ab	0.59 ± 0.14 b	0.34 ± 0.09 b	68%	78%	88%
SH	1.28 ± 0.11 a	0.62 ± 0.13 b	0.40 ± 0.08 bc	0.20 ± 0.07 c	52%	69%	85%

Note: Different lowercase letters in the same line indicate a significant difference between different treatments of the same variety ($p < 0.05$). I: mild drought; II: moderate drought; III: severe drought.

3.2. The Biomass and Root–Crown Ratio

After the drought stress treatment, destructive sampling was used to determine the dry weight of the aboveground and underground parts, as well as the root–crown ratio of the seedlings. It can be shown that when the stress level increased, the aboveground biomass of the eight provenances declined, while the underground part exhibited a general rising trend. Under severe drought stress, the aboveground biomass of the JR, as well as the underground biomass of the SH, were considerably different from that of the control group ($p < 0.05$). In terms of the decrease in aboveground biomass, that of the XH had the highest decline (13.24%), while that of the SY had the lowest decline (2.32%). The SH had the largest increase (12.14%) in underground biomass, while the XW had the smallest increase in underground biomass (1.68%) (Table 4).

Table 4. The biomass distribution of the *Melia azedarach* seedlings among different provenances under drought stress.

Provenance	Biomass/g·Plant-1							
	The Aboveground Part/g·Plant-1				The Underground Part/g·Plant-1			
	CK	I	II	II	CK	I	II	III
KL	16.76 ± 3.04 a	16.59 ± 0.99 a	16.49 ± 3.14 a	11.98 ± 1.34 a	13.81 ± 3.79 a	14.51 ± 2.30 a	14.33 ± 3.02 a	16.27 ± 1.79 a
PZ	16.40 ± 1.47 a	13.94 ± 2.78 a	13.91 ± 2.73 a	12.14 ± 3.10 a	17.10 ± 3.03 a	17.56 ± 4.44 a	23.51 ± 2.68 a	21.98 ± 1.97 a
SY	17.03 ± 2.54 a	16.62 ± 3.04 a	15.32 ± 0.87 a	14.71 ± 0.39 a	10.47 ± 0.81 a	12.35 ± 0.32 a	13.67 ± 3.26 a	15.09 ± 1.64 a
XW	19.31 ± 5.24 a	15.76 ± 0.74 a	14.24 ± 2.43 a	13.49 ± 1.21 a	15.75 ± 0.61 a	16.61 ± 2.94 a	16.63 ± 2.39 a	17.43 ± 1.70 a
XH	27.49 ± 11.15 a	20.02 ± 2.71 a	16.37 ± 4.98 a	14.25 ± 0.18 a	12.74 ± 2.83 a	13.44 ± 2.57 a	16.11 ± 1.65 a	17.79 ± 6.65 a
JR	28.72 ± 0.98 a	25.62 ± 1.23 a	23.00 ± 2.47 ab	18.85 ± 0.17 b	17.33 ± 1.29 a	19.70 ± 2.05 a	20.04 ± 2.12 a	23.16 ± 0.56 a
LG	26.85 ± 6.07 a	26.40 ± 5.61 a	22.48 ± 0.05 a	21.13 ± 1.65 a	18.25 ± 3.56 a	13.91 ± 2.37 a	17.17 ± 3.78 a	21.00 ± 2.01 a
SH	21.26 ± 3.20 a	16.04 ± 1.01 a	15.58 ± 2.30 a	15.53 ± 3.16 a	11.41 ± 0.88 b	14.46 ± 1.44 b	15.18 ± 1.89 b	23.55 ± 1.86 a

Note: Different lowercase letters in the same line indicate a significant difference between different treatments of the same variety ($p < 0.05$). I: mild drought; II: moderate drought; III: severe drought.

The root–crown ratio of the eight provenances tended to rise in general. The severe drought stress in the JR differed significantly from the other treatments, which were similar to the severe drought stress in the SH and the control group. Numerically, the PZ had a larger root–crown ratio than the other seven provenances under any treatment, while the LG had a smaller root-crown ratio than the other provenances following drought stress (Table 5).

Table 5. The root–shoot ratio of the *Melia azedarach* seedlings among different provenances under drought stress.

Provenance	Root Crown Ratio			
	CK	I	II	III
KL	0.81 ± 0.08 a	0.89 ± 0.19 a	0.87 ± 0.02 a	1.39 ± 0.31 a
PZ	1.03 ± 0.09 a	1.38 ± 0.59 a	1.80 ± 0.55 a	1.89 ± 0.32 a
SY	0.62 ± 0.05 a	0.77 ± 0.12 a	0.92 ± 0.20 a	0.98 ± 0.05 a
XW	0.89 ± 0.27 a	1.16 ± 0.15 a	1.06 ± 0.01 a	1.31 ± 0.24 a
XH	0.60 ± 0.35 a	0.67 ± 0.04 a	1.05 ± 0.22 a	1.25 ± 0.48 a
JR	0.61 ± 0.07 b	0.77 ± 0.12 b	0.87 ± 0.00 b	1.23 ± 0.02 a
LG	0.76 ± 0.31 a	0.52 ± 0.02 a	0.81 ± 0.18 a	0.93 ± 0.02 a
SH	0.54 ± 0.18 b	0.90 ± 0.28 b	0.97 ± 0.19 ab	1.52 ± 0.25 a

Note: Different lowercase letters in the same line indicate a significant difference between different treatments of the same variety ($p < 0.05$). I: mild drought; II: moderate drought; III: severe drought.

3.3. Leaf Water Status

The relative water content of the seedling leaves decreased as the stress increased in the three sample periods (Figure 1). The average relative water content of the leaves of the JR was the highest in the early and middle stages of the experiment (67.64%, 83.06%) but the lowest in the late stage of stress (68.57%). The LG was also affected by this change, and the average value over time was only slightly lower than that of the JR. When compared to the control group, that of the PZ had the smallest fall in all three drought stress treatments, while that of the JR had the largest decrease under mild and moderate drought stress, and that of the KL had the largest decrease under severe drought stress.

With the escalation in stress, the leaf water saturation deficit of the eight provenances gradually rose (Figure 2). When compared to the control group, the increase in the water saturation deficit of the SY was always the largest under drought stress, while that of the KL was the lowest under mild and severe drought stress, and the PZ saturation deficit increased the least under moderate drought stress. Numerically, the water saturation deficit of the SH was higher in the late stage of stress (153.66%, 171.19%, 192.21%). The longer the duration of stress and the greater the degree of stress, the more significant differences were shown between the eight provenances ($p < 0.05$).

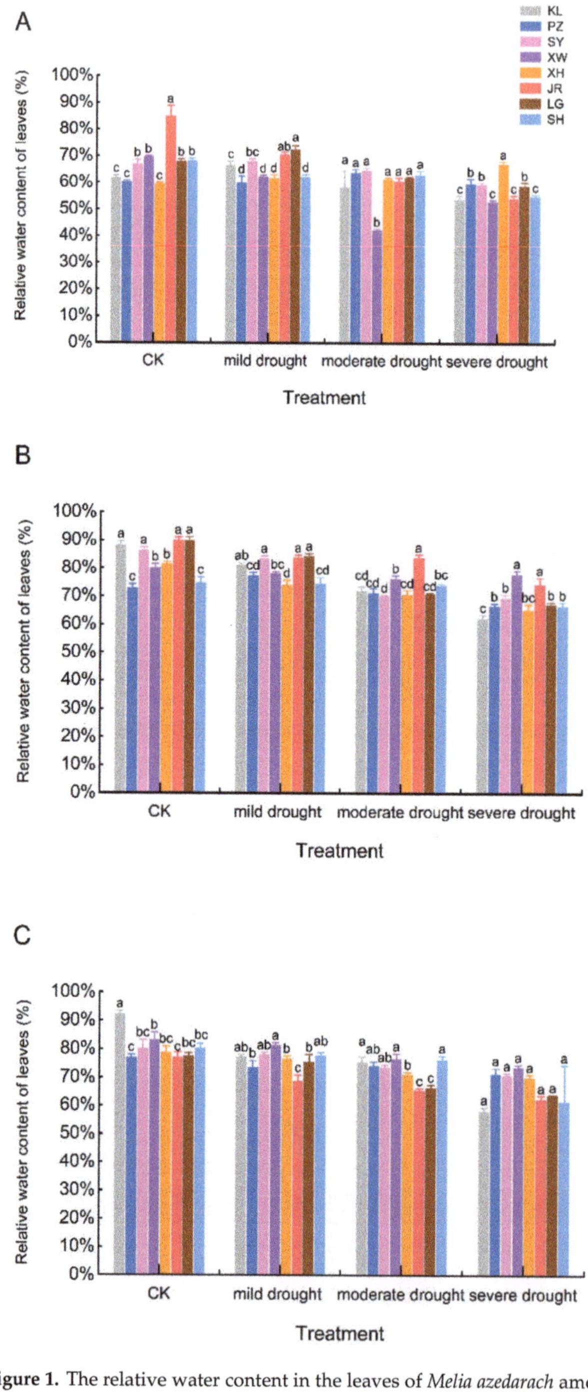

Figure 1. The relative water content in the leaves of *Melia azedarach* among different provenances: (**A**): early stage, (**B**): middle stage, and (**C**): late stage. Note: The error line in the figure is the mean ± standard error, and different lowercase letters in the same treatment indicate significant differences at the 0.05 level.

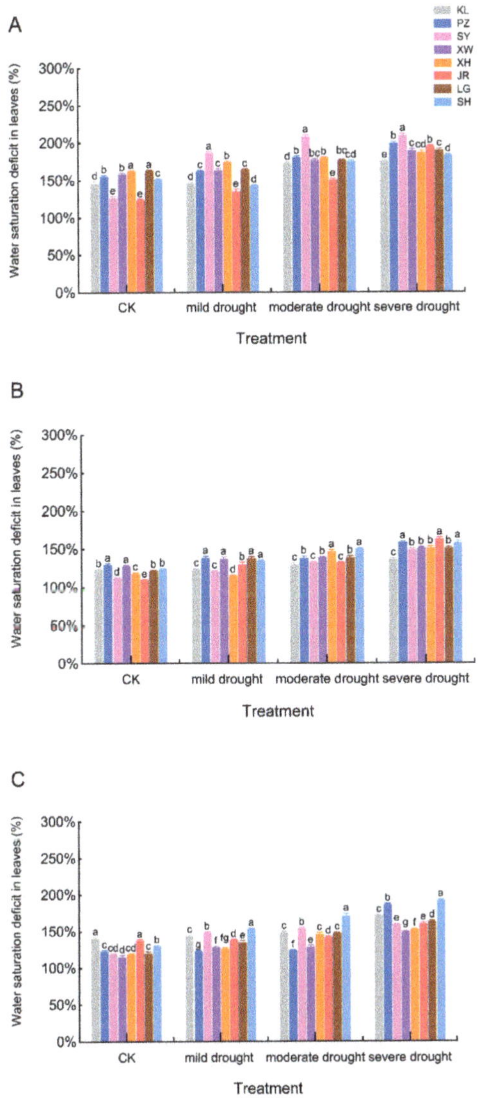

Figure 2. The water saturation deficit in the leaves of *Melia azedarach* among different provenances: (**A**): early stage, (**B**): middle stage, and (**C**): late stage. Note: The error line in the figure is the mean ± standard error, and different lowercase letters in the same treatment indicate significant differences at the 0.05 level.

3.4. Response of Photosynthesis and Photosynthetic Pigment

The changes in the photosynthetic parameters were distinct in each of the eight provenances. At each treatment level, the average *Pn* and *Tr* of the SH were relatively high throughout the experimental period. In the early stages of treatment, the *Pn* of the SY and JR under drought stress was significantly different from that of the control group. The *Pn* and *Tr* of the SH under severe drought stress were considerably different from those of the control group in the middle of the treatment, while the *Pn* and *Tr* of the XW under severe drought stress were likewise significantly different from those of the control group in the late stage of treatment ($p < 0.05$) (Figures A1 and A2). It is obvious that both the *Pn* and *Tr* of the SH were higher in the middle period of the stress (Figures 3 and 4).

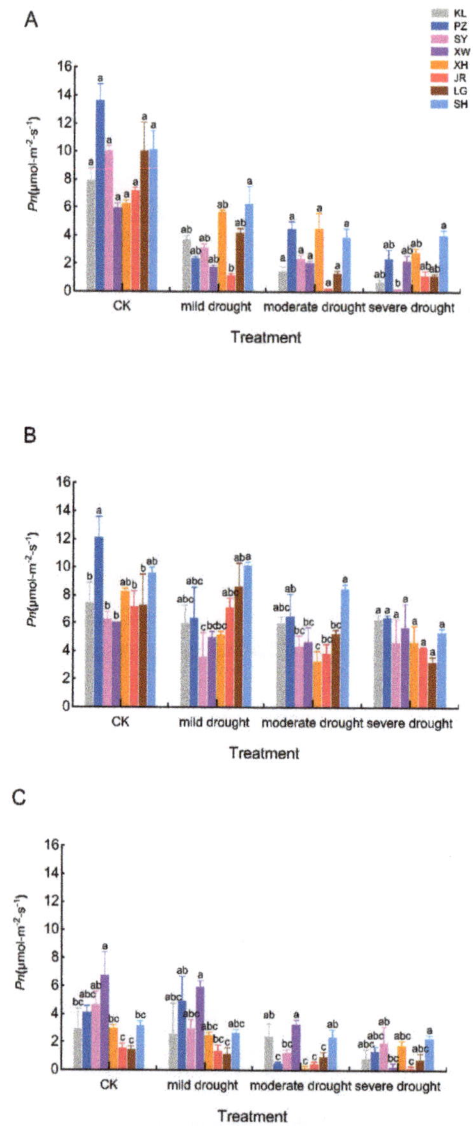

Figure 3. The net photosynthetic rate of *Melia azedarach* among different provenances: (**A**): early stage (**B**): middle stage, and (**C**): late stage. Note: The error line in the figure is the mean ± standard error, and different lowercase letters in the same treatment indicate significant differences at the 0.05 level.

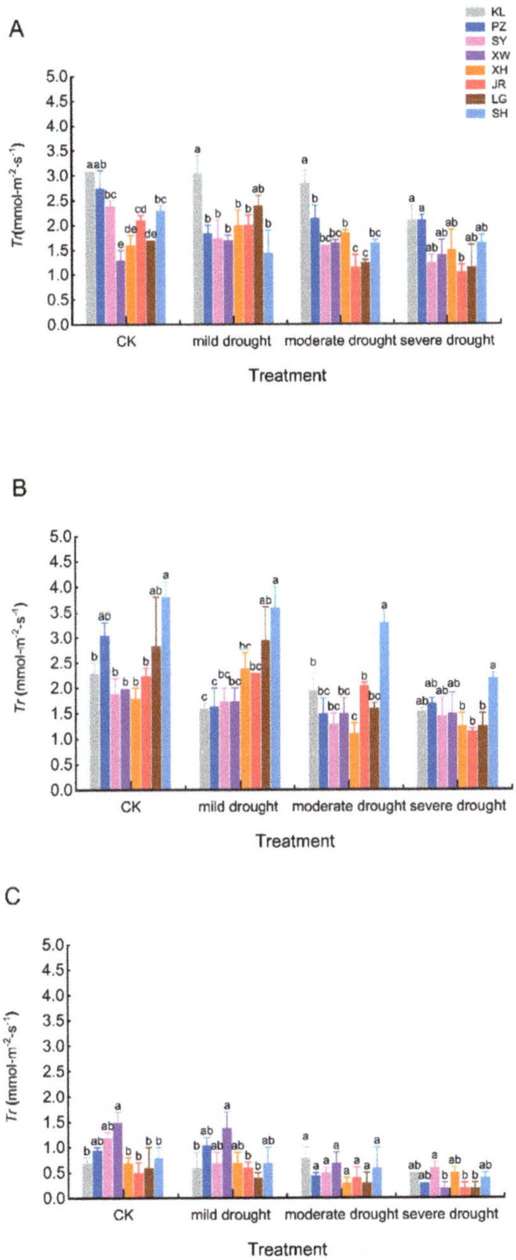

Figure 4. The transpiration rate of *Melia azedarach* among different provenances: (**A**): early stage, (**B**): middle stage, and (**C**): late stage. Note: The error line in the figure is the mean ± standard error, and different lowercase letters in the same treatment indicate significant differences at the 0.05 level.

Under the control group and mild drought, the *Ci* of the KL was higher, whereas that of the SY was higher under moderate and severe drought (Figure 5). The average values of the *Ci* in the LG were the lowest under the three drought stress treatments (226.17, 207, and

209.17 μmol·mol^{-1}). The average of the KL's *Gs* in three periods was the largest among the eight provenances under the four treatment levels; whereas, the that of the XH was the smallest under mild and moderate drought stress, and that of the JR was the smallest under the control and severe drought stress (Figure 6).

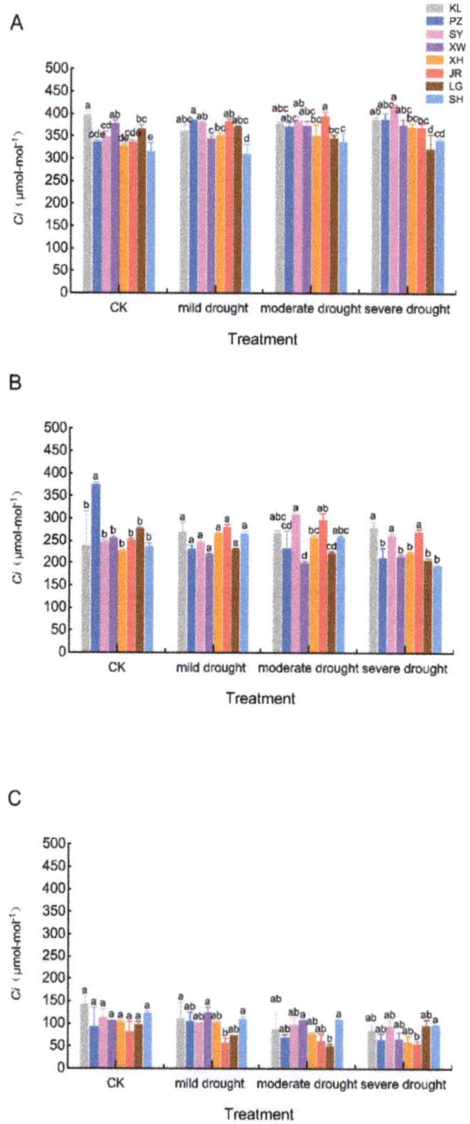

Figure 5. The intercellular CO_2 concentration of *Melia azedarach* among different provenances: (**A**): early stage, (**B**): middle stage, and (**C**): late stage. Note: The error line in the figure is the mean ± standard error, and different lowercase letters in the same treatment indicate significant differences at the 0.05 level.

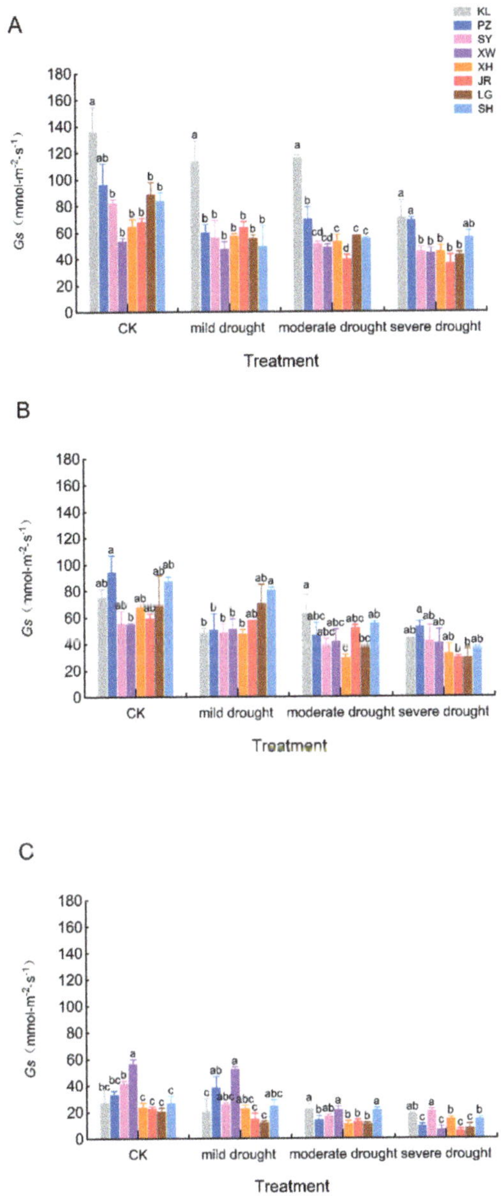

Figure 6. The stomatal conductance of *Melia azedarach* among different provenances: (**A**): early stage, (**B**): middle stage, and (**C**): late stage. Note: The error line in the figure is the mean ± standard error, and different lowercase letters in the same treatment indicate significant differences at the 0.05 level.

As the stress degree and duration increased, the chlorophyll A content of the eight provenances dropped or increased initially and subsequently decreased (Figure 7). The average chlorophyll A content of the SH was the highest in all four treatment levels (2.636, 2.662, 2.656, and 2.582 mg/g), while that of the PZ was the lowest (1.776, 1.727, 1.634, 1.640 mg/g). In the early stages of the treatment, the average chlorophyll A content of the severe drought stress groups of PZ, SY, and LG were significantly different from that of the control group, as were those of the severe drought stress groups of SY and XH in the late stages ($p < 0.05$). In the middle of treatment, the average chlorophyll A content of both the LG and SH under mild drought stress was significantly different from that of the control group ($p < 0.05$). The chlorophyll A content of the JR, LG, and SH under mild and moderate drought stress was considerably different from that of the control group in the later treatment period ($p < 0.05$) (Figure A3). It is apparent that the chlorophyll content of SY was the lowest in each period and under each treatment.

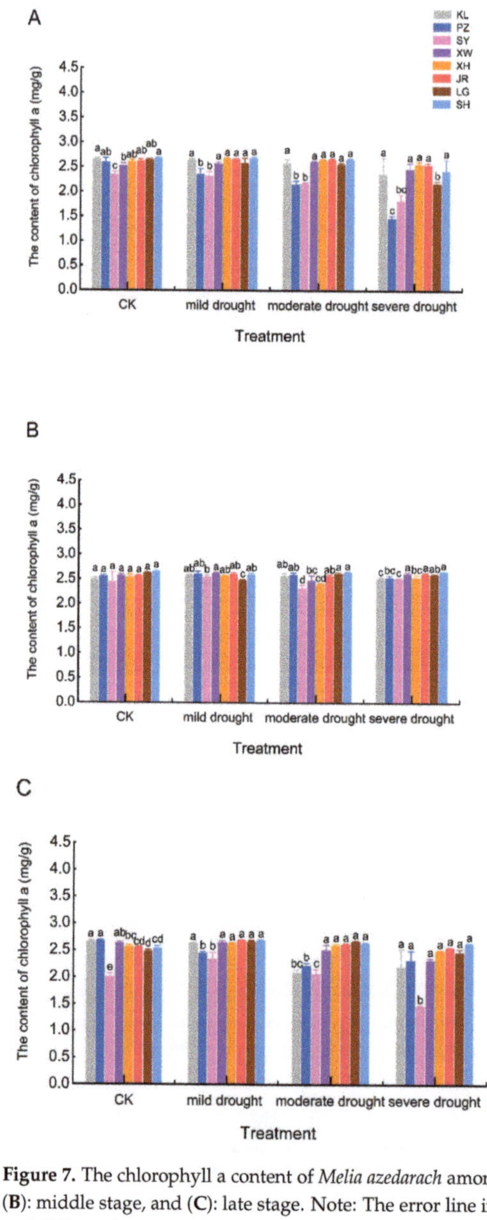

Figure 7. The chlorophyll a content of *Melia azedarach* among different provenances: (**A**): early stage, (**B**): middle stage, and (**C**): late stage. Note: The error line in the figure is the mean ± standard error, and different lowercase letters in the same treatment indicate significant differences at the 0.05 level.

3.5. Response of MDA and SOD

The change in the MDA rose gradually as the level and duration of the stress increased (Figure 8). In all four treatment levels, the KL had the highest average MDA (0.0308, 0.0355, 0.0390, and 0.0433 μmol/g). Under control and severe drought stress, the XH MDA was the lowest (0.0223 and 0.0311 μmol/g), whereas that of the LG was the lowest (0.0006 and 0.0276 μmol/g) under mild and moderate drought stress. All provenances, with the exception of the JR, demonstrated a substantial difference in MDA in severe drought stress in the early stages of treatment as compared to that of the control group ($p < 0.05$). In the middle of the treatment, only the MDA in the severe drought group of the SH was significantly different from that of the control group ($p < 0.05$). Except for the KL and SH, there was a significant difference between the MDA in the severe drought and the control group at the late stages of treatment in six provenances ($p < 0.05$) (Figure A4).

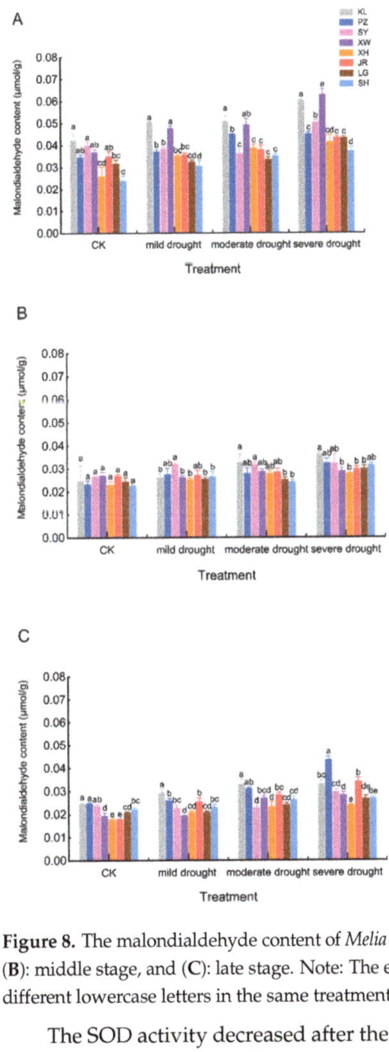

Figure 8. The malondialdehyde content of *Melia azedarach* among different provenances: (**A**): early stage, (**B**): middle stage, and (**C**): late stage. Note: The error line in the figure is the mean ± standard error, and different lowercase letters in the same treatment indicate significant differences at the 0.05 level.

The SOD activity decreased after the initial increase with the deepening of the stress degree and the extension of the stress time (Figure 9). The average SOD activity of the KL under the four treatment levels was the highest (42.128, 90.948, 104.59, and 82.879 U/g), while that of LG was the lowest in the control group (12.444 U/g), and that of XW was the lowest in the three drought stress treatment groups (24.406, 40.158, and 23.920 U/g). In the early and middle stages of treatment, the SOD of the KL, PZ, SY and LG under the three drought stresses were significantly different from that of the control group ($p < 0.05$), as was that of the JR in the middle stage of treatment and that of the KL and the XW in the late stage of treatment (Figure A5).

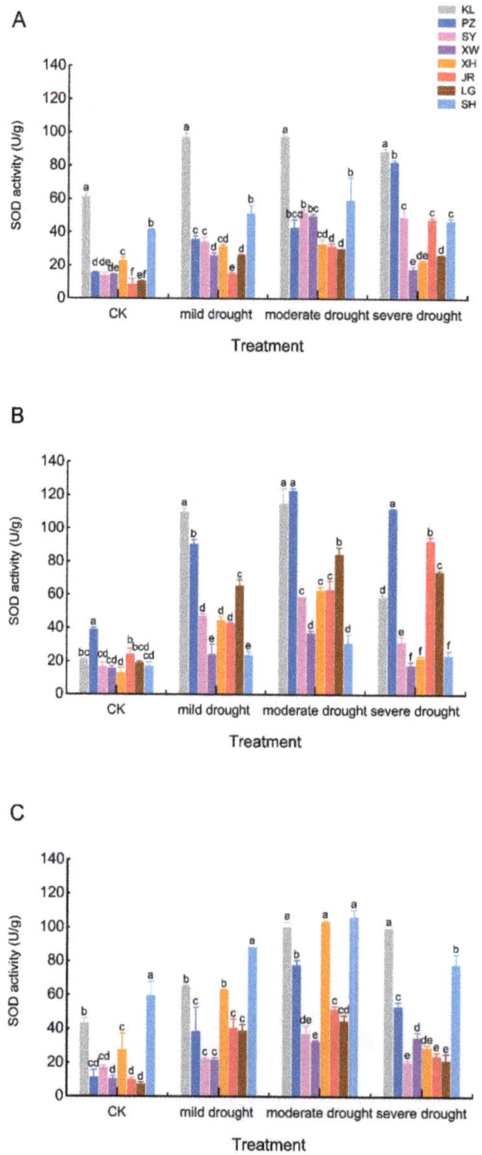

Figure 9. The superoxide dismutase activity of *Melia azedarach* among different provenances: (**A**): early stage, (**B**): middle stage, and (**C**): late stage. Note: The error line in the figure is the mean ± standard error, and different lowercase letters in the same treatment indicate significant differences at the 0.05 level.

3.6. Comprehensive Evaluation of the Drought Resistance of Melia azedarach in Different Provenances

The correlation analysis of the drought resistance coefficients of the 12 indicators of *Melia azedarach* shows that there was a correlation between all the measured indicators (Tables A1, A4 and A7). Individual indicators or some of them could not fully evaluate the drought resistance of each provenance. In addition, the weight of the individual indicators for the evaluation of the drought resistance in *Melia azedarach* was inconsistent. Therefore, the drought resistance of *Melia azedarach* should be analyzed through the screening of the combined indicators.

The results of the principal component analysis of the 12 indicators showed that there were four principal components with eigenvalues greater than 1 under the three drought stresses, with cumulative contributions of 86.539%, 91.285%, and 85.526%, respectively (Tables 6–8). This indicates that these four principal components reflected most of the information of the original data, and a comprehensive evaluation of *Melia azedarach* drought resistance can be carried out by these four integrated indicators.

Table 6. The eigenvalues, contribution ratio, and cumulative contribution ratio of the principal components under mild drought stress.

Principal Component	Eigenvalues	Contribution Ratio/%	Cumulative Contribution Rate/%
1	4.404	36.699	36.699
2	2.729	22.744	59.442
3	1.665	13.877	73.320
4	1.586	13.219	86.539

Table 7. The eigenvalues, contribution ratio, and cumulative contribution ratio of the principal components under moderate drought stress.

Principal Component	Eigenvalues	Contribution Ratio/%	Cumulative Contribution Rate/%
1	4.285	35.707	35.707
2	3.005	25.040	60.748
3	2.164	18.034	78.782
4	1.500	12.503	91.285

Table 8. The eigenvalues, contribution ratio, and cumulative contribution ratio of the principal components under severe drought stress.

Principal Component	Eigenvalues	Contribution Ratio/%	Cumulative Contribution Rate/%
1	4.285	35.707	35.707
2	3.005	25.040	60.748
3	2.164	18.034	78.782
4	1.500	12.503	91.285

The D value represents the comprehensive evaluation of the drought resistance of each provenance under drought stress; the larger the value, the more drought resistant it was.

As can be seen from Tables 9–11, the drought resistance of the eight provenances of *Melia azedarach* under mild drought stress was ranked from largest to smallest: XH, LG, JR, XW, KL, SH, PZ, and SY. The ranking under moderate drought stress was: SH, XH, KL, JR, XW, PZ, SY, and LG, and under severe drought stress, it was: SH, XH, XW, KL, PZ, JR, SY, and LG. The drought resistance of the XH was stronger, while that of the SY was poorer, under all three drought stresses. The SH ranked lower under mild drought, but was the most resistant to drought under moderate and severe stress. The LG, in contrast, had the weakest drought resistance under moderate and severe stress.

Table 9. The function value of the comprehensive indexes ($U(Y_{ik})$), weight (W_k), and the comprehensive evaluation value (D) of the different provenances under mild drought stress.

Provenance	$U(Y_{ik})$				D Value	Rank
	$U(Y_{i1})$	$U(Y_{i2})$	$U(Y_{i3})$	$U(Y_{i4})$		
KL	0.429	0.561	0.549	0.511	0.495	5
PZ	0.469	0.357	0.540	0.044	0.386	7
SY	0.544	0	0.496	0.311	0.358	8
XW	0.776	0.276	0.053	0.933	0.553	4
XH	0.714	0.551	0.788	0.4448	0.642	1
JR	0.633	0.429	0.6288	0.5338	0.563	3
LG	0	1	1	1	0.576	2
SH	1	0.010	0	0	0.427	6
Weights	0.424	0.263	0.160	0.153		

Table 10. The function value of the comprehensive indexes ($U(Y_{ik})$), weight (W_k), and the comprehensive evaluation value (D) of the different provenances under moderate drought stress.

Provenance	$U(Y_{ik})$				D Value	Rank
	$U(Y_{i1})$	$U(Y_{i2})$	$U(Y_{i3})$	$U(Y_{i4})$		
KL	0.731	0.280	0.192	0.617	0.485	3
PZ	0.059	0.354	1	0.808	0.429	6
SY	0.333	0.648	0	0.524	0.380	7
XW	0.628	0.383	0.129	0.538	0.450	5
XH	0.456	0.650	0.688	0.757	0.596	2
JR	0.311	0.431	0.375	1	0.451	4
LG	0	0	0.190	0.823	0.150	8
SH	1	1	0.274	0	0.719	1
Weights	0.391	0.274	0.198	0.137		

Table 11. The function value of the comprehensive indexes ($U(Y_{ik})$), weight (W_k), and the comprehensive evaluation value (D) of the different provenances under severe drought stress.

Provenance	$U(Y_{ik})$				D Value	Rank
	$U(Y_{i1})$	$U(Y_{i2})$	$U(Y_{i3})$	$U(Y_{i4})$		
KL	0.633	0.310	0.296	0.268	0.444	4
PZ	0	1	0.172	0.483	0.346	5
SY	0.548	0.044	0	0.311	0.296	7
XW	0.782	0.612	0.132	0.144	0.541	3
XH	1	0	0.435	0.577	0.595	2
JR	0.031	0.498	0.591	0.564	0.316	6
LG	0.229	0.468	0.079	0	0.232	8
SH	0.998	0.045	1	1	0.759	1
Weights	0.442	0.251	0.170	0.137		

4. Discussion

4.1. Morphological Growth

The analysis of plant growth under water deficit conditions is important for understanding plant responses to drought stress at the whole plant level [20]. Drought can have a devastating impact on plant growth [21]. It inhibits seedling growth by lowering carbon domestication and allocation as well as slowing cell expansion [22]. The reduction in plant height is mainly due to reduced cell expansion, increased leaf abscission, and impaired mitosis under drought conditions [23]. Morphological changes in plants under drought stress include a decrease in the size, area, and number of leaves and an increase in the length of roots and shoots. This is due to the stimulation of the ABA precursor ACC, which prevents the growth and early maturation of the roots. The rapid root growth and low number of stomata increase leaf thickness, curling, folding, and wax formation, thus preventing water loss from leaves and roots [24]. Under drought stress, the leaf area, seedling height, ground diameter, biomass, root–shoot ratio and other indexes have been used to evaluate the drought resistance of the test materials. Most studies have shown that the seedling height, ground diameter, and biomass were all inhibited to varying degrees, but the root–crown ratio increased slightly or significantly with the increase in drought stress [25–27]. This has been explained by studies in that when plants are stressed by drought, the biomass is distributed more underground in order to absorb more water and nutrients, while minimizing water loss due to transpiration [28]. It also can be concluded from the studies of Ying, Narayan Bhusal, and Lei et al. that drought stress had distinct impacts on the morphological growth of various tree species, as well as on different variations or provenances of the same tree species [29–31]. In this study, the seedling height growth, ground diameter growth, and the aboveground biomass of different provenances declined as the stress levels increased, which was consistent with the above results. In addition, the underground biomass increased with the deepening of the stress degree, and the root–crown ratio also exhibited an upward trend in this study. The underground biomass and root–crown ratio of the SH increased significantly under severe drought stress compared to that of the control group, which showed that the SH was more sensitive to drought stress and utilized the water in the soil by increasing the biomass of the belowground part.

4.2. Water Status

When plants are stressed by drought, the rate of water loss by transpiration from leaves exceeds the rate of water uptake through roots [32]. The tissue water content and water potential decrease dramatically, and the cells lose tension, which makes the leaves appear curled, yellowed, scorched, and permanently wilted [33]. The relative water content, leaf water potential, stomatal resistance, transpiration rate, leaf temperature, and canopy temperature are important characteristics that affect plant–water relations [34]. The leaf relative water content and leaf water saturation deficit are key markers of plant water status and are frequently used by researchers to assess plant drought tolerance. The exposure of plants to drought stress substantially decreased the leaf water potential, relative water content, and transpiration rate, with a concomitant increase in leaf temperature but increased leaf water saturation deficit. [35]. This study yielded the same findings. In addition, the relative water content of the leaves in the PZ declined the least in comparison to that of the control group, but the water saturation deficit of the leaves in the SY rose the highest, indicating that the PZ and SY had greater water retention abilities under drought stress.

4.3. Photosynthesis and Photosynthetic Pigment

Photosynthesis is "the most important chemical reaction on Earth". The main force behind plant growth and biomass production is photosynthesis, which provides the energy and carbon needed for the biosynthesis of the organic compounds needed for development [36]. Drought stress causes photosynthetic inhibition, which happens as a result of stomatal closure, chlorophyll degradation, and damage to the photochemical equipment, leading to a reduction in the intercellular CO_2 concentration (*Ci*) [37]. According to the intercellular carbon dioxide concentration, stomatal conductance, transpiration rate, and net photosynthetic rate, researchers frequently evaluate and explore plant stress resistance and the factors that cause a drop in the photosynthetic rate. Previous studies have offered an explanation for the *Pn* decline: when *Gs* decreases while *Ci* increases or remains steady, it is non-stomatal limitation. When both *Gs* and *Ci* decrease, it is stomatal limitation [38]. During drought, plants tend to reduce water loss due to transpiration, which is consistent with our findings [39]. The *Pn* and *Gs* exhibited a declining trend as the stress levels increased in studies by Silva, Wu, Gao et al. [40–42], which was consistent with the results of previous studies. In this study, the KL had the highest *Gs* under all treatments, and the highest *Ci* under the control and mild drought. The *Gs* of the XH was the lowest under mild and moderate drought, while the *Gs* of the JR was the lowest under the control and severe drought. The *Ci* of the LG was the lowest under drought stress. It was found, as shown in the figure, that in the early stages of drought stress, the decrease in *Pn* in several provenances might be caused by some damage to the chloroplast structure to a certain extent, the intensification of membrane lipid peroxidation, and the generation of superoxide free radicals and other non-stomatal restrictions, which corresponded to the decrease in chlorophyll A and the increase in MDA. In the middle stage of drought stress, stomatal restriction was the major reason for the *Pn* decline in most provenances, which might be due to the adaptation of the test materials to water stress. At the late stage of drought stress, the XW, JR, and LG were affected by non-stomatal restriction, and not only did the chlorophyll A decrease, but the MDA increased, and the SOD activity also decreased, indicating that these three provenances were severely impacted by drought stress and had poor drought resistance.

Chlorophyll plays an indispensable role in the absorption, transmission, and transformation of light energy in photosynthesis. Plants accumulate a considerable quantity of reactive oxygen species in their bodies during drought stress, leading to changes in membrane structure and consequently altering chlorophyll concentration [43]. Among them, chlorophyll A is a component of the photosynthetic reaction center and LHC (the most abundant pigment protein complex in thylakoid membrane) [44]. According to Ghobadi, M., et al. [45,46], the value of chlorophyll A under sufficient water stress was higher than that under moderate and severe drought stress. On the other hand, drought stress has been shown in certain experiments to increase the chlorophyll content [47,48]. The chlorophyll A content of the eight provenances in this study was not always highest in the control group, which could be related to the enrichment effect of leaf water loss and extended leaf growth. It is also conceivably due to other elements, such as light. The chlorophyll A content of the SH was the highest under the four treatments, indicating that it had superior photosynthetic efficiency, could maintain plant growth, and was more drought resistant.

4.4. Membrane Lipid Peroxidation and Protective Enzyme Activity

Under the normal growth of plants, the production and elimination of free radicals in cells are balanced. However, under drought stress, the cell structure is disrupted, the relative permeability of plasma membrane of cells increases, and the internal oxides increase, leading to cell membrane peroxidation [49]. Malondialdehyde is the final decomposition product of the membrane lipid peroxidation reaction in plants, which has been used as a lipid peroxidation marker in studies related to oxidative stress and redox signaling [50]. Its content will increase when plants suffer from adversity. Khaleghi and Marcia Carvalho et al. [51,52] found in drought stress experiments that malondialdehyde content increased as the depth of the drought stress increased, supporting the results of our study, and indicating that drought stress induced or exacerbated the membrane lipid peroxidation. It was found that under moderate and severe drought stress, the MDA content in the SY increased the least compared to that of the control group, indicating that the damage degree of the cell membrane in the SY was the least. It is worth noting that the MDA content of the KL was the highest under the four treatments, and the rate of increase in the drought stress treatment group was relatively higher than that of the control group. At the same time, the SOD activity of the KL was the highest among the four treatments, and the increase in SOD activity in the drought stress group was also larger than that in the control group. As a result, plants with increased SOD activity may be able to defend against reactive oxygen species produced by membrane lipid peroxidation, making them more drought-resistant.

One of the most essential defensive enzymes for scavenging superoxide anions and H_2O_2 is superoxide dismutase (SOD). When plants are exposed to drought, they will scavenge accumulated superoxide radicals by increasing the superoxide dismutase activity, in order to maintain the balance of the reactive oxygen metabolism, and slow down the damage caused by free radicals to the cell membrane system, allowing plants to resist drought to some extent. In this study, the SOD activity of different provenances increased initially and, then, subsequently decreased with increasing stress levels, which was consistent with the results of Wang, Zhang et al. [53,54]. Among them, the SOD activity of the KL was the highest in all treatments. The SOD activity of the LG was the lowest in the control group. Under drought stress, the SOD activity of the XW was the lowest. In the study of stress, catalase (CAT), peroxidase (POD), ascorbate peroxidase (APX), and other antioxidant enzymes are commonly utilized as physiological indicators to identify plant stress resistance. Multiple indicators can help to strengthen a conclusion.

5. Conclusions

The relative water content of the leaves in the eight provenances declined as the drought stress became more severe; whereas, the water saturation deficit increased. In different stress treatment periods, the changes in the intercellular CO_2 concentration in the eight provenances were not consistent. With the deepening of the drought stress, the transpiration rate, net photosynthetic rate, and stomatal conductance of the eight provenances showed a general trend of decline, and the drop in the net photosynthetic rate was mainly due to stomatal and non-stomatal factors. In general, chlorophyll A content decreased. With the severity of the drought stress, the malondialdehyde content showed an increasing trend, and the SOD activity increased at first and, then, decreased. A comprehensive evaluation of drought resistance was performed for eight provenances of *Melia azedarach*. The results showed that the SH and XH were more drought resistant, while the LG and SY were less drought resistant.

The physiological mechanism of the drought stress response in the eight different provenances was preliminarily identified from the growth traits and physiological responses in this study, and the drought resistance ability of the eight provenances was ranked. If the molecular response mechanism of *Melia azedarach* to drought stress can be analyzed at the transcriptome level, it will help to identify prospective drought-resistant genes in *Melia azedarach*. These results may be used to build a drought resistance gene database for *Melia azedarach* and provide a theoretical basis for studying the molecular mechanism of *Melia azedarach*'s drought stress response.

Author Contributions: Conceptualization, W.Y. and C.H.; methodology, W.Y. and C.H.; software, J.C. and Z.L.; validation, F.Y. and W.Y.; investigation, C.H. and W.Y.; resources, W.Y.; writing—original draft preparation, C.H.; writing—review and editing, F.Y., W.Y. and H.C.; project administration, F.Y.; funding acquisition, F.Y. All authors have read and agreed to the published version of the manuscript.

Funding: The research was funded by the National Important Research and Development Program Subjects, grant number 2017YFD0600701, the National Natural Science Foundation of China, grant number C161101, and the Forestry Science and Technology Innovation and Promotion Project in Jiangsu Province, grant number LYKJ(2021)30.

Institutional Review Board Statement: Not applicable.

Informed Consent Statement: Not applicable.

Conflicts of Interest: The authors declare no conflict of interest.

Appendix A

Table A1. Correlation analysis of the drought resistance coefficients of each index under mild drought stress.

Indicator	RWC	WSD	ChlA	*Tr*	*Pn*	*Gs*	*Ci*	MDA	SOD	Seedling Height	Ground Diameter	Root–Crown Ratio
RWC	1.000											
WSD	0.142	1.000										
ChlA	−0.329	0.717	1.000									
Tr	−0.349	−0.354	0.075	1.000								
Pn	−0.087	−0.650	−0.304	0.460	1.000							
Gs	−0.510	−0.324	0.007	0.601	0.199	1.000						
Ci	−0.118	0.034	0.315	0.199	−0.028	0.453	1.000					
MDA	−0.443	−0.546	−0.046	0.333	0.469	0.447	0.706	1.000				
SOD	0.245	−0.129	−0.297	0.211	0.070	−0.524	−0.659	−0.422	1.000			
Seedling height	−0.170	−0.185	0.014	0.016	0.152	0.407	−0.268	−0.190	−0.340	1.000		
Ground diameter	−0.326	−0.103	0.324	0.257	0.190	0.484	0.948	0.857	−0.673	−0.204	1.000	
Root–crown ratio	−0.230	0.024	0.027	−0.413	−0.074	0.377	0.531	0.439	−0.910	0.162	0.567	1.000

Table A2. The eigenvectors of the principal components under mild drought stress.

Traits	Components			
	P1	P2	P3	P4
Ground diameter	0.44	−0.08	0.23	−0.12
MDA	0.41	0.18	0.17	−0.22
Ci	0.40	−0.17	0.26	−0.16
SOD	−0.37	0.28	0.32	0.04
Gs	0.35	0.15	−0.19	0.31
Root–crown ratio	0.32	−0.25	−0.36	−0.25
WSD	−0.12	−0.52	0.16	0.23
Pn	0.13	0.45	−0.02	−0.08
Tr	0.15	0.38	0.32	0.40
Chl A	0.10	−0.37	0.26	0.47
Seedling height	0.04	0.08	−0.63	0.36
RWC	−0.24	−0.07	0.01	−0.43

Table A3. The value of the comprehensive indexes of the different provenances under mild drought stress.

Provenance	*Yik*			
	Yi1	*Yi2*	*Yi3*	*Yi4*
KL	0.84	0.37	0.93	0.57
PZ	0.90	0.17	0.92	0.36
SY	1.01	−0.18	0.87	0.48
XW	1.35	0.09	0.37	0.76
XH	1.26	0.36	1.20	0.54
JR	1.14	0.24	1.02	0.58
LG	0.21	0.80	1.44	0.79
SH	1.68	−0.17	0.31	0.34

Table A4. Correlation analysis of the drought resistance coefficients of each index under moderate drought stress.

Indicator	RWC	WSD	ChlA	*Tr*	*Pn*	*Gs*	*Ci*	MDA	SOD	Seedling Height	Ground Diameter	Root–Crown Ratio
RWC	1.000											
WSD	0.105	1.000										
ChlA	−0.479	0.366	1.000									
Tr	−0.412	−0.372	0.005	1.000								
Pn	−0.105	−0.024	0.162	0.550	1.000							
Gs	−0.413	−0.283	−0.048	0.929	0.600	1.000						
Ci	−0.021	0.663	0.312	0.163	−0.118	0.156	1.000					
MDA	0.156	−0.604	−0.175	0.620	0.070	0.384	0.034	1.000				
SOD	−0.130	−0.366	−0.225	−0.485	−0.647	−0.480	−0.613	−0.230	1.000			
Seedling height	0.005	−0.230	−0.002	0.792	0.857	0.708	0.019	0.546	−0.711	1.000		
Ground diameter	0.079	0.199	0.203	0.436	−0.125	0.274	0.727	0.598	−0.442	0.262	1.000	
Root–crown ratio	0.815	0.305	−0.044	−0.300	−0.184	−0.358	0.441	0.273	−0.398	0.001	0.464	1.000

Table A5. The eigenvectors of the principal components under moderate drought stress.

Traits	Components			
	P1	P2	P3	P4
Tr	0.4527	−0.1419	−0.0007	0.1780
Seedling height	0.4391	−0.0490	0.0931	−0.2727
Gs	0.4193	−0.1644	−0.0761	0.0629
SOD	−0.3560	−0.2988	0.0972	0.2899
Pn	0.3367	−0.1160	−0.1373	−0.5226
Root–crown ratio	−0.0198	0.4834	0.3202	−0.1168
Ci	0.1406	0.4644	−0.2162	0.2286
WSD	−0.0908	0.4102	−0.3861	−0.1886
Ground diameter	0.2420	0.3623	0.0823	0.4376
RWC	−0.1024	0.2850	0.4826	−0.3560
Chl A	0.0531	0.1321	−0.4765	0.1208
MDA	0.2928	0.0104	0.4385	0.3135

Table A6. The value of the comprehensive indexes of the different provenances under moderate drought stress.

Provenance	*Yik*			
	*Yi*1	*Yi*2	*Yi*3	*Yi*4
KL	1.002	0.778	0.607	0.993
PZ	0.072	0.885	1.040	1.185
SY	0.451	1.306	0.504	0.900
XW	0.859	0.926	0.573	0.914
XH	0.621	1.308	0.873	1.134
JR	0.421	0.995	0.705	1.377
LG	−0.009	0.378	0.606	1.200
SH	1.374	1.809	0.651	0.375

Table A7. Correlation analysis of the drought resistance coefficients of each index under severe drought stress.

Indicator	RWC	WSD	ChlA	*Tr*	*Pn*	*Gs*	*Ci*	MDA	SOD	Seedling Height	Ground Diameter	Root–Crown Ratio
RWC	1.000											
WSD	0.051	1.000										
Chl A	−0.339	−0.252	1.000									
Tr	0.332	−0.575	0.145	1.000								
Pn	0.296	−0.401	0.373	0.728	1.000							
Gs	0.325	−0.259	−0.222	0.864	0.610	1.000						
Ci	−0.388	0.522	0.021	−0.027	−0.092	0.194	1.000					
MDA	0.134	−0.219	0.068	0.065	−0.036	−0.098	−0.660	1.000				
SOD	−0.076	0.405	−0.416	−0.759	−0.913	−0.686	−0.174	0.301	1.000			
Seedling height	0.241	−0.344	−0.122	0.113	0.162	0.157	−0.711	0.447	−0.077	1.000		
Ground diameter	0.135	−0.446	0.564	0.670	0.741	0.503	−0.008	−0.226	−0.870	0.286	1.000	
Root–crown ratio	−0.004	0.193	0.370	0.167	0.549	0.145	0.066	0.377	−0.338	−0.095	0.085	1.000

Table A8. The eigenvectors of the principal components under severe drought stress.

Traits	Components			
	P1	P2	P3	P4
SOD	−0.438	0.193	−0.024	0.039
Pn	0.434	−0.039	0.074	0.185
Tr	0.419	0.010	−0.179	0.022
Ground diameter	0.407	−0.077	0.076	−0.242
Gs	0.349	−0.068	−0.400	0.146
WSD	−0.265	−0.258	−0.098	0.454
Ci	−0.048	−0.592	−0.104	0.070
Seedling height	0.124	0.478	−0.041	−0.117
MDA	−0.004	0.473	0.269	0.323
Chl A	0.175	−0.129	0.626	−0.139
RWC	0.125	0.244	−0.411	0.321
Root–crown ratio	0.152	−0.067	0.376	0.657

Table A9. The value of the comprehensive indexes of the different provenances under severe drought stress.

Provenance	*Yik*			
	*Yi*1	*Yi*2	*Yi*3	*Yi*4
KL	0.376	0.574	0.583	2.649
PZ	−0.669	1.192	0.497	2.891
SY	0.236	0.335	0.377	2.697
XW	0.623	0.844	0.469	2.509
XH	0.983	0.296	0.680	2.997
JR	−0.618	0.742	0.789	2.983
LG	−0.291	0.715	0.432	2.347
SH	0.979	0.336	1.074	3.474

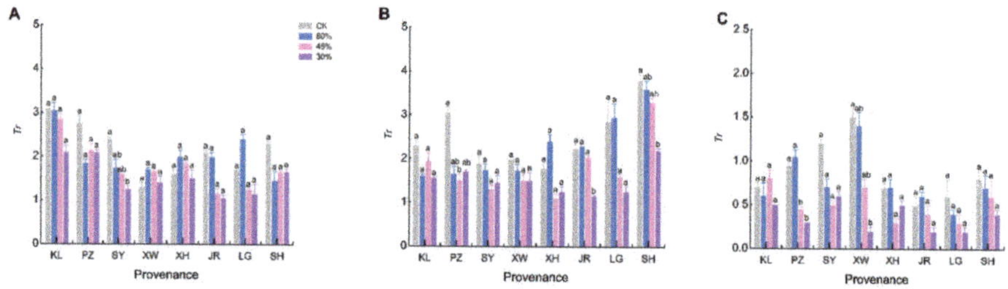

Figure A1. The transpiration rate of *Melia azedarach* among the different provenances: (**A**): early stage, (**B**): middle stage, and (**C**): late stage. Note: The error line in the figure is the mean ± standard error, and different lowercase letters in the same provenance indicate significant differences at the 0.05 level, as in the following figures.

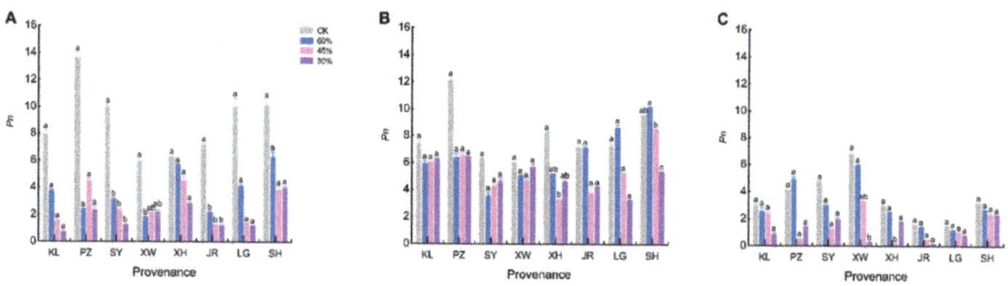

Figure A2. The net photosynthetic rate of *Melia azedarach* among the different provenances: (**A**): early stage, (**B**): middle stage, and (**C**): late stage. Note: The error line in the figure is the mean ± standard error, and different lowercase letters in the same treatment indicate significant differences at the 0.05 level.

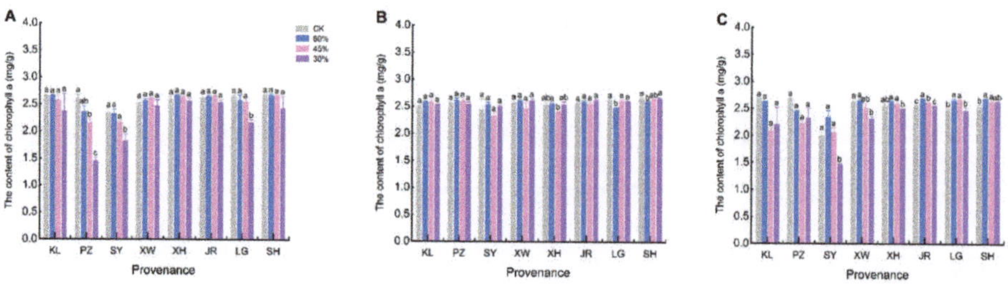

Figure A3. The chlorophyll a content of *Melia azedarach* among the different provenances: (**A**): early stage, (**B**): middle stage, and (**C**): late stage. Note: The error line in the figure is the mean ± standard error, and different lowercase letters in the same treatment indicate significant differences at the 0.05 level.

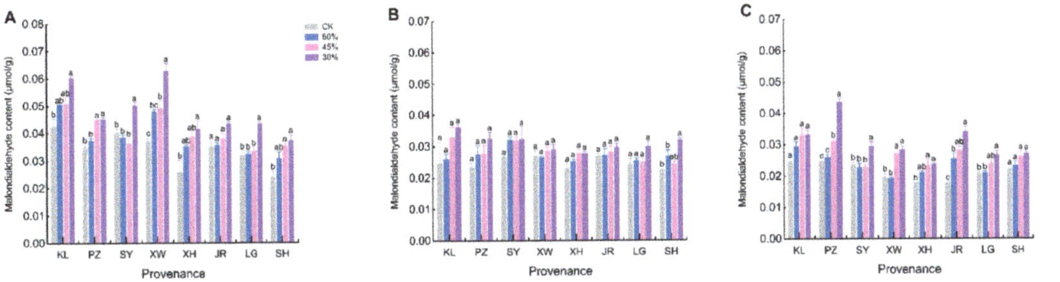

Figure A4. The malondialdehyde content of *Melia azedarach* among the different provenances: (**A**): early stage, (**B**): middle stage, and (**C**): late stage. Note: The error line in the figure is the mean ± standard error, and different lowercase letters in the same treatment indicate significant differences at the 0.05 level.

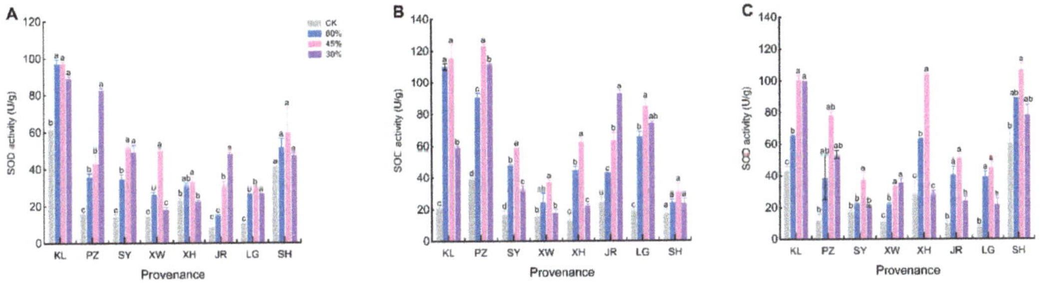

Figure A5. The superoxide dismutase activity of *Melia azedarach* among the different provenances: (**A**): early stage, (**B**): middle stage, and (**C**): late stage. Note: The error line in the figure is the mean ± standard error, and different lowercase letters in the same treatment indicate significant differences at the 0.05 level.

References

1. Ayaz, A.; Huang, H.; Zheng, M.; Zaman, W.; Li, D.; Saqib, S.; Zhao, H.; Lü, S. Molecular Cloning and Functional Analysis of *GmLACS2-3* Reveals Its Involvement in Cutin and Suberin Biosynthesis along with Abiotic Stress Tolerance. *Int. J. Mol. Sci.* **2021**, *22*, 9175. [CrossRef] [PubMed]
2. Rane, J.; Singh, A.K.; Tiwari, M.; Prasad, P.V.V.; Jagadish, S.V.K. Effective Use of Water in Crop Plants in Dryland Agriculture: Implications of Reactive Oxygen Species and Antioxidative System. *Front. Plant Sci.* **2022**, *12*, 778270. [CrossRef] [PubMed]
3. Prăvălie, R. Drylands extent and environmental issues. A global approach. *Earth-Sci. Rev.* **2016**, *161*, 259–278. [CrossRef]
4. Mao, D.; Wang, Z.; Wu, B.; Zeng, Y.; Luo, L.; Zhang, B. Land degradation and restoration in the arid and semiarid zones of China: Quantified evidence and implications from satellites. *Land Degrad. Dev.* **2018**, *29*, 3841–3851. [CrossRef]
5. Sakadevan, K.; Nguyen, M.-L. Extent, Impact, and Response to Soil and Water Salinity in Arid and Semiarid Regions. *Adv. Agron.* **2010**, *109*, 55–74. [CrossRef]
6. Editorial Committee of the Flora of China, Chinese Academy of Sciences. *Flora of China*; Science Press: Beijing, China, 1997; Volume 143, ISBN 9787030012050. (In Chinese)
7. Sen, A.; Batra, A. Melia Azedarach L.-A paradise tree. *Indian J.* **2011**, *1*, 59–69.
8. Adan, Y.; Rubea, A.L. The Potential Uses of Melia *Azedarach* L. as pesticidal and Medicinal Plant, Review. *Am.-Eurasian J. Sustain. Agric.* **2009**, *3*, 185–194.
9. Cheng, S.M.; Gu, W.C. The phenological Division of Distribution area in China for Melia Azedarach. *Sci. Silvae Scinicae* **2005**, *41*, 186–191. (In Chinese)
10. Vishnukanta, A.R. Melia azedarach: A phytopharmacological review. *Pharmacogn. Rev.* **2008**, *2*, 173–179.
11. Sultana, S.; Asif, H.M.; Akhtar, N.; Waqas, M.; Rehman, S.U. Comprehensive review on ethanobotanical use, phytochemistry and pharmacological properties of Melia Azederach Linn. *Asian J. Pharm. Res. Health Care* **2014**, *16*, 26–32.
12. Kumar, R.; Singh, R.; Meera, P.S.; Kalidhar, S.B. Chemical components and insecticidal properties of Bakain (*Melia azedarach* L.)–A review. *Agric. Rev.* **2003**, *24*, 101–115.

13. Silva, S.; Pinto, G.; Correia, B.; Pinto-Carnide, O.; Santos, C. Rye oxidative stress under long term Al exposure. *J. Plant Physiol.* **2013**, *170*, 879–889. [CrossRef] [PubMed]
14. Dias, M.C.; Pinto, G.; Santos, C. Acclimatization of micropropagated plantlets induces an antioxidative burst: A case study with *Ulmus minor* Mill. *Photosynthetica* **2011**, *49*, 259–266. [CrossRef]
15. Jhou, H.-C.; Wang, Y.-N.; Wu, C.-S.; Yu, J.-C.; Chen, C.-I. Photosynthetic gas exchange responses of Swietenia macrophylla King and *Melia azedarach* L. plantations under drought conditions. *Bot. Stud.* **2017**, *58*, 57. [CrossRef]
16. Dias, M.C.; Azevedo, C.; Costa, M.; Pinto, G.; Santos, C. Melia azedarach plants show tolerance properties to water shortage treatment: An ecophysiological study. *Plant Physiol. Biochem.* **2014**, *75*, 123–127. [CrossRef]
17. *DB11/T 770-2010*; Growth media for ornamental. Beijing Quality and Technology Supervision Bureau: Beijing, China, 2010.
18. Li, H.S. *Principles and Techniques of Plant Physiological and Biochemical Experiments*; Higher Education Press: Beijing, China, 2000; ISBN 9787040080766. (In Chinese)
19. Wang, X.K. *Principles and Techniques of Plant Physiological and Biochemical Experiments*, 3rd ed.; Higher Education Press: Beijing, China, 2006; ISBN 9787040396461. (In Chinese)
20. Kuromori, T.; Fujita, M.; Takahashi, F.; Shinozaki, K.Y.; Shinozaki, K. Inter-tissue and inter-organ signaling in drought stress response and phenotyping of drought tolerance. *Plant J.* **2021**, *109*, 342–358. [CrossRef]
21. Suzuki, N.; Rivero, R.M.; Shulaev, V.; Blumwald, E.; Mittler, R. Abiotic and biotic stress combinations. *New Phytol.* **2014**, *203*, 32–43. [CrossRef]
22. Jabbari, H.; Akbari, G.A.; Sima, N.A.K.K.; Rad, A.H.S.; Alahdadi, I.; Hamed, A.; Shariatpanahi, M.E. Relationships between seedling establishment and soil moisture content for winter and spring rapeseed genotypes. *Ind. Crop. Prod.* **2013**, *49*, 177–187. [CrossRef]
23. Yang, X.; Lu, M.; Wang, Y.; Wang, Y.; Liu, Z.; Chen, S. Response Mechanism of Plants to Drought Stress. *Horticulturae* **2021**, *7*, 50. [CrossRef]
24. Bogati, K.; Walczak, M. The Impact of Drought Stress on Soil Microbial Community, Enzyme Activities and Plants. *Agronomy* **2022**, *12*, 189. [CrossRef]
25. Zhang, X.; Zang, R.; Li, C. Population differences in physiological and morphological adaptations of Populus davidiana seedlings in response to progressive drought stress. *Plant Sci.* **2004**, *166*, 791–797. [CrossRef]
26. Li, F.; Bao, W.; Wu, N.; You, C. Growth, biomass partitioning, and water-use efficiency of a leguminous shrub (Bauhinia faberi var. microphylla) in response to various water availabilities. *New For.* **2008**, *36*, 53–65. [CrossRef]
27. Lei, Y. Physiological responses of Populus przewalskii to oxidative burst caused by drought stress. *Russ. J. Plant Physiol.* **2008**, *55*, 857–864. [CrossRef]
28. Brunner, I.; Herzog, C.; Dawes, M.A.; Arend, M.; Sperisen, C. How tree roots respond to drought. *Front. Plant Sci.* **2015**, *6*, 547. [CrossRef] [PubMed]
29. Eying, Y.; Song, L.L.; Jacobs, D.; Emei, L.; Eliu, P.; Ejin, S.; Ewu, J. Physiological response to drought stress in Camptotheca acuminata seedlings from two provenances. *Front. Plant Sci.* **2015**, *6*, 361. [CrossRef] [PubMed]
30. Bhusal, N.; Lee, M.; Lee, H.; Adhikari, A.; Han, A.R.; Han, A.; Kim, H.S. Evaluation of morphological, physiological, and biochemical traits for assessing drought re-sistance in eleven tree species. *Sci. Total Environ.* **2021**, *779*, 146466. [CrossRef]
31. Lei, Y.; Yin, C.; Li, C. Differences in some morphological, physiological, and biochemical responses to drought stress in two contrasting populations of Populus przewalskii. *Physiol. Plant.* **2006**, *127*, 182–191. [CrossRef]
32. Goche, T.; Shargie, N.G.; Cummins, I.; Brown, A.P.; Chivasa, S.; Ngara, R. Comparative physiological and root proteome analyses of two sorghum varieties responding to water limitation. *Sci. Rep.* **2020**, *10*, 11835. [CrossRef]
33. Corso, D.; Delzon, S.; Lamarque, L.J.; Cochard, H.; Torres-Ruiz, J.M.; King, A.; Brodribb, T. Neither xylem collapse, cavitation, or changing leaf conductance drive stomatal closure in wheat. *Plant Cell Environ.* **2020**, *43*, 854–865. [CrossRef]
34. Anjum, S.A.; Xie, X.Y.; Wang, L.C.; Saleem, M.F.; Man, C.; Lei, W. Morphological, physiological and biochemical responses of plants to drought stress. *Afr. J. Agric. Res.* **2011**, *6*, 2026–2032.
35. Siddique, M.R.B.; Hamid, A.; Islam, M.S. Drought stress effects on water relations of wheat. *Bot. Bull. Acad. Sin.* **2000**, *41*, 35–39.
36. Nowicka, B.; Ciura, J.; Szymanska, R.; Kruk, J. Improving photosynthesis, plant productivity and abiotic stress tolerance current trends and future perspectiv-es. *J. Plant Physiol.* **2018**, *231*, 415–433. [CrossRef] [PubMed]
37. Bohnert, H.J.; Jensen, R.G. Strategies for engineering water-stress tolerance in plants. *Trends Biotechnol.* **1996**, *14*, 89–97. [CrossRef]
38. Farquhar, G.D.; Sharkey, T.D. Stomatal conductance and photosynthesis. *Annu. Rev. Plant Physiol.* **1982**, *33*, 317–345. [CrossRef]
39. Rawat, J.M.; Rawat, B.; Tewari, A.; Joshi, S.C.; Nandi, S.K.; Palni, L.M.S.; Prakash, A. Alterations in growth, photosynthetic activity and tissue-water relations of tea clones in response to different soil moisture content. *Trees* **2017**, *31*, 941–952. [CrossRef]
40. Silva, E.N.; Ribeiro, R.V.; Ferreira-Silva, S.L.; Vieira, S.A.; Ponte, L.F.; Silveira, J.A. Coordinate changes in photosynthesis, sugar accumulation and antioxidative enzymes improve the performance of Jatropha curcas plants under drought stress. *Biomass-Bioenergy* **2012**, *45*, 270–279. [CrossRef]
41. Wu, F.Z.; Bao, W.K.; Li, F.L.; Wu, N. Effects of water stress and nitrogen supply on leaf gas exchange and fluorescence parameters of Sophora davidii seedlings. *Photosynthetica* **2008**, *46*, 40–48. [CrossRef]
42. Gao, L.; Caldwell, C.D.; Jiang, Y. Photosynthesis and Growth of Camelina and Canola in Response to Water Deficit and Applied Nitrogen. *Crop Sci.* **2018**, *58*, 393–401. [CrossRef]

43. Smirnoff, N. Antioxidant systems and plant response to the environment. In *Environment and Plant Metab-Olism: Flexibility and Acclimation*; Smirnoff, N., Ed.; BIOS Scientific Publishers: Oxford, UK, 1995; pp. 217–243.
44. Oster, U.; Tanaka, R.; Tanaka, A.; Rüdiger, W. Cloning and functional expression of the gene encoding the key enzyme for chlorophyll b biosynthesis (CAO) from Arabidopsis thaliana. *Plant J.* **2000**, *21*, 305–310. [CrossRef]
45. Ghobadi, M.; Taherabadi, S.; Ghobadi, M.-E.; Mohammadi, G.-R.; Jalali-Honarmand, S. Antioxidant capacity, photosynthetic characteristics and water relations of sunflower (*Helianthus annuus* L.) cultivars in response to drought stress. *Ind. Crop. Prod.* **2013**, *50*, 29–38. [CrossRef]
46. Efeoğlu, B.; Ekmekçi, Y.; Çiçek, N. Physiological responses of three maize cultivars to drought stress and recovery. *S. Afr. J. Bot.* **2009**, *75*, 34–42. [CrossRef]
47. Cui, L.; Zheng, F.; Zhang, D.; Li, C.; Li, M.; Ye, J.; Zhang, Y.; Wang, T.; Ouyang, B.; Hong, Z.; et al. Tomato methionine sulfoxide reductase B2 functions in drought tolerance by promoting ROS scavenging and chlorophyll accumulation through interaction with Catalase 2 and RBCS3B. *Plant Sci.* **2022**, *318*, 111206. [CrossRef] [PubMed]
48. Hao, S.R.; Guo, X.P.; Wang, W.M.; Zhang, L.J.; Wang, Q. Effect of water stress and rehydration on leaf chloroplast pigments during rice nodulation. *J. Hohai Univ.-Y (Nat. Sci. Ed.)* **2006**, *4*, 397–400.
49. Golldack, D.; Li, C.; Mohan, H.; Probst, N. Tolerance to drought and salt stress in plants: Unraveling the signaling networks. *Front. Plant Sci.* **2014**, *5*, 151. [CrossRef] [PubMed]
50. Morales, M.; Munné-Bosch, S. Malondialdehyde: Facts and Artifacts. *Plant Physiol.* **2019**, *180*, 1246–1250. [CrossRef] [PubMed]
51. Khaleghi, A.; Naderi, R.; Brunetti, C.; Maserti, B.; Salami, S.A.; Babalar, M. Morphological, physiochemical and antioxidant responses of Maclura pomifera to drought stress. *Sci. Rep.* **2019**, *9*, 19250. [CrossRef]
52. Carvalho, M.; Castro, I.; Moutinho-Pereira, J.; Correia, C.; Egea-Cortines, M.; Matos, M.; Rosa, E.; Carnide, V.; Lino-Neto, T. Evaluating stress responses in cowpea under drought stress. *J. Plant Physiol.* **2019**, *241*, 153001. [CrossRef]
53. Wang, L.-F. Physiological and molecular responses to drought stress in rubber tree (*Hevea brasiliensis Muell. Arg.*). *Plant Physiol. Biochem.* **2014**, *83*, 243–249. [CrossRef]
54. Zhang, M.; Jin, Z.-Q.; Zhao, J.; Zhang, G.; Wu, F. Physiological and biochemical responses to drought stress in cultivated and Tibetan wild barley. *Plant Growth Regul.* **2015**, *75*, 567–574. [CrossRef]

MDPI
St. Alban-Anlage 66
4052 Basel
Switzerland
Tel. +41 61 683 77 34
Fax +41 61 302 89 18
www.mdpi.com

Agronomy Editorial Office
E-mail: agronomy@mdpi.com
www.mdpi.com/journal/agronomy

www.ingramcontent.com/pod-product-compliance
Lightning Source LLC
Chambersburg PA
CBHW061847210926
51850CB00006B/995
* 9 7 8 3 0 3 6 5 5 9 9 9 5 *